Fungal Pathogenesis in Plants and Crops

Fungal Pathogenesis in Plants and Crops

Contributors

Chia-Lin Chung, Joy M Longfellow et al.

AURIS
Reference

www.aurisreference.com

Fungal Pathogenesis in Plants and Crops

Contributors: Chia-Lin Chung, Joy M Longfellow et al.

Published by Auris Reference Limited

www.aurisreference.com

United Kingdom

Fungal Pathogenesis in Plants and Crops

ISBN: 978-1-78154-959-9

British Library Cataloguing in Publication Data
A CIP record for this book is available from the British Library

Printed in the United Kingdom

Exclusively distributed by CBS Publishers & Distributors Pvt. Ltd.

Sales & Distribution Rights only for India, Pakistan, Bangladesh, Sri Lanka, Nepal and Bhutan.This book is not to be sold outside these territories.

Contents

List of Abbreviations

AUDPC	Area under the Disease Progress Curve
AWF	Apoplastic Washing Fluid
BCA	Biological Control Agent
CWDE	Cell Wall Degrading Enzymes
DD	Differential Display
DLA	Diseased Leaf Area
DTA	Days to Anthesis
EST	Expressed Sequence Tags
FWER	Family Wise Error Rate
GO	Gene Ontology
GS	Growth Stages
HILIC	Hydrophilic Interaction Liquid Chromatography
IP	Incubation Period
ISR	Induced Systemic Resistance
JA	Jasmonic Acid
LC	Liquid Chromatography
LE	Lesion Expansion
LSD	Least Significant Difference
LSP	Leaderless Secreted Proteins
MAPU	Max Planck Unified Database
MDH	Malate Dehydrogenase
MS	Mass Spectrometry
NIL	Near Isogenic Lines
NLB	Northern Leaf Blight
OPD	Open Proteomic Database
PCA	Principle Component Analysis
PCER	Percomparison Error Rate
PDA	Potato Dextrose Agar
PGPR	Plant Growth Promoting Rhizobacteria
PRIDE	Proteomics Identifications Database
QTL	Quantitative Trait Loci
RFLP	Restriction Fragment Length Polymorphism
ROS	Reactive Oxygen Species
ROS	Reactive Oxygen Species
SA	Salicylic Acid
SAGE	Serial Analyze of Gene Expression
SAR	Systemic Acquired Resistance
SAXS	Small Angle X-ray Scattering
SSR	Simple Sequence Repeat
TCA	Trichloroacetic Acid
TE	Transposable Elements
TILLING	Targeting Induced Local Lesions in Genomes
YRCPDR	Yeast Resource Center Public Data Repository

List of Contributors

Chia-Lin Chung
Dept. of Plant Pathology and Plant-Microbe Biology, Cornell University, Ithaca, NY 14853, USA

Joy M Longfellow
Dept. of Plant Pathology and Plant-Microbe Biology, Cornell University, Ithaca, NY 14853, USA

Ellie K Walsh
Dept. of Plant Pathology and Plant-Microbe Biology, Cornell University, Ithaca, NY 14853, USA

Zura Kerdieh
Dept. of Biology, West Virginia State University, Institute, WV 25112, USA

George Van Esbroeck
Dept. of Crop Science, North Carolina State University, Raleigh, NC 27695, USA

Peter Balint-Kurti
USDA-ARS, Plant Science Research Unit; Dept. of Plant Pathology, North Carolina State University, Raleigh, NC 27695, USA

Rebecca J Nelson
Dept. of Plant Pathology and Plant-Microbe Biology, Cornell University, Ithaca, NY 14853, USA
Dept. of Plant Breeding and Genetics, Cornell University, Ithaca, NY 14853, USA

Raquel González-Fernández
Agricultural and Plant Biochemistry and Proteomics Research Group, Department of Biochemistry and Molecular Biology, University of Córdoba, 14071 Córdoba, Spain

Elena Prats
CSIC, Institute of Sustainable Agriculture, 14080 Córdoba, Spain

Jesús V. Jorrín-Novo
1Agricultural and Plant Biochemistry and Proteomics Research Group, Department of Biochemistry and Molecular Biology, University of Córdoba, 14071 Córdoba, Spain

Richard J O'Connell
Department of Plant Microbe Interactions, Max Planck Institute for Plant Breeding Research, Cologne, Germany
These authors contributed equally to this work. Correspondence should be addressed to R.J.O.

Michael R Thon
Centro Hispano-Luso de Investigaciones Agrarias, Departamento de Microbiología y Genética, Universidad de Salamanca, Villamayor, Spain
These authors contributed equally to this work. Correspondence should be addressed to R.J.O.

Stéphane Hacquard
Department of Plant Microbe Interactions, Max Planck Institute for Plant Breeding Research, Cologne, Germany

Stefan G Amyotte
Department of Plant Pathology, University of Kentucky, Lexington, Kentucky, USA

Jochen Kleemann
Department of Plant Microbe Interactions, Max Planck Institute for Plant Breeding Research, Cologne, Germany

Maria F Torres
Department of Plant Pathology, University of Kentucky, Lexington, Kentucky, USA

Ulrike Damm
Centraalbureau voor Schimmelcultures, Koninklijke Nederlandse Akademie van Wetenschappen, Fungal Biodiversity Centre, Utrecht, The Netherlands

Ester A Buiate
Department of Plant Pathology, University of Kentucky, Lexington, Kentucky, USA

Lynn Epstein
Department of Plant Pathology, University of California, Davis, California, USA

Noam Alkan
Department of Postharvest Science of Fresh Produce, Agricultural Research Organization, The Volcani Center, Bet Dagan, Israel

Janine Altmüller
Cologne Center for Genomics, University of Cologne, Cologne, Germany

Lucia Alvarado-Balderrama
Broad Institute, Cambridge, Massachusetts, USA

Christopher A Bauser
GATC Biotech AG, Konstanz, Germany

Christian Becker
Cologne Center for Genomics, University of Cologne, Cologne, Germany

Bruce W Birren
Broad Institute, Cambridge, Massachusetts, USA

Zehua Chen
Broad Institute, Cambridge, Massachusetts, USA

Jaeyoung Choi
Department of Agricultural Biotechnology, Center for Fungal Genetic Resources, Seoul National University, Seoul, Korea

Jo Anne Crouch
Systematic Mycology and Microbiology Laboratory, US Department of Agriculture, Agricultural Research Service, Beltsville, Maryland, USA

Jonathan P Duvick
Pioneer Hi-Bred International, DuPont Agricultural Biotechnology, Wilmington, Delaware, USA
Department of Genetics, Development and Cell Biology, Iowa State University, Ames, Iowa, USA (J.P.D.), Department of Bioproduction Science, Faculty of Bioresources and Environmental Sciences, Ishikawa Prefectural University, Ishikawa, Japan (H.T.), Department of Plant Pathology, Swammerdam Institute for Life Sciences, University of Amsterdam, Amsterdam, The Netherlands (H.C.v.d.D.) and The College of Natural Sciences, University of Massachu-

setts Amherst, Amherst, Massachusetts, USA (L.-J.M.)
Frances Trail28

Mark A Farman
Department of Plant Pathology, University of Kentucky, Lexington, Kentucky, USA

Pamela Gan
Plant Immunity Research Group, RIKEN Plant Science Center, Yokohama, Japan

David Heiman
Broad Institute, Cambridge, Massachusetts, USA

Bernard Henrissat
Laboratoire Architecture et Fonction des Macromolécules Biologiques, Centre National de la Recherche Scientifique, Unité Mixte de Recherche 7257, Université Aix-Marseille, Marseille, France

Richard J Howard
Pioneer Hi-Bred International, DuPont Agricultural Biotechnology, Wilmington, Delaware, USA

Mehdi Kabbage
Department of Plant Pathology and Microbiology, Institute for Plant Genomics and Biotechnology, Borlaug Genomics and Bioinformatics Center, Texas A&M University, College Station, Texas, USA

Christian Koch
Department of Biology, Division of Biochemistry, Friedrich-Alexander-University ErlangenNuremberg, Erlangen, Germany

Barbara Kracher
Department of Plant Microbe Interactions, Max Planck Institute for Plant Breeding Research, Cologne, Germany

Yasuyuki Kubo
Laboratory of Plant Pathology, Graduate School of Life and Environmental Sciences, Kyoto Prefectural University, Kyoto, Japan

Audrey D Law
Department of Plant Pathology, University of Kentucky, Lexington, Kentucky, USA

Marc-Henri Lebrun
Institut National de la Recherche Agronomique, Biologie et Gestion des Risques en Agriculture—Champignons Pathogènes des Plantes, Thiverval-Grignon, France

Yong-Hwan Lee
Department of Agricultural Biotechnology, Center for Fungal Genetic Resources, Seoul National University, Seoul, Korea

Itay Miyara
Department of Postharvest Science of Fresh Produce, Agricultural Research Organization, The Volcani Center, Bet Dagan, Israel

Neil Moore
Department of Computer Science, University of Kentucky, Lexington, Kentucky, USA

Ulla Neumann
Central Microscopy, Max Planck Institute for Plant Breeding Research, Cologne, Germany

Karl Nordström
Department of Plant Developmental Biology, Max Planck Institute for Plant Breeding Research, Cologne, Germany

Daniel G Panaccione
Division of Plant and Soil Sciences, West Virginia University, Morgantown, West Virginia, USA

Ralph Panstruga
Department of Plant Microbe Interactions, Max Planck Institute for Plant Breeding Research, Cologne, Germany
Institute for Biology I, Unit of Plant Molecular Cell Biology, Rheinisch-Westfälische Technische Hochschule Aachen University, Aachen, Germany

Michael Place
Laboratory for Molecular and Computational Genomics, University of Wisconsin-Madison, Madison, Wisconsin, USA

Robert H Proctor
US Department of Agriculture, Agriculture Research Service, National Center for Agricultural Utilization Research, Peoria, Illinois, USA

Dov Prusky
Department of Postharvest Science of Fresh Produce, Agricultural Research Organization, The Volcani Center, Bet Dagan, Israel

Gabriel Rech
Centro Hispano-Luso de Investigaciones Agrarias, Departamento de Microbiología y Genética, Universidad de Salamanca, Villamayor, Spain

Richard Reinhardt
Max Planck Genome Centre Cologne, Cologne, Germany

Jeffrey A Rollins
Department of Plant Pathology, University of Florida, Gainesville, Florida, USA

Steve Rounsley
Broad Institute, Cambridge, Massachusetts, USA

Christopher L Schard
Plant Immunity Research Group, RIKEN Plant Science Center, Yokohama, Japan

David C Schwartz
Laboratory for Molecular and Computational Genomics, University of Wisconsin-Madison, Madison, Wisconsin, USA

Narmada Shenoy
Broad Institute, Cambridge, Massachusetts, USA

Ken Shirasu
Plant Immunity Research Group, RIKEN Plant Science Center, Yokohama, Japan

Usha R Sikhakolli
Department of Plant Biology, Michigan State University, East Lansing, Michigan, USA

Kurt Stüber
Max Planck Genome Centre Cologne, Cologne, Germany

Serenella A Sukno
Centro Hispano-Luso de Investigaciones Agrarias, Departamento de Microbiología y Genética, Universidad de Salamanca, Villamayor, Spain

James A Sweigard
Pioneer Hi-Bred International, DuPont Agricultural Biotechnology, Wilmington, Delaware, USA

Yoshitaka Takano
Laboratory of Plant Pathology, Graduate School of Agriculture, Kyoto University, Kyoto, Japan

Hiroyuki Takahara
Department of Plant Microbe Interactions, Max Planck Institute for Plant Breeding Research, Cologne, Germany
Department of Genetics, Development and Cell Biology, Iowa State University, Ames, Iowa, USA (J.P.D.), Department of Bioproduction Science, Faculty of Bioresources and Environmental Sciences, Ishikawa Prefectural University, Ishikawa, Japan (H.T.), Department of Plant Pathology, Swammerdam Institute for Life Sciences, University of Amsterdam, Amsterdam, The Netherlands (H.C.v.d.D.) and The College of Natural Sciences, University of Massachusetts Amherst, Amherst, Massachusetts, USA (L.-J.M.)
Frances Trail28

H Charlotte van der Does
Department of Plant Microbe Interactions, Max Planck Institute for Plant Breeding Research, Cologne, Germany
Department of Genetics, Development and Cell Biology, Iowa State University, Ames, Iowa, USA (J.P.D.), Department of Bioproduction Science, Faculty of Bioresources and Environmental Sciences, Ishikawa Prefectural University, Ishikawa, Japan (H.T.), Department of Plant Pathology, Swammerdam Institute for Life Sciences, University of Amsterdam, Amsterdam, The Netherlands (H.C.v.d.D.) and The College of Natural Sciences, University of Massachusetts Amherst, Amherst, Massachusetts, USA (L.-J.M.)

Lars M Voll
Department of Biology, Division of Biochemistry, Friedrich-Alexander-University ErlangenNuremberg, Erlangen, Germany

Isa Will
Department of Plant Microbe Interactions, Max Planck Institute for Plant Breeding Research, Cologne, Germany

Sarah Young
Broad Institute, Cambridge, Massachusetts, USA

Qiandong Zeng
Broad Institute, Cambridge, Massachusetts, USA

Jingze Zhang
Broad Institute, Cambridge, Massachusetts, USA

Shiguo Zhou
Laboratory for Molecular and Computational Genomics, University of Wisconsin-Madison, Madison, Wisconsin, USA

Martin B Dickman
Department of Plant Pathology and Microbiology, Institute for Plant Genomics and Biotechnology, Borlaug Genomics and Bioinformatics Center, Texas A&M University, College Station, Texas, USA

Paul Schulze-Lefert
Department of Plant Microbe Interactions, Max Planck Institute for Plant Breeding Research, Cologne, Germany

Emiel Ver Loren van Themaat
Department of Plant Microbe Interactions, Max Planck Institute for Plant Breeding Research, Cologne, Germany

Li-Jun Ma
Broad Institute, Cambridge, Massachusetts, USA
Department of Genetics, Development and Cell Biology, Iowa State University, Ames, Iowa, USA (J.P.D.), Department of Bioproduction Science, Faculty of Bioresources and Environmental Sciences, Ishikawa Prefectural University, Ishikawa, Japan (H.T.), Department of Plant Pathology, Swammerdam Institute for Life Sciences, University of Amsterdam, Amsterdam, The Netherlands (H.C.v.d.D.) and The College of Natural Sciences, University of Massachusetts Amherst, Amherst, Massachusetts, USA (L.-J.M.)

Lisa J Vaillancourt
Department of Plant Pathology, University of Kentucky, Lexington, Kentucky, USA

Dale R. Walters
Crop and Soil Systems Research Group, Scotland's Rural College, Edinburgh, UK

Neil D. Havis
Crop and Soil Systems Research Group, Scotland's Rural College, Edinburgh,

UK

Linda Paterson
Crop and Soil Systems Research Group, Scotland's Rural College, Edinburgh, UK

Jeanette Taylor
Crop and Soil Systems Research Group, Scotland's Rural College, Edinburgh, UK

David J. Walsh
Engineering, Science and Technology Department, Scotland's Rural College, Edinburgh, UK

Cecile Sablou
Crop and Soil Systems Research Group, Scotland's Rural College, Edinburgh, UK

Asghar Heydari
Department of Plant Pathology, Iranian Research Institute of Plant Protection, Tehran, Iran

Mohammad Pessarakli
Department of Plant Sciences, University of Arizona, Tucson, AZ, 85721, USA

Shaowu Meng
Fungal Genomics Laboratory, Center for Integrated Fungal Research, North Carolina State University, Raleigh, NC 27695, USA
Hayes Laboratory, Lineberger Comprehensive Cancer Center, School of Medicine, CB# 7295, University of North Carolina at Chapel Hill, Chapel Hill, NC 27599-7295, USA

Trudy Torto-Alalibo
Virginia Bioinformatics Institute, Virginia Polytechnic and State University, Blacksburg, VA 24061, USA

Marcus C Chibucos
Virginia Bioinformatics Institute, Virginia Polytechnic and State University, Blacksburg, VA 24061, USA
Institute for Genome Sciences, University of Maryland School of Medicine, Baltimore, MD 21201, USA

Brett M Tyler

xvii

Virginia Bioinformatics Institute, Virginia Polytechnic and State University, Blacksburg, VA 24061, USA

Ralph A Dean
Fungal Genomics Laboratory, Center for Integrated Fungal Research, North Carolina State University, Raleigh, NC 27695, USA

Raviraj M. Kalunke
Dipartimento di Scienze e Tecnologie per l'Agricoltura, le Foreste, la Natura e l'Energia, Università della Tuscia, Viterbo, Italy

Silvio Tundo
Dipartimento di Scienze e Tecnologie per l'Agricoltura, le Foreste, la Natura e l'Energia, Università della Tuscia, Viterbo, Italy

Manuel Benedetti
Dipartimento di Biologia e Biotecnologie "Charles Darwin", Sapienza Università di Roma, Roma, Italy

Felice Cervone
Dipartimento di Biologia e Biotecnologie "Charles Darwin", Sapienza Università di Roma, Roma, Italy

Giulia De Lorenzo
Dipartimento di Biologia e Biotecnologie "Charles Darwin", Sapienza Università di Roma, Roma, Italy

Renato D'Ovidio
Dipartimento di Scienze e Tecnologie per l'Agricoltura, le Foreste, la Natura e l'Energia, Università della Tuscia, Viterbo, Italy

Fen Yang
Department of Plant and Environmental Sciences, University of Copenhagen, 1871 Frederiksberg C, Denmark

Wanshun Li
BGI-tech, BGI, 518083 Shenzhen, China

Mark Derbyshire
Department of Plant Biology and Crop Science, Rothamsted Research, Harpenden, Hertfordshire AL5 2JQ, United Kingdom

Martin R Larsen

Department of Biochemistry and Molecular Biology, University of Southern Denmark, 5230 Odense M, Denmark

Jason J Rudd
Department of Plant Biology and Crop Science, Rothamsted Research, Harpenden, Hertfordshire AL5 2JQ, United Kingdom

Giuseppe Palmisano
Department of Biochemistry and Molecular Biology, University of Southern Denmark, 5230 Odense M, Denmark
Institute of Biomedical Science, Department of Parasitology, University of São Paulo, 05508-900 São Paulo, Brazil.

Preface

Fungal pathogenesis is the process by which fungi infect and cause disease in a host. From the molecular basis of host defense mechanisms and molecular events leading to the suppression of defense mechanisms by fungal pathogens to fungal infection processes, the text *Fungal Pathogenesis in Plants and Crops* covers various aspects of molecular plant pathology. First chapter discusses the use of introgression lines for quantitative trait loci (QTL) mapping and characterization in the maize. Second chapter deals with proteomics of plant pathogenic. Lifestyle transitions in plant pathogenic colletotrichum fungi deciphered by genome and transcriptome analysis have been focused in third chapter. In fourth chapter, we report the results of field experiments over 3 consecutive years, undertaken to determine the potential for use of an elicitor combination to control foliar pathogens of spring barley. The objective of fifth chapter is to present an advanced survey of the nature and practice of biological control as it is applied to the suppression of plant diseases. In sixth chapter, we summarize common mechanisms of pathogenesis displayed by oomycetes and fungi. An update on polygalacturonase-inhibiting protein (PGIP), a leucine-rich repeat protein that protects crop plants against pathogens has been presented in seventh chapter. Last chapter reveals a similarity in fungal protein profiles between two interactions, possibly due to the fact that analysis of whole inoculated leaves resulted in the dominance of most abundant plant and fungal proteins, which largely diluted the information about low abundant fungal proteins likely essential for pathogenicity.

Chapter 1

RESISTANCE LOCI AFFECTING DISTINCT STAGES OF FUNGAL PATHOGENESIS: USE OF INTROGRESSION LINES FOR QTL MAPPING AND CHARACTERIZATION IN THE MAIZE - SETOSPHAERIA TURCICA PATHOSYSTEM

Chia-Lin Chung[1], Joy M Longfellow[1], Ellie K Walsh[1], Zura Kerdieh[3], George Van Esbroeck[4], Peter Balint-Kurti[5] and Rebecca J Nelson[1,2]

[1]Dept. of Plant Pathology and Plant-Microbe Biology, Cornell University, Ithaca, NY 14853, USA

[2]Dept. of Plant Breeding and Genetics, Cornell University, Ithaca, NY 14853, USA

[3]Dept. of Biology, West Virginia State University, Institute, WV 25112, USA

[4]Dept. of Crop Science, North Carolina State University, Raleigh, NC 27695, USA and

[5]USDA-ARS, Plant Science Research Unit; Dept. of Plant Pathology, North Carolina State University, Raleigh, NC 27695, USA

ABSTRACT

Background

Studies on host-pathogen interactions in a range of pathosystems have revealed an array of mechanisms by which plants reduce the efficiency of pathogenesis. While R-gene mediated resistance confers highly effective defense responses against pathogen invasion, quantitative resistance is associated with intermediate levels of resistance that reduces disease progress. To test the hypothesis that specific loci affect distinct stages of fungal pathogenesis, a set of maize introgression lines was used for mapping and characterization of quantitative trait loci (QTL) conditioning resistance to *Setosphaeria turcica*, the causal agent of northern leaf blight (NLB). To better understand the nature of quantitative resistance, the identified QTL were further tested for three secondary hypotheses: (1) that disease QTL differ by host developmental stage; (2) that their performance changes across environments; and (3) that they condition broad-spectrum resistance.

Results

Among a set of 82 introgression lines, seven lines were confirmed as more resistant or susceptible than B73. Two NLB QTL were validated in BC_4F_2 segregating populations and advanced introgression lines. These loci, designated *qNLB1.02* and *qNLB1.06*, were investigated in detail by comparing the introgression lines with B73 for a series of macroscopic and microscopic disease components targeting different stages of NLB development. Repeated greenhouse and field trials revealed that *qNLB1.06* $_{Tx303}$ (the Tx303 allele at bin 1.06) reduces the efficiency of fungal penetration, while *qNLB1.02* $_{B73}$ (the B73 allele at bin 1.02) enhances the accumulation of callose and phenolics surrounding infection sites, reduces hyphal growth into the vascular bundle and impairs the subsequent necrotrophic colonization in the leaves. The QTL were equally effective in both juvenile and adult plants; *qNLB1.06* $_{Tx303}$ showed greater effectiveness in the field than in the greenhouse. In addition to NLB resistance, *qNLB1.02* $_{B73}$ was associated with resistance to Stewart›s wilt and common rust, while *qNLB1.06* $_{Tx303}$ conferred resistance to Stewart›s wilt. The non-specific resistance may be attributed to pleiotropy or linkage.

Conclusions

Our research has led to successful identification of two reliably-expressed QTL that can potentially be utilized to protect maize from *S. turcica* in different environments. This approach to identifying and dissecting quantitative resistance in plants will facilitate the application of quantitative resistance in crop protection.

BACKGROUND

Pathogenesis is the series of events that occurs in a host-pathogen interaction, including infection and colonization of the host, and reproduction and dissemination of the pathogen. Genetic variation in host and/or pathogen can have quantitative or qualitative effects on the extent of disease. Many plant genetic factors that modulate pathogenesis have been discovered. The best known group is the R-genes, which provide high levels of resistance or even complete immunity. R-gene mediated resistance is initiated through a gene-for-gene interaction; the recognition of a pathogen effector by a host protein encoded by the R-gene leads to the induction of the hypersensitive response

(HR), the production of antimicrobial metabolites such as phytoalexins, and the expression of pathogenesis-related (PR) proteins [1]. This type of interaction, typically resulting in a highly effective but race-specific defense response against pathogenic invasion, is sometimes known as qualitative resistance. Quantitative resistance, on the other hand, confers intermediate levels of resistance and is believed to be controlled by a set of genes distinct from, or partially overlapping with, those involved in qualitative resistance [2–7].

Although each quantitative resistance locus conditions a relatively small effect on pathogenesis, this type of resistance is of agricultural interest because qualitative resistance tends to be ephemeral in many pathosystems and is unavailable in others. Quantitative resistance is presumably more durable because multiple genes with minor effects lead to lower selection pressure and greater complexity to overcome [8]. A large number of quantitative trait loci (QTL) for disease resistance have been mapped in plants [6, 9], but little is known about the underlying genetic basis or defense mechanisms involved. A range of genetic mechanisms controlling basal resistance, defense signalling pathways, detoxification, morphology, and development in the plant host, is hypothesized to be associated with reducing disease progress [6]. A small number of quantitative resistance genes have recently been cloned [3–5, 7, 10], implicating diverse host functions in quantitative resistance.

Given that diverse host functions affect quantitative resistance, it is likely that QTL act at different stages of pathogenesis. The ways in which quantitative resistance affects different stages of pathogenesis has been addressed, to a limited extent, by comparing trait values obtained using distinct (usually macroscopic) disease components. In most (or probably all) of the phytopathosystems analyzed to date, differences in various disease parameters can be observed among plant genotypes. Previous QTL studies for foliar diseases have mapped distinct loci associated with incubation period, lesion number, lesion size, or diseased leaf area, with results suggesting that defense genes affecting lesion formation and lesion expansion may not be the same. In breeding programs, selection for decreased lesion length or lesion numbers can have insignificant effects on incubation period or disease severity (eg. [11]). These observations suggest that distinct resistance mechanisms govern different macroscopic components of resistance.

More insights into the role of a given disease QTL in limiting pathogenesis can be gained through histopathological analysis. While biochemical and microscopic analyses have been applied to investigate major gene resistance and fungal pathogenicity factors (reviewed by Vidhyasekaran [12]), few studies have reported the effect of individual QTL on distinct stages of pathogenesis from a microscopic view (exceptions include [5, 13]). If QTL effective at specific stages of pathogenesis can be identified, combining favorable alleles for complementary QTL (eg. for infection and colonization) will likely provide greater levels of resistance.

Northern leaf blight (NLB; also known as turcicum blight and northern corn leaf blight) of maize was used as a model system to identify and characterize disease QTL at the macroscopic and microscopic levels. NLB, caused by *Setosphaeria turcica* (anamorph *Exserohilum turcicum*, syn. *Helminthosporium turcicum*), is one of the most prevalent foliar diseases in most maize-growing regions of the world. The disease causes periodic epidemics associated with significant yield losses [14–17], particularly under conditions of moderate temperature and high humidity [18]. Qualitative and quantitative forms of resistance against *S. turcica* are available in maize germplasm [19, 20], and have been widely utilized alone or in combination in resistance breeding programs [21]. A few histological studies have revealed the pathogenesis of *S. turcica*on maize leaves by staining, whole mount and serial dissection [22–25]. Marked phenotypic variation in symptom development has been observed among diverse maize lines in our multiple field and greenhouse trials. How macroscopic and microscopic phenotypes relate to specific QTL remains to be determined.

To answer questions concerning individual QTL effects, such as testing the hypothesis that distinct QTL act at different stages of pathogenesis, well-defined genetic stocks that differ only at specific loci are required. Introgression lines have been successfully used to study QTL in maize [26, 27], rice [28], barley [29, 30], tomato [31], and Arabidopsis [32]. While QTL analysis using recombinant inbred lines (RILs) provides greater statistical power in detecting QTL [33], RIL-based approaches have limitations in estimating QTL effects [31, 32, 34]. Introgression lines can be efficiently used to produce near-isogenic lines (NILs), which permit careful analysis of phenotypic effects associated with introgressed segments [27,31].

NILs allow many long-standing questions about quantitative disease resistance to be addressed, such as the relationship between disease QTL and plant maturity, the interaction of QTL and environmental factors, and the specificity of resistance conditioned by QTL. The interplay between disease resistance and plant development has been widely recognized [35] yet remains poorly understood. In general, the resistance in adult plants or older leaves is greater than in juvenile plants or younger leaves (eg. [36–38]), and a correlation between resistance and flowering time has been found [[39], R. Wisser, J. Kolkman, and P. Balint-Kurti, unpublished]. Some QTL effects may thus be specific to certain plant developmental stages. In addition, the expression and effectiveness of many genes/QTL have been observed to be regulated by environmental conditions [9, 40, 41]. Another issue of fundamental and practical interest is whether a disease QTL confers specific or broad-spectrum resistance. A single locus can condition resistance to more than one disease, if it encompasses linked QTL effective against different diseases, or its underlying genes are involved in broad-spectrum resistance pathways.

Here, we describe the use of introgression lines and derived NILs for QTL mapping and macro-/microscopic characterization in the maize - *S. turcica* pathosystem. We used an available population of introgression lines named TBBC3 [27]. This population is composed of introgression lines, each of which carries one or a few chromosomal segments of the donor genotype Tx303 in the genetic background of the recurrent parent B73. To better understand the nature of quantitative resistance, we assembled a panel of conventional and novel disease components targeting different stages of disease development. These were used to demonstrate that two QTL affect different stages of pathogenesis. The QTL were further characterized to shed light on three secondary hypotheses: (1) that disease QTL differ by host developmental stage (young versus adult plants); (2) that their performance changes across environments (field versus greenhouse); and (3) that they condition broad-spectrum resistance. This approach to identifying and dissecting quantitative resistance in plants will facilitate more effective and efficient application of quantitative resistance in crop protection.

METHODS

Plant Materials

A set of 82 TBBC3 introgression lines was provided by J. Holland of the USDA-ARS unit at North Carolina State University. The TBBC3 (for Tx303 by B73 Backcross 3) population, originally developed by C. Stuber at North Carolina State University [26], was the most extensively developed set of introgression lines available at the time for public use in maize. The population was derived from an initial cross of Tx303 and B73, followed by backcrossing to B73 for three generations. Each line was then selfed for several generations to attain homozygosity. Genotypic information was publicly available for each line, consisting of 14 restriction fragment length polymorphism (RFLP) and 116 simple sequence repeat (SSR) markers across the genome. Each line was known to carry one or more Tx303 introgressions, covering on average 2.5% of the genome, in the background of the sequenced reference maize line B73. Taken together, the set of introgression lines collectively carries ~89% of Tx303 genome [27]. To validate and characterize the effects of Tx303 introgressions, several BC_4F_2 populations were developed by crossing selected TBBC3 lines to B73. Sets of BC_4F_3 and BC_4F_4 lines carrying different introgression(s) were subsequently derived by single-seed descent. After four generations of marker-assisted backcrossing, the BC_4F_3 and BC_4F_4 lines were designated as NILs.

Assessments of Northern Leaf Blight

A single isolate of *S. turcica* (NY001, race 1) was used in the experiments carried out in New York (NY), and a mixture of isolates representing race 1, race 23, and race 23N of *S. turcica* was used in North Carolina (NC). For preparation of liquid inoculum, *S. turcica* was cultured for two to three weeks on lactose - casein hydrolysate agar (LCA) plates under a 12 hr/12 hr normal light-dark cycle at room temperature. The conidia were then dislodged with sterile ddH_2O and a glass rod, filtered through four layers of cheesecloth, and adjusted to the final concentration with the aid of a haemocytometer. Solid inoculum was prepared by culturing *S. turcica* on sorghum grains in plastic milk jugs for two to three weeks under the same condition. For each jug, 900 ml of sorghum grains were soaked overnight in 600 ml of water in a milk jug. The jug was then autoclaved twice at 121°C, 15 lb/cm², for 25 minutes per run. The jugs were inoculated by dividing the spore suspension produced from one heavily colonized LCA plate (10 cm diameter) among five jugs. Jugs were shaken daily to prevent caking and accelerate fungal colonization.

In this study, the "juvenile" phase refers to earlier vegetative stages, and the "adult" phase refers to later vegetative stages and reproductive stages. Juvenile plants (at the five- to six-leaf stage) and adult plants (late vegetative stage, around two weeks before tasselling) were used for inoculation. The inoculation technique utilized depended on the specific objectives of the experiment. In the field trials in NY, plants were inoculated with both liquid (0.5 ml of spore suspension, 4×10^3 conidia per ml, 0.02% Tween 20) and solid inoculum (1/4 teaspoon, ~1.25 ml of colonized sorghum grains) placed in the whorl. This was done to ensure the viability of inoculum across a range of conditions (under optimal conditions, the liquid inoculum was considered most effective, while the solid inoculum was considered to perform more effectively under dry conditions). In the field trials in NC, ~20 grains of sorghum colonized with *S. turcica* were placed in the whorl.

In greenhouse trials, whorl inoculation with aforementioned liquid inoculum was carried out for assessing individual plants in the segregating populations. Spray inoculation was performed for detailed QTL characterization using NILs, as it provides significantly better differentiation for NLB evaluation (data not shown). The spray method was preferred for microscopic examination and real-time PCR quantification. A higher number of spores could be evenly distributed on leaf surface with spraying, making the subsequent sampling more effective and accurate. About 0.5 ml of concentrated spore suspension (5×10^4 conidia per ml, 0.02% Tween 20) was evenly sprayed on the first fully expanded leaf with an airbrush (Badger® Model 150) at 20 psi. After inoculation, the plants were kept overnight in a mist chamber at > 85% RH, then maintained at 22°C day/18°C night temperature with a 14 hr-light/10 hr-dark cycle.

Phenotypic Characterization of Resistance to NLB

Field experiments were conducted at Cornell's Robert Musgrave Research Farm in Aurora, NY and Central Crops Field Station in Clayton, NC. Plants were evaluated for different disease parameters and for days to anthesis (DTA). DTA, which was only assessed for field-grown plants, was scored on a row basis when > 50% of the plants in a row started to shed pollen. An overview of various disease components used in this study and their corresponding stages during NLB development is summarized in Table 1. The evaluation method for each parameter is illustrated as below.

Table 1: Overview of disease components used to target different stages of northern leaf blight (NLB) development

Disease component	Description (unit)	Targeted disease development stage(s)	Evaluation	Literature
Incidence of multiple appressoria[a]	The incidence of > 1 appressorium developed from each germinated conidium (%)	Pre-penetration	Trypan blue staining and microscopy	[25, 42]
Infection efficiency	The incidence of successful infection per germinated conidium (%)	Penetration into the epidermal cell	Trypan blue staining and microscopy, KOH-aniline blue fluorescence microscopy	[25, 42, 94]

Accumulation of callose and phenolics[a]	Diameter of enhanced fluorescing area surrounding the infection site (μm)	Intercellular and intracellular hyphal growth from primary infected cell to surrounding mesophyll cells	KOH-aniline blue fluorescence microscopy	[43]
Vascular invasion efficiency[a]	The incidence of hyphae entering vascular bundles per infection site (%)	Hyphal growth into the vasculature	KOH-aniline blue fluorescence microscopy	[43]
Fungal biomass ratio[a]	The percentage of fungal DNA divided by the total DNA in the infected leaf tissues (%)	Overall fungal growth in leaves before the appearance of necrotic lesions	DNA-based real-time quantitative PCR	[44]
Incubation period (IP)	The number of days from inoculation to the appearance of the first lesion on a plant (days)	Xylem plugging due to extensive hyphal growth in the vascular veins	Visual examination	[73, 74, 76–78,95]
Lesion expansion (LE)	The longitudinal expansion of a lesion per day (mm)	Destructive hyphal growth in primary inoculated leaves	Digital caliper measurement	[78, 96]

Diseased leaf area (DLA)	The percentage of infected leaf area of the entire plant, disregarding decayed bottom leaves (%)	Destructive hyphal growth on the leaves of a entire plant, caused by primary and secondary inoculum	Visual examination	[73, 76, 77, 94,97]
Disease severity[b]	The severity of infected leaf area of the entire plant (scale 1-10, 1: little diseased area)	Destructive hyphal growth on the leaves of a entire plant, caused by primary and secondary inoculum	Visual examination	[98]
Area under the disease progress curve (AUDPC)	Total area under the graph of DLA or other disease severity rating (area unit)	An overall destructive hyphal growth on a plant throughout the season	Calculated from visual examination scores	[76, 77]

[a] To authors' knowledge, not previously utilized as a disease component for evaluating NLB resistance.

[b] Only applied in the 2006 trial in North Carolina.

Microscopic Analysis

Two microscopy techniques were applied to investigate differential development of *S. turcica* in the NILs: trypan blue staining and KOH-aniline blue staining. In greenhouse trials, infected leaf samples were harvested from individual plants. In the field trial, for the purpose of obtaining a sufficient number of infection sites for examination, samples (per genotype per block) were collected from four plants in a row and pooled for subsequent treatments.

Trypan blue staining was performed as previously described [25, 42] with some modifications. Infected tissues were collected at two days post inoculation (dpi) from plants in the greenhouse, and at 3 dpi from plants in the field trials. Leaf samples were cut into 1 × 1 cm² segments, cleared first in an acetic acid: ethanol (1:3, v/v) solution overnight, then in an acetic acid: ethanol: glycerol (1:5:1, v/v/v) solution for at least 3 hours. The samples were subsequently incubated overnight in a staining solution of 0.01% (w/v) trypan blue in lactophenol, and rinsed then stored in 60% glycerol until examination. Specimens were transferred onto microscopic slides and examined under a compound microscope. Fifty to 60 germinated conidia were assessed per individual plant (greenhouse) or per row (field).

A modified KOH-aniline blue fluorescence technique [43] was used to visualize the growth of fungal hyphae inside the infected leaves and the accumulation of (plant-produced) callose and phenotypic compounds around the infection sites. Infected leaves were sampled at 4 dpi and 7 dpi in greenhouse trials, and at 6 dpi in the field trial. The samples were cut into 1×1 cm^2 segments, incubated in 1 M KOH at room temperature for 24 hours, then autoclaved at 121°C, 15 lb/cm^2 for 2-5 min. Autoclaving time was adjusted according to the rigidity of leaves, which varied with plant genotype and maturity. The autoclaved specimens were rinsed in ddH$_2$O three times, then stored in autoclaved ddH$_2$O until examination. Specimens were carefully placed on microscopic slides and mounted in a staining solution of 0.05% aniline blue in 0.067 M K$_2$HPO$_4$ (prepared at least 2 hrs prior to use). Thirty five to 40 germinated conidia were checked per individual plant (greenhouse) or per row (field) under a Zeiss fluorescence microscope with a G365 excitation filter, a FT395 dichromatic beam splitter, and an LP420 barrier filter.

Quantitative Real-Time PCR for Quantifying Fungal Colonization

DNA-based quantitative PCR (qPCR) was performed as described by Qi and Yang [44] with some modifications. The specific pair of primers for *S. turcica*: forward: 5'-TCTTTTGCGCACTTGTTGTT and reverse: 5'-CGATGCCAGAACCAAGAGAT, was designed based on the internal transcribed spacer 1 (ITS1) of ribosomal DNA gene in *S. turcica*. The ITS1 sequence (GenBank: AF163067.1) was obtained from the nucleotide database of the National Center for Biotechnology Information (NCBI). PCR amplification resulted in a specific fragment of 170 base pairs.

Inoculation experiments were performed three times in the greenhouse. Five plants per genotype were spray-inoculated. The same amounts of infected tissue (0.12 g per plant) were collected at 9 dpi from the middle part of each leaf. Leaf samples were ground with liquid nitrogen and DNA was extracted following the protocol described later. The extracted DNA from each individual plant was dissolved in 100 µl TE buffer. Total DNA concentration was determined using the PicoGreen® dsDNA quantitation assay kit (Invitrogen,

Eugene, Oregon, USA). Fungal DNA was quantified using qPCR. The ratio of fungal biomass in maize leaves was computed from the amount of fungal DNA divided by total DNA. Each qPCR reaction was performed in a total volume of 25 µl, containing 12.5 µl of iTaq SYBR® Green Supermix with ROX (Bio-Rad Laboratories, Hercules, CA, USA), 3 µl of 7.5-fold diluted DNA from an infected plant and 300 nM of each forward and reverse primer. PCR samples were incubated in an ABI Prism 7000 Sequence Detection System (Applied Biosystems, Foster City, CA, USA) with thermal cycling parameters of 95°C for 2 min followed by 40 cycles of 15 sec at 95°C and 30 sec at 56°C. Two standard curves were constructed by mixing a series of *S. turcica* DNA (0, 1, 10, 50, 100, 500 and 1000 pg) with 50 ng of maize DNA extracted from non-inoculated B73 and Tx303 plants, respectively. The quantification of fungal biomass in infected B73 and derived NILs was based on the first standard curve (created from mixing with a constant amount of B73 DNA), while the quantification for DNA from infected Tx303 was based on the second standard curve. Three technical replicates were carried out in individual plates for both Picogreen quantification and iTaq SYBR Green PCR, with the samples for standard curves repeated twice in the same plates.

Incubation Period (IP)

Individual plants were checked every day after 7 dpi for the appearance of the first wilted lesions. The number of dpi when the first lesions were observed was scored as the IP. In the trials at Aurora NY, IP scores were rated for individual plants, then averaged for the rows. In the trial at Clayton NC, IP was recorded on a row basis when > 50% of the plants in a row started showing lesions.

Lesion Expansion (LE)

LE was scored in the trials in NY but not NC. Around two to three weeks after inoculation, three lesions per plant were randomly chosen for measurement. Lesion margins were marked and then measured 10-14 days later for the longitudinal extension with a digital caliper. The expansion measurements taken from three lesions were averaged, and divided by the number of days from the marking until measurement of the lesions.

Primary Diseased Leaf Area (PrimDLA)

Primary DLA was rated as the percentage of infected leaf area of the inoculated leaves in the 2006 trial in NY. It was scored once on a row basis at around three to four weeks after inoculation.

Diseased Leaf Area (DLA)

DLA was rated as the percentage of infected leaf area of the entire plant, disregarding decayed bottom leaves. DLA was rated on a row basis for TBBC3 lines and derived NILs, and on individual plants for testing trait-marker association in segregating populations. DLA was rated three to four times per season at 10-14 day intervals. The first DLA score was recorded at one to two weeks after observing the onset of secondary infection.

Disease Severity

Disease severity was rated on a row basis four times through the season in the 2006 trial in NC. The severity score was based on a 1 to 9 scale corresponding to the percentage of infected leaf area on primarily the ear leaf as well as the leaves above and below the ear leaf (severity 1: 0%, 2: 12.5%, 3: 25%, ..., 9: 100%).

Area under the Disease Progress Curve (AUDPC)

The AUDPC was calculated as

$$\sum_{i=1}^{n-1} \frac{(y_i + y_{i+1})(t_{i+1} - t_i)}{2},$$

where y_i = DLA or disease severity at time i, t_{i+1} - t_i = day interval between two ratings, n = number of ratings [45].

Evaluation for Multiple Disease Resistance

Stewart's Wilt

Pantoea stewartii (syn. *Erwinia stewartii*) strain PsNY003, originally collected in NY in 1991, was obtained from Helene Dillard of Cornell University. Inoculum was prepared as previously described [46] with the use of nutrient agar plates and nutrient broth as media. Plants at the five- to six-leaf stage were inoculated with *P. stewartii* following the pinprick method [47, 48] with a modified inoculator. The multiple-pin inoculator was made with 30 T-pins (1.5 inch long), pieces of 5.5 cm × 6.5 cm sponge and cork board (3/8 inch thick) fastened on two arms of a tong with rubber bands. Primary diseased leaf area was rated as the percentage of infected area of the inoculated leaves. It was scored twice (two and three weeks after inoculation) on a row basis and the scores were averaged.

Anthracnose Stalk Rot

A New York isolate of *Colletotrichum graminicola* (teleomorph: *Glomerella graminicola*) (isolate Cg151) was obtained from Gary Bergstrom of Cornell University. Each plant was inoculated with 1 ml of 10^6 conidia per ml (0.02% Tween 20) when more than 50% of the plants in every row were tasseling [49]. Inoculum preparation and inoculation were conducted as described by Muimba-Kankolongo and Bergstrom [50], with the replacement of plastic straw with 1 ml pipette tips. Four weeks after inoculation, stalks were split longitudinally and the percentages of discolored area of individual internodes were visually rated [49] and summed for analysis. In 2007, eight consecutive internodes were scored from four plants per row (inoculation was conducted in NLB plot); in 2008, six consecutive internodes were scored from eight plants per row.

Common Smut

In the 2007 trial, plants in NLB plots were evaluated for the development of ear galls and stalk galls resulting from natural infection. Artificial inoculation was conducted in the 2008 trial using six compatible strains of *Ustilago maydis* (UmNY001, UmNY002, UmNY003, UmNY004, UmNY008 and UmNY009), which were isolated from naturally infected smut galls collected at Aurora NY in 2007. *U. maydis* strains were cultured as described by du Toit and Pataky [51]. Inoculum was prepared by adjusting sporidial suspension (in potato dextrose broth) to a final concentration of 10^6 sporidia per ml (0.02% Tween 20) with sterilized distilled water. Equal amounts of sporidia from the six compatible strains were mixed prior to inoculation. Non-pollinated (shoot-bagged) primary ear of each plant was injected with 2 ml of mixed sporidial suspension at the time that the silks of most primary ears had emerged 1-5 cm. Every plant in the row was rated from four weeks after inoculation for ear galls on a 0-10 scale, corresponding to the number and size of galls and the disease severity of the entire plant (0 = no smut galls and 10 = a dead smut-infected plant). Stalk galls resulting from natural infection were also scored.

Common Rust

Urediniospores of *Puccinia sorghi* were collected from naturally infected leaves at Aurora NY in 2007. Inoculum was increased on three- to four-leaf stage seedlings of susceptible sweet corn in the greenhouse. About 300 mg of stock urediniospores (preserved at -80°C) were suspended in 100 ml of Sortrol oil (Chevron Phillips Chemical Company, Phillips, TX, USA) [52] and evenly applied on leaves with a spray gun (Preval, Yonkers, NY, USA). Plants were kept overnight in a mist chamber at > 85% RH, then grown for two more weeks until *Puccinia sorghi* sporulated vigorously. The urediniospores were collected by agitating infected leaves with mature rust pustules in distilled water and filtering through four layers of cheesecloth. Spore suspension was adjusted to a final concentration of 2×10^5 urediniospores per ml with the aid of a haemocytometer. Field plants were inoculated at the six- to eight-leaf stage by adding 1 ml of spore suspension (0.02% Tween 20) in the whorl [53]. Evaluation for disease severity was based on a 0-10 scale with 0.5 increments, corresponding to the percentage of infected leaf area of the entire plant (0 = no disease, 1 = 10%, ..., 10 = 100%). Disease severity was scored three times at 9-day intervals from four weeks after inoculation in 2008, and at 15-20 day intervals from 10 days after inoculation in 2009. AUDPC was calculated from the three severity scores as described above.

DNA Extraction and Genotyping

About 0.1 g of fresh or lyophilized leaf tissue and a stainless steel ball (5/32 inch diameter, OPS Diagnostics, NJ, USA) were loaded in each well of a 96-well plate (Corning® Costar 96 Well Polypropylene Cluster Tubes). Tissues were frozen at -80°C and pulverized at 450 strokes/min for 50-120 sec using Genogrinder 2000 (SPEX CertiPrep Inc., Metuchen, NJ, USA). Genomic DNA was extracted following a standard CTAB extraction protocol [54, 55] with 500 μl of CTAB extraction buffer, 400 μl of chlorophorm/isoamyl alcohol (24:1, v/v), and 300 μl of isopropanol. Fungal DNA was extracted as described, except that *S. turcica* mycelium was ground in liquid nitrogen with a pestle and mortar, and transferred to a 1.5 ml eppendorf tube for later steps.

Simple sequence repeat (SSR) markers were used for genotypic analysis following a single-reaction nested PCR method [56]. Each PCR reaction was performed as described by Wisser *et al.* [57] in a total volume of 13 μl, with the same thermal cycling parameters as described by Schuelke [56]. Amplicons labeled with different fluorescent dyes were multiplexed (combining up to four PCR reactions, 0.7 μl PCR product per specific primer pair), mixed with 9 μl formamide and 0.05-0.1 μl GeneScan-500 LIZ size standard (Applied Biosystems), and analyzed on the Applied BioSystems 3730xl DNA Analyzer at Biotechnology Resource Center at Cornell University. The sizes of amplicons were scored using GeneMapper v. 3.0 (Applied Biosystems).

Experimental Design

The full set of 82 TBBC3 lines, a subset of 15 TBBC3 lines, and derived sets of BC_4F_3 and BC_4F_4 NILs were evaluated in the field following the resolvable incomplete block design (also known as an alpha design [58]). The 82 TBBC3 lines were evaluated at Aurora NY for IP, PrimDLA, DLA, and DTA. To precisely estimate the effects of Tx303 introgressions, every experimental row was grown next to a row of B73 (B73 in every third row). The whole design had two replicates with 14 blocks per replicate, and six experimental rows plus three B73 rows per block. A parallel experiment was conducted on the 82 TBBC3 lines at Clayton NC with two replicates, nine blocks per replicate and 10 rows per block. The B73 rows were included in every other block for the control (five B73 rows per replicate). Disease severity and DTA were evaluated. The field trials on the 15 selected TBBC3 lines were conducted at Aurora NY and Clayton NC following the alpha design with B73 in every third row as described above. The 15 lines were selected from those significantly different from B73 in NLB resistance (based on the first-year results), and/or from the lines carrying Tx303 introgression(s) corresponding to previously identified NLB QTL. In each location, there were four replicates with four blocks per replicate, and four experimental rows plus two B73 rows per block. The IP, DLA, and DTA were evaluated. An additional disease component, lesion expansion (LE), was evaluated at Aurora NY as well. During the same field season, to understand the effectiveness of identified QTL in adult plants, the same 15 selected TBBC3 lines were grown in a separate field plot at Aurora NY, following identical arrangement and design with three replicates. The plants were inoculated two weeks before flowering (rather than the usual five- to six-leaf stage), and evaluated for IP and DLA. To associate resistance with specific introgressions, several BC_4F_2 populations were genotyped and phenotyped for IP, LE, and DLA in either greenhouse or in the field at Aurora NY. Individual plants of each population were grown within a single block with B73 rows as the border. For QTL confirmation, derived BC_4F_3 and BC_4F_4 NILs were evaluated at Aurora NY for IP, LE, DLA, and DTA, following the previously described alpha design with B73 rows in every third row. The experiment was planted in two replicates, with five blocks per replicate, and four experimental rows plus two B73 rows per block. To test whether the identified NLB QTL confer resistance to other important maize diseases, derived BC_4F_3 and BC_4F_4 NILs were evaluated at Aurora NY for anthracnose stalk rot (ASR) and common smut in 2007 and 2008, and for Stewart's wilt and common rust in 2008 and 2009. In 2007, the assessments of artificially inoculated ASR and naturally occurring common smut were conducted on the plants in NLB trial. In 2008 and 2009, derived NILs were grown in separate field plots for different disease evaluations. Trials for Stewart's wilt and common rust were planted following the previously described alpha design, with two replicates, five blocks per replicate, and four experimental rows plus two B73 rows per block. Trials for ASR were planted following the same alpha design, with two replicates, two blocks per replicate, and seven experimental rows plus four B73 rows per block.

Plants in the trials of common smut were randomized in two replicates. Four genotypes were used for microscopic analysis and DNA-based real time PCR quantification: B73, Tx303, TBBC3-38-05F (the NIL with $qNLB1.06_{Tx303}$), and TBBC3-42-10E-02 (the NIL with $qNLB1.02_{Tx303}$). Three greenhouse trials (each with five plants per genotype) and one field trial (five blocks, one row per genotype per block) were conducted from December 2007 to July 2008. Real-time PCR quantification was not performed in the 2008 field trial. In the greenhouse, plants subject to the same treatment were randomized within a block in order to eliminate the variance due to environmental factors.

Data Analysis

Mixed model analyses were performed in JMP 7.0 for the field trails following an alpha design. For trials conducted in one field location, "maize lines" was specified as a fixed factor, whereas "replicates" and "blocks within replicates" were specified as random. Data from the 2006 trials in NY and NC were analyzed separately because of the different field arrangements and disease rating scales. A combined analysis across two locations (NY and NC) was conducted for the 2007 trials by fitting a mixed model with "maize lines", "locations" and "maize lines by locations" as fixed factors, and "replicates within locations" and "blocks within replicates within locations" as random effects. Phenotypic differences between TBBC3 lines/NILs and B73, or between any two NILs, were determined by pair-wise comparisons of least squares means using two-tailed Student's t-test at $P < 0.05$. Correction for multiple comparisons was not made since all the tests were independent. The TBBC3 lines and derived NILs were each compared to the recurrent parent B73, and pairs of NILs were only compared to each other when sharing identical genotypes except for the target introgression(s) under testing. Overall comparisons of TBBC3 lines or NILs were not intended in this analysis. The QTL effect of each marker locus was investigated in the full TBBC3 population. Considering the presence of unlinked, potentially unrecognized introgressed segments in the same lines, a series of statistical tests was conducted as described by Szalma et al. [27] with some modification. Markers likely associated with resistance ($P < 0.05$) were identified by: 1) comparing the least squares mean of TBBC3 lines homozygous for Tx303 alleles at each locus to the least squares mean of B73 rows, and 2) performing one-way analysis of variance (ANOVA) on an individual marker - trait basis in the TBBC3 population. Significant markers were grouped into linked introgression blocks, and the markers with the greatest significance in each linked block were considered to reflect the most likely QTL position. Correlations between the most significant markers were checked. For each pair of unlinked but correlated markers, TBBC3 lines fixed for one marker were grouped and analyzed for the co-segregation of the second marker and disease traits, using two-tailed Student's t test at $P < 0.05$. A putative QTL was declared if the mean of all the lines with Tx303 alleles at this locus was significantly different from B73, and significant difference was detected for lines contrasting for this locus under the condition of another potential QTL locus being fixed.

In segregating populations, the phenotypic and genotypic data were first analyzed by mixed stepwise regression, with a significance probability of $P < 0.05$ for each parameter to enter/leave the model [59]. Once the markers significantly associated with the traits were identified, ANOVA was conducted on a single marker basis at $P < 0.05$ to estimate QTL effects. For the populations with more than one marker showing significant effect on traits, the effective markers were further tested for epistatic interactions by examining marker pairs by two-way ANOVA. Allele effects were determined by pair-wise two-tailed Student›s t test according to the least significant difference (LSD) at $P = 0.05$. The proportion data from microscopic analysis and qPCR were transformed to arcsine of the square roots prior to statistical analysis. Arcsine transformed data were analyzed by fitting a linear least squares model with «genotype» and «replicate» as two independent variables, and with the number in the denominator of each proportion as a weighting factor. For qPCR, considering the biological and technical replicates, the fungal biomass ratios were also arcsine square-root transformed, but were analyzed with a different linear least squares model of $y_{ijm} = \beta_0 + \beta_1 (genotype_i) + \beta_2 (rep_j) + \beta_3 (plate_m (rep_j)) + e_{ijm}$ (genotype i = B73, Tx303 and two derived NILs; rep j = 1, 2, 3; plate m = 1, 2, 3; e_{ijm}: random experimental error). Least squares means and 95% confidence intervals were back-transformed to percentages. Confidence intervals were larger than significance levels due to asymmetry resulting from back transformation of arcsine scale. The significant QTL effects were estimated by pair-wise comparisons of least squares means of introgression lines/NILs and B73 using two-tailed Student's t-test at $P < 0.05$. All the statistical analyses in the study were conducted using JMP 7.0.

RESULTS

A stepwise strategy (Figure 1) was conducted to map and characterize QTL using the TBBC3 introgression lines and derived NILs.

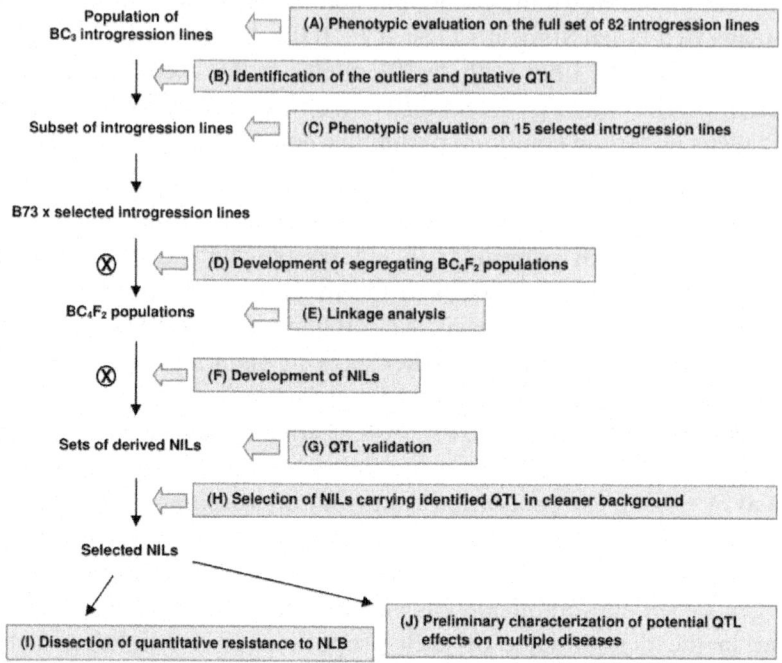

Figure 1: Strategy for mapping and characterizing QTL using introgression lines. (A) Initial screening was conducted at Aurora NY and Clayton NC. Conventional disease components, including incubation period (IP), diseased leaf area (DLA), and disease severity were evaluated. (B) Lines that differed from B73, and lines showed extreme phenotypes in the population were determined. Putative QTL were also identified. (C) Lines carrying putative NLB QTL were evaluated to confirm their differential resistance/susceptibility relative to B73. Plants were tested at the juvenile and adult stages for IP, LE, and DLA. (D) Selected lines were backcrossed to B73, then selfed to generate F_2 populations segregating for the introgressed regions within the populations. (E) Individual plants were tested for markers targeting introgressed regions, and co-segregation was assessed with the disease components measured in the greenhouse or field in NY. Candidate NLB QTL were determined if trait-marker association was detected in more than one F_2population. (F) The BC_3F_3 and/or BC_3F_4 near-isogenic lines (NILs) carrying different Tx303 introgressions were derived from F_2populations. (G) The effects of different introgressions on conventional disease components were further tested. The NLB QTL were declared if the QTL effects were validated in the NILs. (H) Selected NILs carrying identified QTL were used for detailed QTL characterization. (I) Selected NILs were evaluated with a series of disease components targeting different stages of disease development (Table 1) in the greenhouse and field. (J) Selected NILs were also evaluated for anthracnose stalk rot, common rust, common smut and Stewart's wilt at Aurora NY.

Identification of Outliers and Putative NLB QTL in the TBBC3 Population

The full set of 82 TBBC3 lines was screened for NLB resistance at field sites in New York and North Carolina in 2006. As shown in Additional file 1, the TBBC3 population was generally more susceptible than B73 at both locations. In the trial at Aurora NY, three lines were scored as significantly more resistant and 20 lines were scored as significantly more susceptible than B73 for primary DLA, while no lines showed significantly more resistance and 39 lines showed more susceptibility than B73 based on AUDPC. In the trial at Clayton NC, eight lines were significantly more resistant and 13 lines were significantly more susceptible than B73 for AUDPC. The only line that showed contrasting phenotypes in the two environments was TBBC3-25, suggesting experimental error or environmental influence on the expression of NLB QTL.

A combination of analytic approaches was used to associate the resistance or susceptibility effects with specific chromosomal segments [27]. Based on this analysis, 13 putative NLB QTL were identified in the TBBC3 population (Additional file 2). Except for bins 1.01 and 5.08-5.09, the putative QTL co-localized with NLB QTL previously detected in four maize mapping populations [20].

Phenotypic Evaluation on the Subset of 15 TBBC3 Lines

A subset of 15 TBBC3 lines was selected for further evaluation based on the degree of difference in NLB resistance between each individual TBBC3 line and B73 at the two field sites (Additional file 1 and Table 2). Lines significantly or marginally different ($P < 0.15$) from B73 in either of the two locations were taken into consideration. Line selection was also influenced by the identity of the Tx303 introgressions carried by the lines; those carrying introgressions corresponding both to putative QTL identified in the TBBC3 population (Additional file 2), and to previously reported NLB QTL (specifically those at bins 1.07-1.08, 5.02, and 6.04-6.05 [20]) were given priority because of our interest in QTL validation and characterization rather than in QTL discovery.

Table 2: NLB resistance of the subset of 15 TBBC3 introgression lines

Maize line	Introgressed region (Bin)[a]	Days to anthesis	Trait (juvenile plants being inoculated; NY and NC)[b]			Trait (adult plants being inoculated; NY)	
			Relative IP	Relative LE	Relative AUDPC	Relative IP	Relatvie AvgDLA

Tx303		8.7***	2.1***	-0.3	-164.1***	7.5***	-8.7***
TBBC3-38	1.03, 1.06, 5.00, 5.03, 5.07-5.09	4.0***	2.1***	-0.1	-110.2***	2.6***	-4.5***
TBBC3-39	1.03, 1.06, 5.00, 5.07	2.4***	0.9*	0.0	-67.2**	1.3*	-4.3***
TBBC3-18	1.11, 7.05, 7.06, 10.03, 10.04	4.2***	0.8*	-0.1	-35.8	1.3*	-2.0**
TBBC3-19[c]	7.04, 7.06, 9.03, 10.04	1.4*	1.2**	0.1	-29.0	1.2	3.2***
TBBC3-26	7.06	1.0	1.4**	0.0	-1.8	-0.1	1.0
TBBC3-75[d]	6.04-6.05	1.2	0.4	0.1	12.2	2.6***	-2.3**
TBBC3-02	1.07-1.08, 5.01-5.02, 8.03-8.05	3.2***	-0.7	0.2	35.7	0.6	1.5*
TBBC3-14[c]	1.09-1.10, 4.07	3.1***	-0.9*	0.0	67.0**	0.3	6.7***
TBBC3-03	1.01, 1.04, 7.03, 8.03-8.05	2.8***	-0.1	-0.1	117.0***	-1.4*	5.7***
TBBC3-61[c]	1.10	1.4*	-0.5	0.6***	126.7***	0.2	7.0***
TBBC3-77	1.01-1.03, 6.04-6.05	4.3***	-1.4**	0.9***	139.4***	-1.4*	4.0***
TBBC3-30	4.02-4.03, 5.04	0.2	-0.9*	0.6***	195.4***	-0.6	4.7***
TBBC3-21	4,07, 9.01-9.04	0.6	-1.7***	0.8***	275.5***	-1.1	7.8***
TBBC3-36	1.01, 1.03-1.05, 4.01, 8.08	1.3	-0.8	0.7***	305.6***	-1.8**	8.2***
TBBC3-42	1.01-1.02, 4.07-4.08, 5.01-5.02, 8.02-8.05	2.2**	-2.1***	0.7**	310.8***	-1.3*	8.7***

In 2007, the same set of plant materials was inoculated at the five- to six-leaf stage in New York (NY) and North Carolina (NC), and were grown in a separate field plot in NY and inoculated 1-2 weeks before anthesis. Trait values shown are the least squares means of maize lines relative to the recurrent parent B73. Phenotypes that were significantly more resistant than B73 are highlighted in bold, while the phenotypes significantly more susceptible than B73 are underscored. The significance level was determined by pair-wise two-tailed Student's t test on the least square difference (denoted as * 0.01 <P< 0.05; ** 0.001 <P< 0.01; *** P < 0.001). Maize lines in bold are the ones that were consistently more resistant or more susceptible than B73 across two field environments.

[a] Introgressed genome regions from Tx303 (heterozygous regions not shown). Bin position was based on genetic map of intermated B73 × Mo17 population (version IBM 2008 neighbors).

[b] Relative least squares means of incubation period (IP) and area under the disease progress curve (AUDPC) shown were from the combined analysis across the trials in NY and NC. Lesion expansion (LE) was only scored in NY.

[c] Individual TBBC3 lines in each field environment were inspected. TBBC3-19 was significantly more resistant than B73 for IP (P = 0.03), DLA1 (P = 0.001), DLA2 (P = 0.035) and AUDPC (P = 0.010) only in NC. TBBC3-14 and TBBC3-61 showed greater susceptibility than B73 for three DLA ratings (P < 0.003) and AUDPC (P < 0.0001) only in NY. The same differential degrees of resistance or susceptibility in two field environments were also observed in 2006 (Additional file 1).

[d] Resistance to NLB was more effective in adult plants.

In 2007, the 15 selected lines were evaluated at the same field locations in NY and NC to confirm their disease phenotypes. Significant «line» effects (P < 0.0001) were found for all disease parameters in each of NY and NC, and across two locations. In the combined analysis across environments, significant effects of «location» (P < 0.002 for IP, DLA and AUDPC) and "line-by-location" (P < 0.0001 for DLA and AUDPC) were also detected. The variations were attributed to dissimilar pathogen strains, environmental conditions and disease ratings conducted by different scorers at two sites. Inspection of individual TBBC3 lines in NY and NC respectively suggested that most of the lines performed similarly across the two locations. Pair-wise comparisons validated the phenotypic differences for seven out of the 15 TBBC3 lines and B73 (Table 2). Consistently in NY and NC, TBBC3-38 and TBBC3-39 were more resistant, whereas TBBC3-42, TBBC3-36, TBBC3-21, TBBC3-30, and TBBC3-77 were more susceptible than B73.

Ten selected TBBC3 lines showed significantly greater DTA than B73 (about 1-4 days of difference, Table 2), indicating the presence of flowering time QTL in those lines. To test the effect of plant development on QTL expression, the 15 selected lines were inoculated at juvenile and adult stages (Table 2) and their disease traits were compared. In all the lines, QTL effects in reducing or increasing resistance to NLB were generally consistent for juvenile and adult plant stages. No lines showed contrasting phenotypes (more resistant and more susceptible relative to B73) when inoculated at the five- to six-leaf stage or just before flowering, which suggested that allele effects of NLB QTL were not altered by developmental stage. There was, however, evidence suggesting that the effectiveness of QTL might change over plant development. The resistance in TBBC3-38, TBBC3-39, TBBC3-18 and TBBC3-75 was more effective in adult than in juvenile individuals. The formation of NLB lesions was delayed by around 0.5-2 days on mature plants of the same genotypes.

Validation of QTL Effects by Linkage Analysis

Each of the TBBC3 lines had more than one identified introgression from Tx303. To determine which of the introgressed regions in a given line was associated with resistance, several segregating BC_4F_2 populations were developed from crosses between selected TBBC3 lines (TBBC3-02, TBBC3-38, TBBC3-39, TBBC3-42, and TBBC3-77) and the recurrent parent B73. The BC_4F_2 individuals from each cross were evaluated for NLB resistance and genotyped using markers for the respective introgressed regions. Significant trait-marker associations were identified in all the BC_4F_2 populations under investigation (Additional file 3).

The QTL at bin 1.02 was found effective in two segregating populations: those derived from TBBC3-42 and TBBC3-77. Although the introgressed region extended from bin 1.01-1.02, higher significance was detected at the marker located at bin 1.02 so the QTL was designated qNLB1.02. In the BC_4F_2 derived from TBBC3-42, ~14% of the IP variation could be attributed to qNLB1.02. The B73 allele at qNLB1.02 was associated with slower disease development. Individuals homozygous for Tx303 alleles in bin 1.02 (replacement of $qNLB1.02_{B73}$) were significantly more susceptible than B73. The Tx303 homozygotes showed lesions ~0.6 days earlier (lower IP; $P <$ 0.0001) and lesions expanded more rapidly by ~0.35 mm/day (greater LE; $P <$ 0.05) in the B73 × TBBC3-42 population. The same QTL acted differently in the B73 × TBBC3-77 population. While qNLB1.02 was effective in reducing DLA and AUDPC, no effect on IP was observed in the B73 × TBBC3-77 population. It is worth noting that TBBC3-42 is a much more susceptible genotype than TBBC3-77, which provided a better background for evaluating minor disease QTL.

The QTL at bin 1.06 was not identified based on previously reported genotype information. Although clear phenotypic variation was observed on

IP and DLA in the BC_4F_2 populations from TBBC3-38 and TBBC3-39, no significant association was found between IP and the known introgressions at bins 1.03, 5.00, 5.01-5.03, and 5.07-5.09. The IP on BC_4F_2 individuals of TBBC3-38 and TBBC3-39 ranged from 11 to 18 dpi, while IP on B73 was generally between 12 and 15 dpi. Since these two populations displayed a greater variation in resistance, we inferred that a disease QTL was present in the population but not associated with a recognized introgression. To search for unrecognized introgressions, an additional 68 SSR markers across the genome were chosen to target chromosomal regions that were not well-covered by the original 130 RFLP and SSR markers on TBBC3 map. Pooled DNA samples from BC_4F_2 individuals of TBBC3-38 and TBBC3-39, along with B73 DNA as a control, were tested with the additional SSR markers for heterozygosity. Among the 68 SSR markers tested, *umc1754* and *umc2234* (both in bin 1.06) were the only polymorphic markers segregating in both BC_4F_2 populations. This indicated the presence of a Tx303 introgression at bin 1.06 in both TBBC3-38 and TBC3-39. Individuals in each population were then genotyped with *umc1754* and *umc2234*, and analyzed for the effects on various disease components. A highly significant association was detected for various parameters in both populations (Additional file 3). In the B73 × TBBC3-38 and B73 × TBBC3-39 populations, around 33% and 53% (respectively) of the variation in AUDPC was explained by this QTL. Compared to the B73 allele, the Tx303 allele at *qNLB1.06* increased the IP by ~2 days, decreased LE by ~0.3 mm per day, and reduced DLA by about 5-10%. Another QTL at bins 5.08-5.09 (Tx303 allele for resistance) was found in the B73 × TBBC3-38 population. The effect was relatively minor ($R^2 ≈ 0.05$) compared to the *qNLB1.06* $_{Tx303}$ identified in the same population.

QTL were also identified in bins 5.00, 5.01-5.02, and 8.03-8.05. Although contributing significant levels of resistance for IP or DLA in the B73 × TBBC3-77 and B73 × TBBC3-02 populations, the same introgressions/markers were not found effective in the populations derived from TBBC3-38, TBBC3-39 and TBBC3-42. The ambiguous results called the association into question, suggesting that the effectiveness of QTL at bins 5.00, 5.01-5.02, and 8.03-8.05 was affected by genetic background and/or environmental conditions.

Validation of *qNLB1.02* $_{B73}$ and *qNLB1.06* $_{Tx303}$ in Selected NIL Sets

To confirm the QTL effects detected in segregating populations, NILs were generated by selfing selected BC_4F_2 lines from the populations B73 × TBBC3-38, B73 × TBBC3-39 and B73 × TBBC3-42. Selected lines were chosen to represent different introgressed regions in the original TBBC3 lines. The

QTL effects were determined by pair-wise comparisons between individual introgression lines and B73, and between the NILs developed from a TBBC3 line. Within each set of NILs, lines contrasting for each marker locus were grouped and analyzed for their phenotypic differences. The NILs derived from different TBBC3 lines were not compared to each other since the phenotypic difference can be attributed to not only introgressed regions but also to minor differences in genetic backgrounds.

Evaluations of the BC_4F_3 NILs validated the effects of $qNLB1.06_{Tx303}$. Five NILs derived from B73 × TBBC3-38 and four NILs derived from B73 × TBBC3-39 were phenotyped in 2008. Individual trait-locus analysis in the two NIL sets revealed that bin 1.06 was the only locus that showed significant effects on resistance. The QTL effect at bin 5.08-5.09 was detected in the B73 × TBBC3-38 population, but was not confirmed in the NILs. As shown in Additional file 4, most of the NILs carrying Tx303 alleles at bin 1.06 were significantly more resistant than B73. The only exception was the line TBBC3-39-11A, which might have lost the resistance gene(s) due to recombination. Significant differences between lines with $qNLB1.06_{Tx303}$ and B73 were observed for IP, DLA and AUDPC, which was in agreement with the results from linkage analysis in BC_4F_2 populations (Additional file 3). The effects of $qNLB1.06_{Tx303}$ relative to $qNLB1.06_{B73}$ on IP (1.5-2.5 days) and DLA (2-10% through the season) were determined by comparing TBBC3-38-05F with B73 (Figure 2A) and TBBC3-39-19E with B73. The significant effect of reducing lesion expansion was not confirmed in the NILs. Lines with $qNLB1.06_{Tx303}$ even showed significantly higher LE scores, indicating some variations from different genetic backgrounds and environments.

(A)

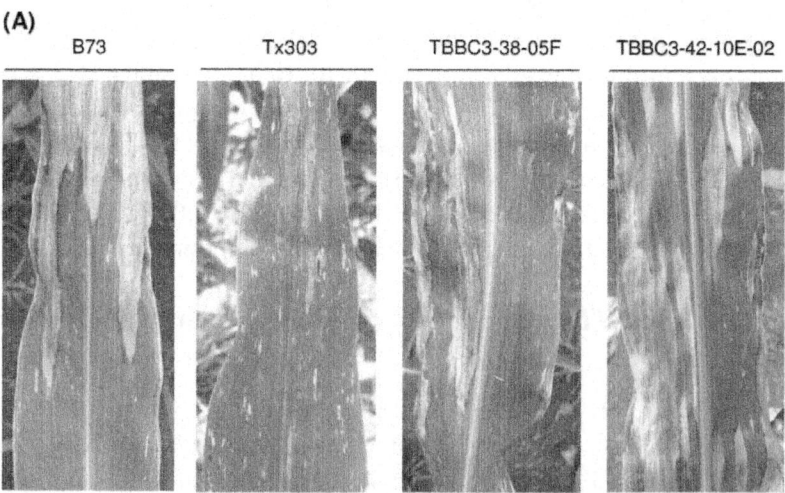

| B73 | Tx303 | TBBC3-38-05F | TBBC3-42-10E-02 |

(B)

B73   TBBC3-42-10E-02
                                          (Replacement of $qNLB1.02_{B73}$)

Figure 2: Effects of $qNLB1.06_{Tx303}$ and $qNLB1.02_{B73}$ in the field. (A) Different levels of NLB resistance in B73, Tx303, TBBC3-38-05F (the NIL carrying $qNLB1.06_{Tx303}$), and TBBC3-42-10E-02 (the NIL carrying $qNLB1.02_{Tx303}$, which is essentially «replacement of $qNLB1.02_{B73}$"). The photographs were taken in Aurora NY, 2008, on the 7th leaves at 28 days after inoculation. (B) Replacement of the B73 allele with the Tx303 allele at bin 1.02 largely increased the susceptibility to NLB. The photograph was taken in Aurora NY, 2008, at 68 days after inoculation. The comparison between B73 and TBBC3-38-05F is not shown because their difference in DLA ($< 10\%$) is not differentiable in photographs.

$qNLB1.02$ was validated by analyzing six BC_4F_3 NILs and seven BC_4F_4 NILs derived from a B73 × TBBC3-42 cross. Evaluations conducted in 2007 and 2008 led to similar results. Since BC_4F_4 NILs had cleaner genetic backgrounds, the data from the 2008 trial was shown to represent the overall result. Individual trait-locus analysis in this NIL set suggested that bin 1.01-1.02 was the only locus affecting resistance. The NILs carrying Tx303 allele(s) at bin 1.01-1.02 were all significantly more susceptible than the ones carrying B73 allele(s) at the same region (Additional file 5), confirming the resistance effect of $qNLB1.02_{B73}$. Highly significant differences between TBBC3-42-10E-02 (the NIL with $qNLB1.02_{Tx303}$ in the cleaner background) and B73 were observed across all the disease components under investigation (Figure 2). The B73 allele(s) at $qNLB1.02$ delayed lesion formation by 1.6 days ($P = 0.004$), inhibited lesion expansion by 0.84 mm per day ($P < 0.0001$), and reduced DLA 18-38% through the season ($P < 0.0001$). Overall, the relative allele effects detected in the NILs were much greater than in the segregating populations, and the resistance of $qNLB1.02_{B73}$ was more effective in the field than in the greenhouse.

Characterization of $qNLB1.02_{B73}$ and $qNLB1.06_{Tx303}$ using Derived NILs

Two derived NILs carrying identified NLB QTL in "cleaner" B73 backgrounds (minimal amount of introgression from Tx303 not associated with NLB resistance) were chosen for detailed QTL characterization. TBBC3-38-05F was a BC_4F_3 NIL with a Tx303 introgression at bin 1.06 (designated as the NIL with $qNLB1.06_{Tx303}$), and TBBC3-42-10E-02 was a BC_4F_4 NIL with a Tx303 introgression at bin 1.02 (designated as the NIL with $qNLB1.02_{Tx303}$). To investigate the effect of QTL during NLB development, a series of disease components (Table 1) was selected or developed and applied on the two derived NILs in three replicated experiments in the greenhouse during December 2007 - April 2008, and at Aurora, NY during summer 2008.

Microscopic Investigation on the Pathogenesis of *S. turcica*

Trypan blue staining and KOH-aniline blue fluorescence microscopy techniques were used for histological examination of NLB development *in planta*. To understand typical resistant and susceptible interaction patterns in the NLB pathosystem, preliminary microscopic analysis was conducted on the following maize lines challenged with *S. turcica*: B73, Tx303, TBBC3-38, TBBC3-42, the resistant maize inbred line CML52 and CML103, and a susceptible recombinant inbred line IBM262. As illustrated in Figures 3 and 4, pathogenesis was successfully visualized and can be summarized as follows. After landing and attaching on the leaf surface of a susceptible maize plant, the conidium of *S. turcica* germinates. It forms an appressorium, which produces a penetration peg that punches through the cell wall and into the epidermal cell. From this cell, infective hyphae are produced, grow towards the vascular bundle, enter the vasculature and ramify within it. Aggressive hyphal growth and extension in the vascular bundles can be seen. The hyphae can grow for several days inside the vascular bundles without causing visible symptoms (incubation period was approximately seven days on susceptible maize lines). After extensive growth within the vascular veins, hyphae grow out to colonize the neighboring bundle sheath cells and then branch out to colonize the rest of leaves. This progression is consistent with several previous studies on *S. turcica* pathogenesis on maize [23–25]. The limited damage *S. turcica* causes to host tissues at early phase of pathogenesis and the extended incubation period suggests the possibility that this is a hemibiotrophic interaction.

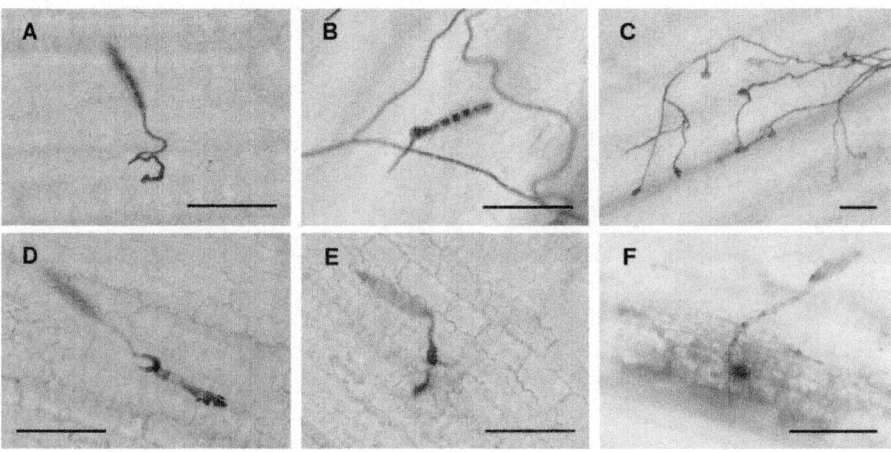

Figure 3: Light micrographs of the infection and early colonization of *Setosphaeria turcica* in corn leaves. Samples from greenhouse trials were stained with trypan blue. (A) A conidium germinated and formed an appressorium. (B) The penetration peg from the appressorium punched through the cuticle and epidermal cell wall. (C) The conidium could continue to produce new germ tubes and appressoria until exhausting its reserves or ultimately gaining entry into a plant cell. This phenomenon was occasionally observed on a resistant inbred line CML52. (D) A subcuticular palm-shaped structure occasionally developed from the penetration peg (likely before hyphae infected the epidermal cell). (E and F) Cytoplasmic depletion of the conidium was seen after the infection process. Infective hyphae spread from primary infected cell to surrounding area, causing host cell death. (Infected leaf samples of A, D and E: CML52, 2dpi; B: B73, 2 dpi; C: CML52, 3 dpi; F: IBM262, 5 dpi) (Scale bars, 100 µm).

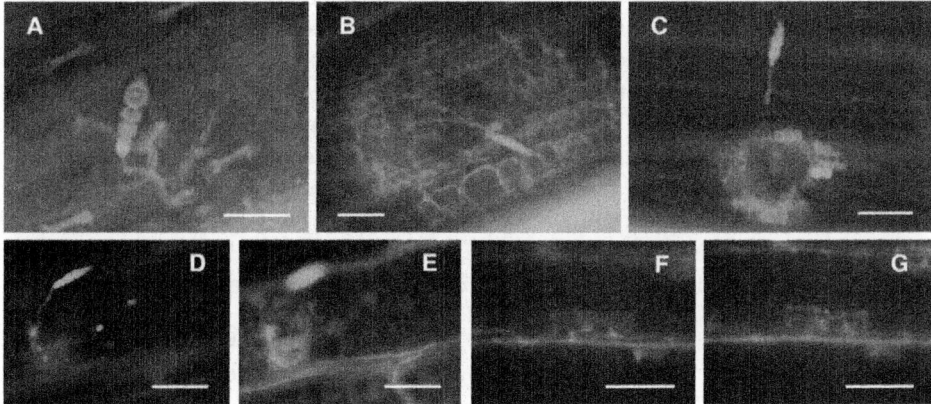

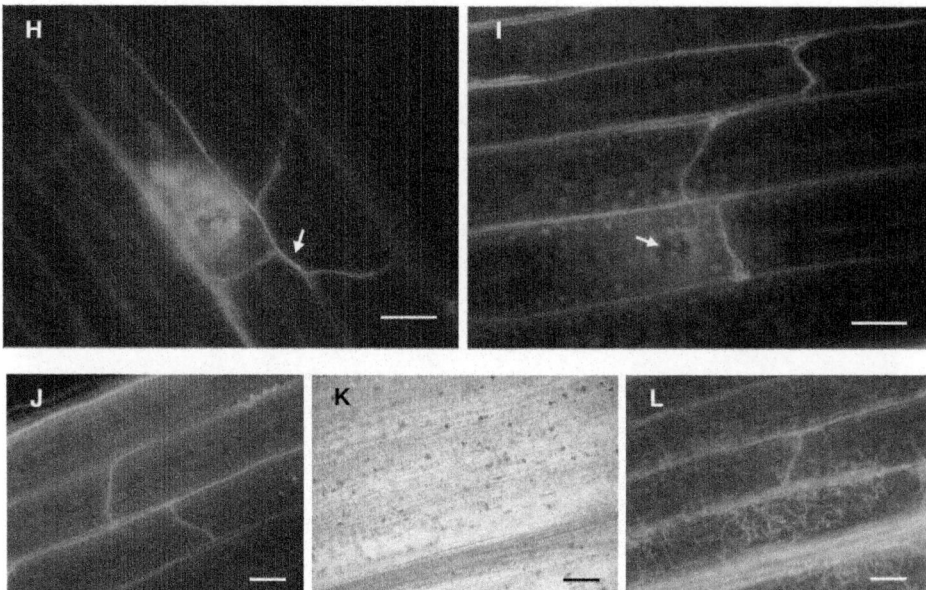

Figure 4: Fluorescence micrographs of the pathogenesis of *Setosphaeria turcica*in corn leaves. Samples from greenhouse trials were treated with KOH-aniline blue. (A) A germinated conidium on leaf surface. (B) Infective hyphae grew into contact with mesophyll cells. (C and H) Defense responses induced around the infection site. The brightly fluorescing area was presumably caused by callose deposition and the accumulation of autofluorescent phenolic compounds. The fluorescing vascular bundles (arrow), possibly due to lignification, could be differentiated from the hyphae growing in the vasculature by the lack of distinguished hyphal coils. (D and E) Infective hyphae grew towards the vascular bundle and invaded it. D and E represent different focal planes. (F and G) Hyphae grew out the vascular vessel to colonize the neighboring bundle sheath cells. F and G represent different focal planes. (I) Hyphae successfully spread through the vascular system. The weak fluorescence and collapsed cells surrounding the infection site (arrow) were typical symptoms of the compatible interaction. (J) Movement of infective hyphae through vascular bundles and cross veins. Leaf tissue remained non-wilted at this stage. (K and L) Colonization of necrotic hyphae in the wilted lesion. Vascular bundles were plugged with aggressively growing hyphae. Hyphae branched out from the colonized vasculature to the rest of the leaf. K and L were viewed in the light- and fluorescent-field, respectively. (Infected leaf samples of A: Tx303, 4 dpi; B: CML52, 4 dpi; C: B73, 4 dpi; D and E: TBBC3-42, 4 dpi; F and G: CML103, 7 dpi; H: B73, 4 dpi; I to L: B73, 10 dpi) (Scale bars, 100 μm).

Among diverse maize genotypes (B73, Tx303, CML52, CML103 and IBM262) and on genotypes carrying different QTL for resistance (B73, TBBC3-38, and TBBC3-42), differential phenotypes were observed for the timing and extent of certain steps in the processes of infection and colonization.

To document these differences, a series of microscopic parameters was chosen for characterizing NLB QTL efficacy at different stages during pathogenesis. The microscopic disease components used in the study included the number of appressoria, infection efficiency, vascular invasion efficiency, the size of strongly fluorescent area surrounding the infection site, and the appearance of necrotic or fortified vascular bundles. The microscopic evaluations on B73, Tx303 and the derived NILs were performed on plants grown in controlled greenhouse conditions as well as in the field.

QTL Effect on the Infection of *S. turcica*

Infection efficiency was examined by trypan blue and KOH-aniline blue staining, followed by scoring the number of conidia that had successfully penetrated the epidermal cell walls. The infection efficiency was defined as the ratio of successful infection sites over the total number of germinated conidia. Non-germinated conidia were excluded because of the difficulty of determining whether a non-germinated conidium was viable, given the fact that a small proportion of damaged conidia was always present in the spore suspension.

(A)

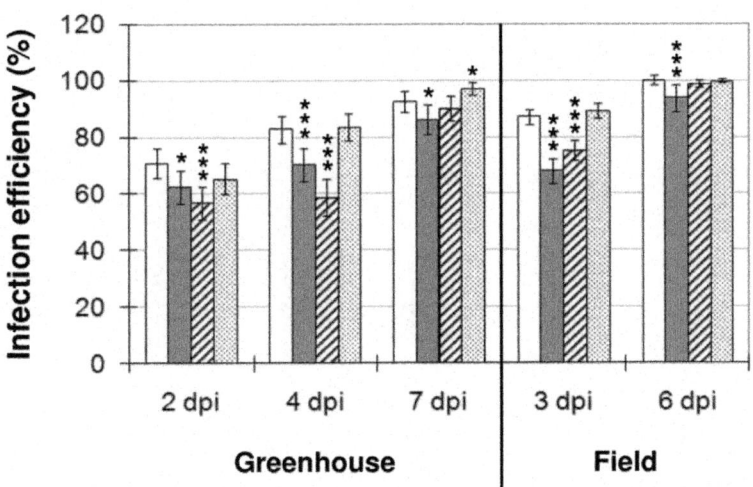

(B)

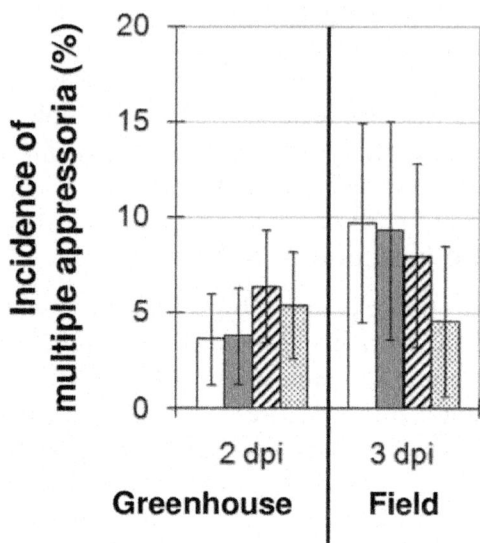

(C)

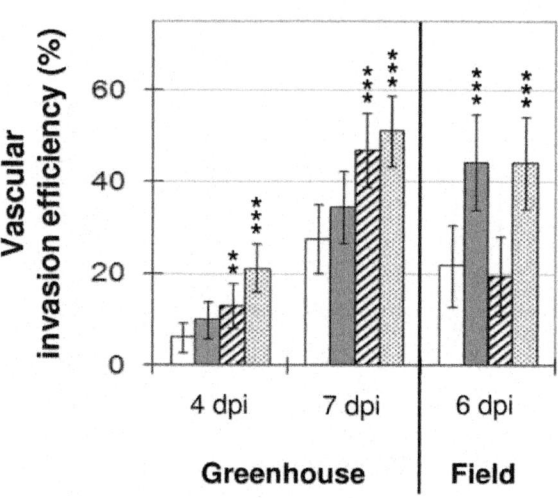

(D)

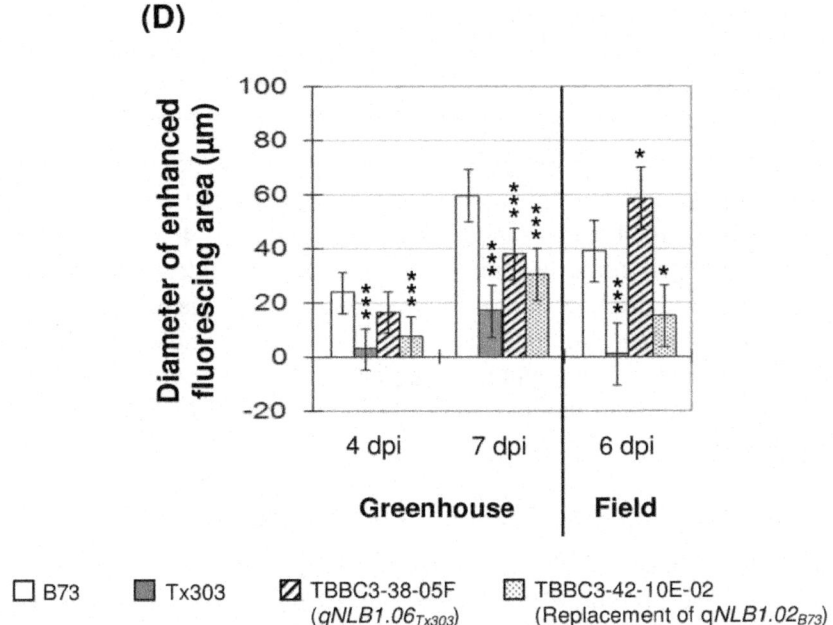

☐ B73   ■ Tx303   ▨ TBBC3-38-05F   ▨ TBBC3-42-10E-02
                                (qNLB1.06$_{Tx303}$)   (Replacement of qNLB1.02$_{B73}$)

Figure 5: Investigation of QTL effects on microscopic disease components. Microscopic disease components including (A) Infection efficiency, (B) incidence of multiple appressoria, (C) vascular invasion efficiency, and (D) size of strongly fluorescing area surrounding the infection site, were used to assess QTL effects in controlled greenhouse condition and field. Infected leaf samples were collected from maize genotypes B73 (open), Tx303 (gray), TBBC3-38-05F (hatched; the NIL carrying $qNLB1.06_{Tx303}$) and TBBC3-42-10E-02 (dotted; the NIL carrying $qNLB1.02_{Tx303}$, which is essentially «replacement of $qNLB1.02_{B73}$»). Samples collected 2 days post inoculation (dpi) from greenhouse and 3 dpi from field were stained with trypan blue, while the samples collected 4 dpi and 7 dpi from greenhouse and 6 dpi from field were treated with KOH-aniline blue fluorescence technique. Differences between least square means of different genotypes relative to B73 were determined by two-tailed Student›s t test (significance level: * 0.01 $<P<$ 0.05; ** 0.001 $<P<$ 0.01; *** $P<$ 0.001). In graphs A - C, the proportion data were arcsine transformed for statistical analysis, and the corresponding least squares means and 95% confidence intervals were back transformed to original scale before plotting. The confidence intervals are bigger than significance levels due to asymmetry resulting from back transformation.

In both the greenhouse and field, lower infection efficiency was consistently observed on the NIL with $qNLB1.06_{Tx303}$ at earlier time points examined, as well as on Tx303 at all time points examined (Figure 5A), compared with B73. In the greenhouse, $qNLB1.06_{Tx303}$ reduced infection efficiency by ~14% at 2 dpi and by ~24% at 4 dpi ($P<0.0001$). Similarly, in the field, infection efficiency was reduced by ~12% at 3 dpi ($P < 0.0001$) relative to B73. The resistant parental genotype Tx303 showed a stronger effect on reducing the infection of *S. turcica* in the field (decreased by ~19% relative to B73, $P < 0.0001$) than in the greenhouse (decreased by 6.5-12.5% relative to B73, $P < 0.05$). No significant effect on pathogen penetration was seen from $qNLB1.02_{Tx303}$.

The incidence of multiple appressoria per germinated conidium was used as another indicator for resistance to fungal penetration. It was observed that when a germinated conidium failed to penetrate the epidermis, the conidium would continue to produce new germ tubes and appressoria until exhausting its reserves or ultimately gaining entry into a plant cell (Figure 5B). Despite the effect of $qNLB1.06_{Tx303}$ on reducing the penetration of *S. turcica*, no difference was detected on the formation of multiple appressoria among B73, Tx303, and the derived NILs.

QTL Effect on the Vascular Invasion of *S. turcica*

Intracellular hyphal growth from the initially infected epidermal cell to surrounding mesophyll cells, and the subsequent invasion of the vascular bundle, were investigated by KOH-aniline blue fluorescence microscopy. Vascular invasion efficiency was defined as the incidence of hyphal growth into the vasculature per infection site. Differences in vascular invasion efficiency were observed between B73 and the other genotypes (Figure 5C). The NIL with $qNLB1.02_{Tx303}$ had greater vascular invasion efficiency, indicating the superiority of the B73 allele(s) at this locus. Replacement of the B73 allele with the Tx303 allele at $qNLB1.02$ (in lines carrying an introgression from Tx303 at this locus) led to 15-24% of increase in the incidence of vascular invasion in both greenhouse and field ($P < 0.0001$). The $qNLB1.06_{Tx303}$ allele(s) was not effective for inhibiting hyphal growth into the vasculature. Under greenhouse conditions, the NIL with $qNLB1.06_{Tx303}$ was observed to have a 7-19% higher susceptibility for vascular invasion ($P < 0.0001$), which is a further indication that this QTL functions at the stage of penetration but not at later stages of pathogenesis. The susceptibility to vascular invasion was also seen in Tx303. In the greenhouse, higher vascular invasion efficiencies were observed in Tx303 than in B73, although with marginal significance levels. In the field, the invasion efficiency in Tx303 was ~22% greater than in B73 ($P < 0.0001$).

QTL Effects on the Accumulation of Defense Compounds around the Infection Site

Infected tissues stained with aniline blue were also used for revealing defense responses induced upon pathogen challenge. The fluorescence emitted from the cells surrounding the infection site is presumably due to callose deposition [60, 61] and the accumulation of autofluorescent phenolic compounds [62–64]. Callose and phenolics are antimicrobial compounds that help restrict the establishment and spread of *S. turcica* in the leaf. In susceptible genotypes, weak fluorescence and collapsed cells were generally observed around the infected sites (Figure 4I), suggesting the lack of induced defense responses. The more resistant genotypes showed enhanced fluorescence from localized cell regions (Figure 4C and 4H).

To assess the degree of defense response around the primary infected cell, the diameter of brightly fluorescing area was measured under a fluorescence microscope with a micrometer (Figure 5D). The average diameter of the fluorescing area in the NIL with *qNLB1.02* $_{Tx303}$ was consistently smaller than that seen for B73 ($P < 0.0001$ in greenhouse, $P = 0.02$ in the field), indicating that the B73 allele(s) at *qNLB1.02* contribute to the regulation, production or accumulation of callose, lignin or other phenolic compounds. Interestingly, compared to B73, the fluorescing area in the NIL with *qNLB1.06* $_{Tx303}$ was slightly larger ($P = 0.052$) in the field, but was significantly smaller ($P < 0.0001$) at 7 dpi in the greenhouse, suggesting that the phenotype conferred by *qNLB1.06* $_{Tx303}$ is vulnerable to environmental influences.

Considering the potential role of vascular resistance in the NLB pathosystem, the incidence of brightly fluorescing vascular bundles (likely caused by lignification) was also evaluated. Although this response was pronounced in CML52 (Figure 4H), this type of resistance was rarely observed in B73, Tx303 and the derived NILs. The induced defense reaction around the infection site was fairly weak at all times examined in either greenhouse or field conditions. It was thus not considered a relevant parameter for testing the effect of *qNLB1.06* $_{Tx303}$ and *qNLB1.02* $_{B73}$.

QTL Effect on Mycelial Growth of *S. turcica* in Planta

DNA-based qPCR was developed to precisely quantify mycelial growth of *S. turcica* in maize leaves. The *R*-square values for the two standard curves based on B73 and Tx303 respectively were both higher than 0.99 (Figure 6A), indicating that the designed ITS primer pair had good sensitivity and specificity for reliably amplifying fungal DNA. To assess the qPCR technique as a tool for analyzing the NLB pathosystem, several time-course experiments were performed on seven maize genotypes with a wide range of differential levels of NLB resistance (C. Chung, unpublished). Preliminary results showed that the levels of fungal DNA ratios determined by qPCR approximately conformed to the performance of resistance in the field. For maize genotypes with low to intermediate levels of resistance, fungal DNA was detected as early as three days after inoculation. The detectable fungal DNA ratios increased over time until the end of the incubation period (when visible lesions formed), and thereafter decreased. This reduction in pathogen DNA was unexpected, and apparently corresponded to a loss of DNA integrity. DNA samples extracted from tissues showing lesions were usually brownish and of poor quality. The decrease in detectable fungal DNA ratio in highly diseased leaves is contradictory to our microscopic observation of abundant mycelial growth in NLB lesions. It is likely that in tissues with actively developing lesions, more DNA-degrading enzymes involving in cell death are present and can digest fungal DNA during the extraction. To better differentiate resistant and susceptible responses, it is critical to apply qPCR on infected tissues taken prior to the appearance of necrotic lesions on the most susceptible genotype in an experimental set.

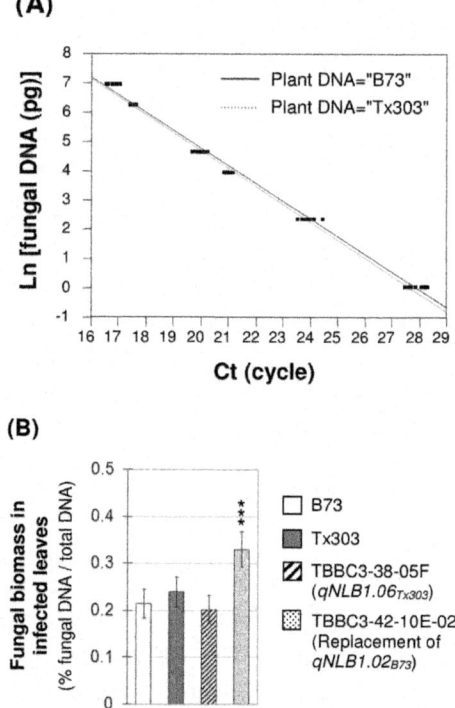

(A)

(B)

Figure 6: Quantifying QTL effect on mycelial growth of *Setosphaeria turcica in planta* using DNA-based real-time PCR. (A) Two standard curves were constructed by mixing a series of *S. turcica* DNA and 50 ng of maize DNA from non-inoculated B73 and Tx303 plants, respectively. With B73 DNA, Ln[fungal DNA] = 16.912 - 0.606*Ct (R^2 = 0.99); with Tx303 DNA, Ln[fungal DNA] = 16.906 - 0.610*Ct (R^2 = 0.99). Ln: natural logarithm. Ct: threshold cycle, the number of PCR amplification cycle at which the exponential increase of the product was detected. (B) Measurement of *S. turcica* DNA ratio in infected leaves collected 9 dpi from maize genotypes B73 (open), Tx303 (gray), TBBC3-38-05F (hatched; the NIL carrying $qNLB1.06_{Tx303}$) and TBBC3-42-10E-02 (dotted; the NIL carrying $qNLB1.02_{Tx303}$, which is essentially «replacement of $qNLB1.02_{B73}$«). Differences between least square means of different genotypes relative to B73 were determined by two-tailed Student›s t test (significance level: * 0.01 <P < 0.05; ** 0.001 <P < 0.01; *** P < 0.001). The proportion data were arcsine transformed for statistical analysis, and the corresponding least squares means and 95% confidence intervals were back-transformed to original scale before plotting. The confidence intervals are bigger than significance levels due to asymmetry resulting from back transformation.

Infected leaves from B73, Tx303, and two derived NILs were collected nine days after inoculation (according to the IP of the susceptible NIL), and measured for their fungal DNA content using qPCR. As shown in Figure 6B, neither Tx303 nor$qNLB1.06_{Tx303}$ showed significant effect on reducing the growth of *S. turcica* in leaves. However, the fungal biomass ratio (% of fungal DNA in the infected leaf) in the NIL with $qNLB1.02_{Tx303}$ was 24% higher than in B73 (P < 0.0001), indicating that the *in planta* development of *S. turcica* is more extensive on plants without $qNLB1.02_{B73}$.

Preliminary Characterization Of Potential QTL Effects for Multiple Disease Resistance using Derived NILs

Significant differences were observed between Tx303 and B73 for resistance to Stewart's wilt, anthracnose stalk rot (ASR), common smut and common rust. In the field, Tx303 was more resistant to Stewart's wilt and ASR, while B73 was more resistant to common smut and common rust. To investigate the resistance spectrum of *qNLB1.02* and *qNLB1.06*, several NILs derived from B73 × TBBC3-38, B73 × TBBC3-39 and B3 × TBBC3-42 were evaluated for the four diseases (Additional files 4 and 5). The choice of NILs was constrained by the availability of seeds.

The evidence suggested that both $qNLB1.06_{Tx303}$ and $qNLB1.02_{B73}$ were effective for resistance to Stewart›s wilt. As shown in Additional file 4, all the NILs carrying Tx303 introgression at bin 1.06 were significantly more resistant than B73. Based on comparing TBBC3-39-19E to B73, $qNLB1.06_{Tx303}$ was inferred to be effective for reducing lesions of Stewart›s wilt on inoculated leaves by as much as 39% (P < 0.0001). Likewise, in Additional file 5, lines with Tx303 alleles at bin 1.01-1.02 were significantly more susceptible than B73. The effect of $qNLB1.02_{B73}$ on Stewart›s wilt was estimated as reducing 19% of primary DLA based on a comparison of TBBC3-42-10E-02 and B73 (P < 0.0001). Significant differences were also observed among the lines fixed (homozygous for Tx303 or B73 alleles) at bins 1.01-1.02, indicating that there are QTL other than $qNLB1.02_{B73}$ segregating in the NIL set. However, none of the rest of the introgressed regions were unambiguously associated with resistance.

$qNLB1.06_{Tx303}$ was ineffective for resistance to ASR, smut and rust. No phenotypic difference in the development of rust and smut galls was detected among B73, TBBC3-38, TBBC3-39 and their derived NILs. Significant difference in discolored internode area was observed between TBBC3-38 and B73 (~32%, $P = 0.001$), but the variation was not associated with $qNLB1.06$, as TBBC3-39 and the NIL carrying a single introgression at bin 1.06 (TBBC3-39-19E) showed the same resistance level as B73 (Additional file 4).

$qNLB1.02_{B73}$ was ineffective for resistance to ASR and common smut, but potentially effective for resistance to common rust (Additional file 5). For ASR, some phenotypic difference was found among B73, TBBC3-42 and derived NILs, but none of the identified introgressed regions were associated with resistance. B73 was more resistant than Tx303 for common smut and common rust. For common smut, there were almost no ear galls or stalk galls observed on B73, TBBC3-42 and TBBC3-42-10E-04, suggesting that introgressed segments in TBBC3-42 did not affect smut development. (More ear galls were scored on TBBC3-42-10E-04 in the 2008 trial, but the significance was marginal ($P = 0.04$), and the finding was not consistent with the 2007 result). For common rust, except for TBBC3-42-06F-2, lines with Tx303 alleles at bin 1.01-1.02 were consistently more susceptible than B73 for all three severity ratings (differed by 1.1-3.1 scales in severity, $P < 0.025$) and AUDPC (differed by about 43-53 scale-day, $P < 0.001$). The result indicated the possibility of rust QTL at bins 1.01-1.02 and/or unidentified introgressed regions in TBBC3-42.

DISCUSSION

Identification of NLB QTL using TBBC3 Introgression Lines

Two QTL for resistance to NLB, $qNLB1.02_{B73}$ and $qNLB1.06_{Tx303}$, were successfully identified, validated and characterized using a population of introgression lines and derived NILs. We applied a stepwise strategy that allowed phenotyping of informative NILs over a series of generations. A special field design, in which each row of CSSL/NIL arranged next to a row of B73, allowed accurate visual comparisons and a relatively uniform epidemic across the field. We detected and sequentially validated QTL using multiple disease components, in a full set of 82 lines, a subset of 15 selected lines, five selected BC_4F_2 populations, and three sets of derived BC_4F_3/BC_4F_4 NILs. Our primary hypothesis that individual QTL affect distinct stage(s) of NLB development was ultimately confirmed using two selected NILs (compared to B73) in repeated greenhouse and field trials, with a panel of conventional and novel components. Our results revealed that $qNLB1.06_{Tx303}$ and $qNLB1.02_{B73}$ are mainly effective against the infection and colonization of *S. turcica*, respectively. We found that the QTL were both effective in both juvenile and adult plants, and that both chromosomal segments were associated with resistance to more than one disease.

The mechanism underlying resistance conferred by $qNLB1.02_{B73}$ and $qNLB1.06_{Tx303}$ is unlikely to be the same as the mechanism underlying qualitative resistance conferred by currently known major genes. Qualitative resistance to NLB is generally characterized by chlorotic-necrotic lesions [65–69], chlorotic halo lesions [70], or extremely prolonged IP [71]. Considerable levels of resistance were seen for $qNLB1.02_{B73}$ and $qNLB1.06_{Tx303}$ in the two field sites, without observing distinct lesion type or much greater IP. Moreover, although co-localized NLB QTL at bins 1.02 and 1.06 have been reported in other populations [72, 73], no major genes have been mapped to these chromosomal regions.

Conventional and Newly-Developed Components of Resistance Targeting Different Stages of NLB Development

QTL effects were analyzed using five conventional macroscopic disease parameters and four microscopic parameters. Each of the macroscopic components reflected different phase(s) of disease development. Incubation period quantifies the time to appearance of wilted lesions, which reflects the speed of xylem plugging due to extensive hyphal growth in the veins. Lesion expansion quantifies the rate of expansion of wilted and necrotic lesions, which reflects the speed of destructive hyphal growth in the leaves. Ratings for diseased leaf area, disease severity and area under the disease progress curve, on the other hand, involve visual quantification of overall disease progress on the entire plant. In our greenhouse and field trials, IP showed a better correlation than LE with DLA, disease severity and AUDPC. This indicates that IP is a more discriminating parameter than LE for measuring NLB severity. This confirms previous reports that IP is a convenient target trait in selection and breeding for resistance to NLB [11, 73–78]. In a recurrent selection study for NLB resistance, selection for prolonged IP resulted in ~7.5% more gain in reducing AUDPC per selection cycle compared to selection for lesion length [11].

QTL effects were also microscopically investigated at the pre-penetration, infection, and colonization phases. In most fungal pathosystems, pre-penetration resistance is associated with specialized physical and chemical features of plant surface which help reduce the incidence of landing, adhesion, germination, appressorium formation and penetration of pathogenic fungi [79]. Post-penetration resistance to microbial attack, on the other hand, is characterized by programmed necrosis, along with the induction of callose, lignin, phenolic compounds and other pathogenesis-related proteins around the primary-infected cell. Four microscopic components, including the incidence of multiple appressoria, infection efficiency, accumulation of defense materials surrounding the infection site, and vascular invasion efficiency, were developed to complement the conventional macroscopic components. Modified trypan blue staining [25, 42] and KOH-aniline blue fluorescence [43] methods were useful in determining the degree and timing of allelic contribution to quantitative resistance to NLB. Previous studies utilizing sectioning and electron microscopy have revealed that xylem plugging is a key stage for the formation of NLB lesions (Jennings and Ullstrup, 1957). Our microscopic examination using KOH-aniline blue fluorescence technique confirmed the finding and allowed the observation of *in planta* colonization of *S. turcica* from a different angle. In all the maize genotypes examined, the mycelium appeared to grow preferentially towards vascular bundle, initially invading the vasculature, then extending through the xylem vessel, and eventually, aggressively growing out to the neighboring mesophyll tissues. To our knowledge, the mechanisms of the post-infection directional growth of vascular fungal pathogens remain to be elucidated.

qPCR was employed for the first time for the measurement of *in planta* biomass of *S. turcica*. Although it has been widely applied in other pathosystems, qPCR can only be used for NLB quantification at early stages of pathogenesis. The accuracy and differentiating power are highly dependent on uniform inoculation and precise sampling at the early stages of pathogenesis, so the method is best suited to work conducted under controlled conditions. The biggest constraint of this technique comes from its not being able to measure the destructive mycelial growth in the blighted leaf areas, possibly due to the poor quality of DNA extracted from necrotic tissues. The cost of qPCR reagents and the required workload make it unfeasible for regular phenotypic screening. However, qPCR serves as a good tool for quantifying the degree of earlier mycelial colonization of *S. turcica* in maize leaves.

Hypothesis #1: Individual QTL Affect Distinct Stages of Disease Development

Detailed macroscopic and microscopic evaluations revealed the distinct features of *qNLB1.06* $_{Tx303}$ and *qNLB1.02* $_{B73}$ in NLB development. $qNLB1.06_{Tx303}$ conditions resistance mainly against the penetration of *S. turcica*. The anti-penetration effect was observed at earlier but not later time points examined, indicating that $qNLB1.06_{Tx303}$ acts in a quantitative manner, which delays rather than prevents the occurrence of infection. In contrast, $qNLB1.02_{B73}$ appears to condition resistance not to infection, but rather by enhancing the induction of defense reactions surrounding infection sites, as well as by inhibiting hyphal growth into the vascular bundle, and the subsequent necrotrophic colonization in the leaves. It is difficult to infer whether the resistance is associated with defense reactions in mesophyll, parenchyma, bundle sheath and/or other cells.

Delaying the invasion and extension of *S. turcica* in the vascular system was shown to be critical for quantitative resistance to NLB. This is consistent with previous microscopic analyses [23, 24] suggested that xylem invasion/plugging is an important pathogenetic phase in NLB progress. Polygenic resistance (in an inbred line CI90A) has been associated with reduced hyphal spread into the xylem [22]. The significance of protecting the vasculature from hyphal invasion is supported by the evidence that the replacement of superior B73 allele with Tx303 allele at $qNLB1.02$ led to an 18-38% higher DLA (relative to B73). Incorporating a QTL effective in slowing down the invasion of the vasculature would thus considerably increase overall resistance.

We speculate that preventing the attachment of conidia and the subsequent infection plays a role in quantitative resistance to NLB in the donor line Tx303. Although this was not quantified, it was visually obvious that the number of spores per leaf segment of Tx303 was lower than the spore numbers on B73 or the NILs carrying $qNLB1.06_{Tx303}$ and $qNLB1.02_{B73}$. Lower infection efficiency of *S. turcica* was observed for Tx303 relative to B73. Interestingly, Tx303 showed a greater susceptibility than B73 for the parameters of vascular invasion efficiency, enhanced fluoresced area, and lesion expansion. Despite the lack of resistance in delaying vascular invasion and extension of *S. turcica* in Tx303, its higher effectiveness in reducing spore attachment and infection may contribute to the overall moderate resistance to NLB.

Hypothesis #2: the Effectiveness of Disease QTL is Affected by Plant Maturity

A number of maize diseases caused by hemibiotrophic fungi (eg. gray leaf spot, anthracnose leaf blight, anthracnose stalk rot) and necrotrophic fungi (eg. southern leaf blight) are known to progress more rapidly on the plants after anthesis [18,49, 80, 81]. In maize, a significant correlation has been detected between disease resistance and flowering time in a panel of 253 diverse lines (R. Wisser, J. Kolkman, P. Balint-Kurti, and R. Nelson, unpublished). The fact that flowering time may account for a considerable proportion of resistance variation leads to the hypotheses that the effects of disease QTL reflect indirect expression of flowering time QTL, and/or that the effects of disease QTL are modulated differently at varied plant developmental stages. In fact, several QTL identified in biparental populations are associated with both NLB resistance and flowering time (J. Poland, pers. comm.).

The association between NLB QTL and plant maturity was investigated in the subset of 15 selected TBBC3 lines. To separate resistance effects from maturity effects, plants in two different fields were inoculated at juvenile and adult stage (two weeks before tasselling) respectively, and evaluated for NLB resistance and flowering time. Generally consistent expression of resistance or susceptibility was observed at the two stages, implying that the effectiveness of these QTL was not altered substantially by plant maturity. While several of the introgression lines (10 out of the 15 TBBC3 lines with differential resistance relative to B73) showed significantly different flowering times relative to B73, the interactions between NLB QTL and flowering time QTL were relatively minor. For the $qNLB1.02_{B73}$ and $qNLB1.06_{Tx303}$, the analysis of advanced NIL sets further provided strong evidence of their independence from flowering time. This agrees with a previous report of flowering-time QTL in TBBC3 introgression lines [27]: bins 1.01-1.02 and 1.06 were not among the QTL affecting days to anthesis, days to silking, or anthesis-silking interval.

Hypothesis #3: the Effectiveness of QTL is Affected by Environmental Conditions

A major limitation for QTL applications is the lack of consistency of QTL effects across environments. The inconsistent detection of QTL has been associated with experimental errors and differential gene expression affected by environmental factors [82]. In view of the widely reported inconsistency, the present study revealed that the introgressions/QTL with relatively large effects (carried in seven introgression lines: TBBC3-38, TBBC3-39, TBBC3-42, TBBC3-36, TBBC3-21, TBBC3-30, and TBBC3-77) were consistent in their performance across field sites and years. Reliable expression of these QTL, including $qNLB1.02_{B73}$ and $qNLB1.06_{Tx303}$, suggests their applicability to resistance breeding. Disease resistance can be expressed differentially, not only in different field environments but also under field versus greenhouse conditions (eg. [80, 83, 84]). Differential efficacy of $qNLB1.02_{B73}$ and $qNLB1.06_{Tx303}$ in controlled greenhouse and field conditions was evaluated using macroscopic and microscopic phenotypes. It should be noted that due to the different inoculation and sampling treatments necessarily employed in different environments, the comparison between the effectiveness of disease QTL in the greenhouse and field should be addressed based on relative observation rather than the absolute values. Our overall results suggest that the resistance of $qNLB1.02_{B73}$ was stably expressed in the greenhouse and field, whereas $qNLB1.06_{Tx303}$ conferred a higher level of resistance in the field than in the greenhouse. Macroscopically, the effects of $qNLB1.06_{Tx303}$ on IP and primary DLA in greenhouse-grown plants were occasionally insignificant. Microscopically, the positive effect of $qNLB1.06_{Tx303}$ on triggering defense response surrounding the initially infected cell was only detected in the field. Greenhouse-grown plants carrying $qNLB1.06_{Tx303}$, however, displayed an opposite effect (smaller fluorescent area), which suggests that this type of resistance is highly affected by environmental factors. The observation that $qNLB1.06_{Tx303}$ increased the efficiency of *S. turcica* invasion into vasculature in the greenhouse but not field, may be an indirect effect of decreased accumulation of host defensive materials at the post-infection stage.

Hypothesis #4: QTL for Resistance to NLB Comprises Genes or Gene Clusters Involving in Broad-Spectrum Resistance

The phenomenon of multiple disease resistance (MDR) has been inferred based on the detection of QTL clusters affecting different diseases [2, 19, 20, 85], genetic correlations in populations [86], non-specific defense mechanisms (eg. SAR [87–89]), and genes (eg. *npr1* [90] and *mlo* [91]). Maize bins 1.02 and 1.06 have been previously associated with a number of disease QTL mapped in diverse maize populations [20]. Our evaluations of resistance to common rust, Stewart›s wilt, anthracnose stalk rot, common smut, and common rust in sets of NILs suggest that both $qNLB1.02_{B73}$ and $qNLB1.06_{Tx303}$ encompass gene(s) contributing nonspecific defense effects. $qNLB1.02_{B73}$ was effective in decreasing NLB, common rust and Stewart›s wilt, while $qNLB1.06_{Tx303}$ was effective in decreasing NLB and Stewart›s wilt. Nevertheless, due to the low resolution of QTL localization in this study, it is unclear whether the nonspecific resistance of $qNLB1.02_{B73}$ and $qNLB1.06_{Tx303}$ is caused by pleiotropy or linkage, or whether their effects are mongenic or polygenic. Although the underlying mechanisms are unknown, the feature of broad-spectrum resistance makes the two identified disease QTL appealing in practical applications.

It is worth noting that the NILs carrying $qNLB1.06_{Tx303}$ showed remarkable resistance against Stewart›s wilt, suggesting the involvement of major gene effect. In fact, a dominant major gene locus for Stewart›s wilt, *Sw1*, has been mapped to bin 1.05-1.06 in inbred line Ki14 [92, 93]. The co-localization of QTL for resistance to NLB and Stewart's wilt has also been observed in the NILs derived from the inbred line CML52 crossed to B73 (CML52 allele for resistance; C. Chung, unpublished).

ELECTRONIC SUPPLEMENTARY MATERIAL

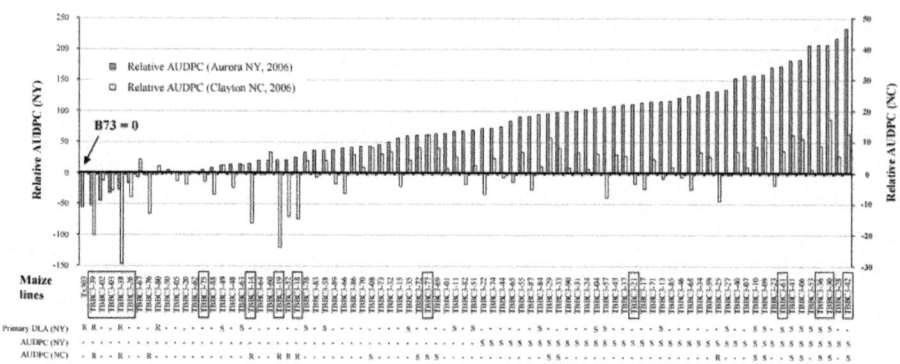

Additional file 1: NLB resistance of the full set of 82 TBBC3 introgression lines.

Additional file 2: Putative NLB QTL identified in the TBBC3 population.

Chr. Bin	Marker	Map Position [a]	Resistance allele	Aurora NY, 2006						Clayton NC, 2006				Previously reported NLB QTL at the locus	
				IP (days)	PrimDLA (%)	DLA1 (%)	DLA2 (%)	DLA3 (%)	AUDPC (%-day)	Severity1 (scale)	Severity2 (scale)	Severity3 (scale)	AUDPC (scale-day)	Mapping population [b]	Reference
Correlated loci															
1.01	umc1071	85.2	B73		6.6*		5.5***	8.0***	110.1***	0.7**				None	
1.02	bnlg1429	143.5	B73		6.9*	6.2***	7.1***	7.4**	134.7***					B52 x Mo17	[1]
	bnlg1953	170.0	B73		7.9**		6.3***	9.3***	120.0***	0.7**				None	
4.07	bnlg1621	349.6	B73		5.0*	3.0**	4.9***	9.4***	91.8***					D32 x D145	[2]
5.02	bnlg565	150.9	B73		4.6*	2.7***	4.9***	10.0***	90.7***					B52 x Mo17	[1, 3]
7.03	bnlg434	323.3	B73		6.6*	3.8***	4.7***	4.1*	84.0***					B52 x Mo17	[1, 3]
8.03 [d]	UMC32B	199.1	B73		8.8***	12.9***	23.5***		246.0***	1.3**			13.2**	None	
Correlated loci															
1.06 [c]	umc2234	529.0	Tx303							-1.1*			-12.2*	D32 x D145	[2]
5.08 [d]	umc1225	641.4	Tx303	1.6**	-18.6**					-2.0**	-1.9***	-1.2**	-31.9***	None	
	bnlg1695	664.2	Tx303	1.6**	-18.6**					-2.0**	-1.9***	-1.2**	-31.9***	None	
5.09 [d]	umc1829	671.5	Tx303	1.6**	-18.6**					-2.0**	-1.9***	-1.2**	-31.9***	None	
Correlated loci															
4.03	umc2082	141.6	B73		11.6***		7.4***	12.4***	132.9***	0.6*	0.4*		7.4*	D32 x D145	[2]
	umc2176	174.6	B73		11.6***		7.4***	12.4***	132.9***	0.6*	0.4*		7.4*		
5.04	bnlg1208	323.1	B73	-0.6*	15.4***	3.3**	6.3***	7.4**	103.8***	0.8*	0.5*	0.4*	10.2**	Lo951 x CML202	[4, 5]
														D32 x D145	[2]
Correlated loci															
1.11 [d]	bnlg131	1065.6	Tx303							-1.5***	-0.9*		-16.5**	IL731a x W6786	[6]
														Lo951 x CML202	[4, 5]
3.06 [d]	bnlg2241	452.7	Tx303							-1.5***	-0.9*		-16.5**	IL731a x W6786	[6]
														B52 x Mo17	[1]
														Lo951 x CML202	[4, 5]

[a] The marker position was based on genetic map of the intermated B73 x Mo17 population (version IBM 2008 neighbors).
[b] The mapping population in which the same locus was detected for resistance to NLB. The resistance donor was underlined.
[c] This marker was first found highly associated with NLB resistance in segregating populations B73 x TBBC3-38 and B73 x TBBC3-39. It was then tested in 82 TBBC3 lines.
[d] Introgressed segment at the locus was only present in a single TBBC line (bin 8.03 in TBBC3-42; bin 5.08-5.09 in TBBC3-38; bin 1.11 and 3.06 in TBBC3-30).

Relative area under the disease progress curve (Relative AUDPC) values shown are the differences of least squares means (from mixed models) between TBBC3 lines and B73 recurrent parent. AUDPC was calculated from three diseased leaf area (DLA) scores in the 2006 trial in NY (solid bars), or three disease severity scores in the 2006 trial in NC (open bars). In NY, primary DLA was also rated for diseased leaf area on inoculated leaves. The letters "R" and "S" below the graph indicate the lines significantly more resistant and more susceptible than B73 at $P < 0.05$, respectively, based on primary DLA and AUDPC. The 15 TBBC3 lines selected for subsequent phenotypic validation are indicated by rectangles highlighting the maize line designation.

Putative QTL for northern leaf blight (NLB QTL) affecting incubation period (IP), primary diseased leaf area (PrimDLA), diseased leaf area (DLA), disease severity (severity), and AUDPC (area under the disease progress curve calculated from DLA or disease severity) were identified using 82 TBBC3 introgression lines. QTL effects for each marker locus are the significant differences of least squares means of Tx303 homozygous genotypes at the locus relative to B73 recurrent parent line (* $0.01 <P < 0.05$, ** $0.001 <P < 0.01$, *** $P < 0.001$). Putative QTL are reported as correlated groups because of the high dependencies among those introgressed segments in TBBC3 lines.

Additional file 3: Validation of NLB QTL in the BC$_4$F$_2$ segregating populations.

Introgressed region (Chr. bin)[a]	Map interval (cM)[b]	SSR marker (map position, cM)[b]	Parental donor lines of the BC₄F₂ population[c]	Environment (location, year)	Resistance allele	R²	Relative allele effect (LSMean Tx303 – LSMean B73)[d]					
							IP (dpi)	LE (mm/day)	DLA1 (%)	DLA2 (%)	DLA3 (%)	AUDPC (%-day)
1.06 [§]	487.5-590.5	umc2234 (529.0)	TBBC3-38	Aurora NY, 2006	Tx303	0.330[f]	1.96**	-0.29*	-6.36***	-6.69***	-5.31***	-119.27***
1.06 [§]	487.5-590.5	umc2234 (529.0)	TBBC3-39	Aurora NY, 2006	Tx303	0.526[f]	2.38***	-0.33**	-5.79***	-8.39***	-9.54***	-151.70***
1.01-1.02 [§]	0-184.2	bnlg1953 (170.0)	TBBC3-42	Greenhouse, 2006	B73	0.139[e]	-0.60***	0.35*	na	na	na	na
1.01-1.02 [§]	0-184.2	bnlg1953 (170.0)	TBBC3-77	Aurora NY, 2006	B73	0.074[f]	-0.58	0.11	6.05***	3.36*	1.42	70.95*
5.00	0-30.0	mmc0151 (20.5)	TBBC3-77	Aurora NY, 2006	B73	0.059[f]	-0.71	0.26	4.30**	4.37**	3.04*	80.37**
5.01-5.02	115.1-153.9	bnlg565 (150.9)	TBBC3-02	Aurora NY, 2006	B73	0.053[f]	-1.30**	0.03	1.98	0.87	0.52	21.13
5.08-5.09	603.0-676.1	bnlg1829 (313.3)	TBBC3-38	Aurora NY, 2006	Tx303	0.046[f]	0.71	0.01	-2.02	-2.45*	-3.67*	-49.85**
8.03-8.05	265.2-344.0	umc1130 (330.1)	TBBC3-02	Aurora NY, 2006	Tx303	0.114[f]	0.41	-0.04	-2.97**	-3.90***	-3.02***	-68.90***

[§] The QTL effects on NLB resistance were further confirmed in derived BC₄F₃ and/or BC₄F₄ NILs.
[a] Introgressed regions that were significantly associated with traits in segregating populations. The genome regions are shown as bin positions (eg. bin 1.06 is the sixth segment in maize chromosome 1).
[b] SSR marker with the most significant effects on traits. The marker position was based on genetic map of the intermated B73 x Mo17 population (version IBM 2008 neighbors). The map interval of each introgression is assumed to extend halfway between two markers around the border of each end of the introgression.
[c] The selected TBBC3 lines used for developing the segregating BC₄F₂ populations. The lines were backcrossed to B73 and selfed. Phenotyping and genotyping were conducted on individual plants in the populations.
[d] The significance level was determined by pair-wise two-tailed Student's t test on the least squares difference (denoted as * $0.01 < P < 0.05$; ** $0.001 < P < 0.01$; *** $P < 0.001$).
[e] R-square from the ANOVA using IP as the dependent variable, referring to the proportion of the variation in IP accounted for by the marker in the third column.
[f] R-square from the ANOVA using AUDPC as the dependent variable, referring to the proportion of the variation in AUDPC accounted for by the marker in the third column.

The BC$_4$F$_2$ populations were genotyped and phenotyped for incubation period (IP), lesion expansion (LE), diseased leaf area (DLA), and area under the disease progress curve (AUDPC). The trait-marker association was tested by ANOVA at $P < 0.05$. Introgressions/markers significantly associated with NLB resistance are listed. The relative allele effects are the differences on the least squares means (LSMean) between Tx303 homozygous genotypes and B73 homozygous genotypes at the locus.

Bin 1.03: bnlg439
Bin 1.06: umc1754, umc1665, umc1949, umc1665
Bin 3.02: bnlp647
Bin 5.00: mmc0151
Bin 5.03: umc1357
Bin 5.07: bnlg1306
Bin 5.08: umc1225
Bin 5.09: bnlg1693, mc1829

		Northern leaf blight							Stewart's wilt	Anthracnose stalk rot	Common smut		Common rust	Anthesis
		IP	LE	DLA1	DLA2	DLA3	DLA4	AUDPC	PrimDLA	Discolored internode area	Ear gall	Stalk gall	AUDPC	Days to anthesis
	TBBC3-38	17 A	1.8 DEFG	10 H	13 I	18 IJ	20 GH	462 J	0 N	174 A	0 BC	0 B	53 H	78 B
	TBBC3-38-11E	15 ABC	1.5 GH	15 DHFG	14 HI	19 I	20 GH	501 IJ	16 JKL				64 DEFGH	77 BCDEFGH
	TBBC3-38-05F	15 BCD	1.4 H	10 H	13 I	19 IJ	25 GH	504 IJ	16 KL				70 DEFGH	75 EFGHI
	TBBC3-38-18A	14 CDEF	1.6 FGH	11 GH	18 FGH	25 EFGH	28 F	637 H	28 IJK				55 GH	78 BCD
	TBBC3-38-07H	15 BCD	1.6 FGH	13 DEFGH	22 DEFG	28 DEFG	38 CD	748 EFG	30 HIJ				59 FGH	76 CDEFGH
	TBBC3-38-16G	13 DEFG	2.0 BCD	21 BC	28 C	29 DEF	34 DE	840 DE	48 DEF	141 CD	1.1 BC	0 B	74 DEF	77 BC
	B73	13 F	1.5 H	15 DEF	21 EF	26 GH	33 E	710 G	44 F	142 D	0 C	0 B	61 GH	74 I
	Tx303	14 BCDE	2.0 BCDE	12 EFGH	14 HI	18 J	16 H	427 J	13 KLM	97 E	2.5 A	1.6 A	83 CD	87 A
	TBBC3-39	15 BCD	1.9 CDEF	12 FGH	18 GHI	24 H	29 F	616 H	3 MN	148 CD	0.1 BC	0 B	57 GH	78 B
	TBBC3-39-19E	16 AB	1.6 FGH	13 EFGH	18 FG	19 IJ	23 G	558 HI	5 LMN	141 CD	0.1 BC	0 B	67 DEFGH	75 FGH
	TBBC3-39-08F	14 BCD	1.6 FGH	13 DEFGH	18 GHI	25 FGH	29 F	634 H	16 JKL				61 FGH	76 CDEFGH
	TBBC3-39-11A	12 FGHIJ	1.7 DEFGH	17 CD	26 CD	29 DE	35 CDE	813 DEF	11 KLMN				61 FGH	76 CDEFGH
	TBBC3-39-09F	12 GHIJ	1.8 DEFG	16 DE	27 C	35 C	38 CD	885 D	35 GH	155 ABCD			64 EFGH	75 H
	B73	13 F	1.5 H	15 DEF	21 EF	26 GH	33 E	710 G	44 F	142 D	0 C	0 B	61 GH	74 I
	Tx303	14 BCDE	2.0 BCDE	12 EFGH	14 HI	15 J	16 H	427 J	13 KLM	97 E	2.5 A	1.6 A	83 CD	87 A

Chr.1   Chr.3   Chr.5

Additional file 4: Genotypes and disease phenotypes for Tx303, B73 and the NIL sets derived from B73 × TBBC3-38 and B73 × TBBC3-39.

Among the target introgressions at bins 1.03, 1.06, 3.02, 5.00, 5.02-5.03 and 5.07-5.09, only $qNLB1.06_{Tx303}$ (Tx303 allele at bin 1.06) was validated for association with resistance to NLB. The open bars and solid bars represent the loci homozygous for B73 alleles and Tx303 alleles, respectively. The gray bars represent heterozygous loci or missing genotypic data. Only the chromosomes with introgressed regions in the two NIL sets are shown. The rest of the genome was assumed fixed for B73 alleles. Trait values are least squares means calculated from the mixed model. Pair-wise Student›s t tests were performed to analyze the differences between each NIL and B73, and between every pair of NILs in each set. Trait values with different letters are significantly different from each other. Disease phenotypes that were significantly more resistant than B73 are highlighted in bold and shaded, while the phenotypes significant more susceptible than B73 are underscored. Lines that showed significantly different days to anthesis are in bold italic. $qNLB1.06_{Tx303}$ was also effective for resistance to Stewart›s wilt. (IP: incubation period; LE: lesion expansion; DLA: diseased leaf area; PrimDLA: primary DLA; AUDPC: area under the disease progress curve) (PDF 91 KB)

	Northern leaf blight							Stewart's wilt	Anthracnose stalk rot	Common smut		Common rust	Anthesis
	IP	LE	DLA1	DLA2	DLA3	DLA4	AUDPC	PrimDLA	Discolored internode area	Ear gall	Stalk gall	AUDPC	Days to anthesis
TBBC3-42	11 HIJ	2.5 A	30 A	45 A	57 A	67 A	1503 A	73 A	154 ABCD	0.5 BC	0 B	106 A	76 CDEFG
TBBC3-42-10E-02	11 HIJ	2.3 AB	33 A	46 A	55 A	71 A	1525 A	63 BC	139 CD			104 AB	77 BCDE
TBBC3-42-02E-01	11 IJ	2.2 ABC	31 A	45 A	56 A	71 A	1524 A	71 AB	167 ABC			115 A	76 CDEFG
TBBC3-42-06F-02	11 J	2.5 A	25 B	39 B	55 A	69 A	1406 B	58 CD	172 AB			84 BCDE	76 CDEFGH
TBBC3-42-12F-01	12 HIJ	2.1 ABCD	25 B	38 B	49 B	60 B	1287 C	52 DE	148 ABCD			100 ABC	76 CDEFGH
TBBC3-42-10E-04										1.2 B	0.1 B	111 A	77 BCDEF
TBBC3-42-05H-04	12 GHIJ	1.5 GH	15 DEFG	24 CDE	30 D	40 C	814 DEF	48 DEF	145 BCD			63 DEFGH	75 FGH
TBBC3-42-05H-01	13 FGHI	1.5 FGH	15 DEFG	21 EFG	31 CD	37 CD	781 EF	30 HI	146 CD			74 DEFG	76 DEFGH
TBBC3-42-09E-01	13 EFGHI	1.6 EFGH	15 DEFG	22 DEFG	27 DEFGH	36 CDE	743 FG	45 EFG	142 CD			65 EFGH	75 GH
B73	13 F	1.5 H	15 DEF	21 EF	26 GH	33 E	710 G	44 F	142 D	0 C	0 B	61 GH	74 I
Tx303	14 BCDE	2.0 BCDE	12 EFGH	14 HI	15 J	16 H	427 J	13 KLM	97 E	2.5 A	1.6 A	83 CD	87 A

Bin 1.01: umc1071
Bin 1.02: umc1568, bnlg1953
Bin 4.06: bnlg2291
Bin 4.06: bnlg1621
Bin 5.02-5.03: bnlg365
Bin 7.01: umc0171
Bin 8.02: bnlg2082
Bin 8.03-8.05: umc1130

Chr.1 Chr.4 Chr.5 Chr.7 Chr.8

Additional file 5: Genotypes and disease phenotypes for Tx303, B73 and the NIL set derived from B73 × TBBC3-42.

Among all target introgressions at bins 1.01-1.02, 4.06-4.07, 5.02-5.03, 7.01, 8.02 and 8.03-8.05, only $qNLB1.02_{B73}$ (B73 allele at bin 1.02) was validated for association with resistance to NLB. The open bars and solid bars represent the loci homozygous for B73 alleles and Tx303 alleles, respectively. The gray bars represent heterozygous loci or missing genotypic data. Only the chromosomes with introgressed regions in the NIL sets are shown. The rest of the genome was assumed fixed for B73 alleles. Trait values are least squares means calculated from the mixed model. Pair-wise Student›s t tests were performed to analyze the differences between each NIL and B73, and between every pair of NILs. Trait values with different letters are significantly different from each other. Disease phenotypes that were significantly more resistant than B73 are highlighted in bold and shaded, while the phenotypes significant more susceptible than B73 are underscored. Lines that showed significantly different days to anthesis are in bold italic. Preliminary evidence also suggested that $qNLB1.02_{B73}$ was effective for resistance to Stewart›s wilt, and QTL at bin 1.01-1.02 and/or 5.02-5.03 were associated with resistance to common rust. (PDF 85 KB)

CONCLUSIONS

Our research has led to successful identification of two reliably-expressed QTL that can potentially be utilized to protect maize from *S. turcica* in different environments. Map-based cloning will reveal more about the genes and mechanisms underlying the distinct features of $qNLB1.02_{B73}$ and $qNLB1.06_{Tx303}$ in the pre-penetration, penetration and post-penetration phases of pathogenesis. Large mapping populations have been generated from the NILs and fine-mapping of $qNLB1.02_{B73}$ and $qNLB1.06_{Tx303}$ is in progress.

ACKNOWLEDGEMENTS

We thank Jim Holland for providing the seeds of TBBC3 lines, Jesse Poland for help with rust inoculation, and Pioneer Hi-Bred International Inc. for the supply of sorghum grains used in production of inoculum. We appreciate help from Judith Kolkman, Oliver Ott and Kristen Kennedy with aspects in the research. The work was funded by The CGIAR Generation Challenge Program, The McKnight Foundation, USDA-ARS, and Ministry of Education, Taiwan.

AUTHORS' CONTRIBUTIONS

CC and RN conceived of the study. CC designed and carried out all the experiments in New York, conducted statistical analysis, and drafted the manuscript. JL and EW contributed to the design of microscopic analysis, participated in phenotypic data collection in the greenhouse and field trials in NY during 2007-2008. ZK participated in the inoculation and phenotypic data collection in the field trials in NY in 2008. GVE carried out the field trials in NC. PBK designed and carried out the field trials in NC, participated in coordination of the study, and contributed to its revision. RN participated in the coordination of the work and the experimental design, helped to draft the manuscript and contributed to its revision. All authors read and approved the final manuscript.

REFERENCES

1.   Jones JDG, Dangl JL: The plant immune system. *Nature* 2006,444(7117):323–329.

2.   Wisser RJ, Sun Q, Hulbert SH, Kresovich S, Nelson RJ: Identification and characterization of regions of the rice genome associated with broad-spectrum, quantitative disease resistance. *Genetics* 2005,169(4):2277–2293.

3.   Manosalva PM, Davidson RM, Liu B, Zhu X, Hulbert SH, Leung H, Leach JE: A germin-Like protein gene family functions as a complex quantitative trait locus conferring broad-spectrum disease resistance in rice. *Plant Physiol* 2009,149(1):286–296.

4.   Krattinger SG, Lagudah ES, Spielmeyer W, Singh RP, Huerta-Espino J, McFadden H, Bossolini E, Selter LL, Keller B: A putative ABC transporter confers durable resistance to multiple fungal pathogens in wheat. *Science* 2009,323(5919):1360–1363.

5. Fukuoka S, Saka N, Koga H, Ono K, Shimizu T, Ebana K, Hayashi N, Takahashi A, Hirochika H, Okuno K, Yano M: Loss of function of a proline-containing protein confers durable disease resistance in rice. *Science* 2009, 325:998–1001.

6. Poland JA, Balint-Kurti PJ, Wisser RJ, Pratt RC, Nelson RJ: Shades of gray: the world of quantitative disease resistance. *Trends Plant Sci*2009,14(1):21–29.

7. Fu D, Uauy C, Distelfeld A, Blechl A, Epstein L, Chen X, Sela H, Fahima T, Dubcovsky J: A Kinase-START gene confers temperature-dependent resistance to wheat stripe rust. *Science* 2009,323(5919):1357–1360.

8. Parlevliet J: Durability of resistance against fungal, bacterial and viral pathogens; present situation. *Euphytica* 2002, 124:147–156.

9. Young ND: QTL mapping and quantitative disease resistance in plants. *Annu Rev Phytopathol* 1996, 479–501.

10. Broglie KE, Butler KH, Butruille MG, da Silva Conceicao A, Frey TJ, Hawk JA, Jaqueth JS, Jones ES, Multani DS, Wolters PJCC, E.I. du Pont de Nemours and Company, Pioneer Hi-Bred International, Inc., University of Delaware: Polynucleotides and methods for making plants resistant to fungal pathogens. United States Patent 20060223102. *United States Patent and Trademark Office* 2006.

11. Carson ML: Response of a maize synthetic to selection for components of partial resistance to *Exserohilum turcicum* . *Plant Dis*2006,90(7):910–914.

12. Vidhyasekaran P: *Fungal Pathogenesis in Plants and Crops: Molecular Biology and Host Defense Mechanisms.* 2nd edition. Boca Raton, FL: CRC Press; 2007.

13. Moldenhauer J, Pretorius ZA, Moerschbacher BM, Prins R, Van Der Westhuizen AJ: Histopathology and PR-protein markers provide insight into adult plant resistance to stripe rust of wheat. *Mol Plant Pathol* 2008,9(2):137–145.

14. Ullstrup AJ, Miles SR: The effects of some leaf blights of corn on grain yield. *Phytopathology* 1957,47(6):331–336.

15. Raymundo AD, Hooker AL: Measuring the relationship between northern corn leaf blight and yield losses. *Plant Dis* 1981,65(4):325–327.

16. Perkins JM, Pedersen WL: Disease development and yield losses associated with northern leaf blight on corn. *Plant Dis*1987,71(10):940–943.

17. Pingali PL, Pandey S: Meeting world maize needs: technological

opportunities and priorities for the public sector. In *CIMMYT 1999/2000 World Maize Facts and Trends*. Edited by: Pingali PL, Mexico DF. CIMMYT; 2001.

18. Carson ML: Helminthosporium leaf spots and blights. In *Compendium of Corn Diseases*. 3rd edition. Edited by: White DG. St. Paul, Minnesota: The American Phytopathology Society; 1999:15–24.

19. Welz HG, Geiger HH: Genes for resistance to northern corn leaf blight in diverse maize populations. *Plant Breed* 2000,119(1):1–14.

20. Wisser RJ, Balint-Kurti PJ, Nelson RJ: The genetic architecture of disease resistance in maize: a synthesis of published studies. *Phytopathology* 2006,96(2):120–129.

21. Pratt RC, Gordon SG: Breeding for resistance to maize foliar pathogens. *Plant Breed Rev* 2006, 27:119–173.

22. Hilu HM, Hooker AL: Host-pathogen relationship of *Helminthosporium turcicum* in resistant and susceptible corn seedlings. *Phytopathology* 1964,54(5):570–575.

23. Hilu HM, Hooker AL: Localized infection by *Helminthosporium turcicum* on corn leaves. *Phytopathology* 1965,55(2):189–192.

24. Jennings PR, Ullstrup AJ: A histological study of three Helminthosporium leaf blights of corn. *Phytopathology* 1957,47(12):707–714.

25. Knox-Davies PS: Penetration of maize leaves by *Helminthosporium turcicum* . *Phytopathology* 1974,64(11):1468–1470.

26. Stuber CW, Polacco M, Senior ML: Synergy of empirical breeding, marker-assisted selection, and genomics to increase crop yield potential. *Crop Sci* 1999,39(6):1571–1583.

27. Szalma SJ, Hostert BM, LeDeaux JR, Stuber CW, Holland JB: QTL mapping with near-isogenic lines in maize. *Theor Appl Genet*2007,114(7):1211–1228.

28. Li Z-K, Fu B-Y, Gao Y-M, Xu J-L, Ali J, Lafitte HR, Jiang Y-Z, Rey JD, Vijayakumar CHM, Maghirang R, Zheng T-Q, Zhu L-H: Genome-wide introgression lines and their use in genetic and molecular dissection of complex phenotypes in rice (*Oryza sativa* L.). *Plant Mol Biol*2005,59(1):33–52.

29. Toojinda T, Baird E, Booth A, Broers L, Hayes P, Powell W, Thomas W, Vivar H, Young G: Introgression of quantitative trait loci (QTLs) determining stripe rust resistance in barley: an example of marker-assisted line development. *Theor Appl Genet* 1998,96(1):123–131.

30. Brown AHD, Munday J, Oram RN: Use of isozyme-marked segments

from wild barley (*Hordeum spontaneum*) in barley breeding.*Plant Breed* 1988,100(4):280–288.

31. Eshed Y, Zamir D: An introgression line population of *Lycopersicon pennellii* in the cultivated tomato enables the identification and fine mapping of yield-associated QTL. *Genetics* 1995,141(3):1147–1162.

32. Keurentjes JJB, Bentsink L, Alonso-Blanco C, Hanhart CJ, Vries HB-D, Effgen S, Vreugdenhil D, Koornneef M: Development of a near-isogenic line population of *Arabidopsis thaliana* and comparison of mapping power with a recombinant inbred line population.*Genetics* 2007,175(2):891–905.

33. Kaeppler SM: Quantitative trait locus mapping using sets of near-isogenic lines: relative power comparisons and technical considerations. *Theor Appl Genet* 1997,95(3):384–392.

34. Remington DL, Purugganan MD: Candidate genes, quantitative trait loci, and functional trait evolution in plants. *Int J Plant Sci*2003,164(3 Supplement):S7-S20.

35. Whalen MC: Host defence in a developmental context. *Mol Plant Pathol* 2005,6(3):347–360.

36. Kim KD, Hwang BK, Koh YJ: Evaluation of rice cultivars under greenhouse conditions for adult-plant resistance to *Pyricularia oryzae*. *J Phytopathol* 1987,120(4):310–316.

37. Century KS, Lagman RA, Adkisson M, Morlan J, Tobias R, Schwartz K, Smith A, Love J, Ronald PC, Whalen MC: Developmental control of*Xa21* -mediated disease resistance in rice. *Plant J* 1999,20(2):231–236.

38. Kus JV, Zaton K, Sarkar R, Cameron RK: Age-related resistance in *Arabidopsis* is a developmentally regulated defense response to*Pseudomonas syringae* . *Plant Cell* 2002,14(2):479–490.

39. Collins A, Milbourne D, Ramsay L, Meyer R, Chatot-Balandras C, Oberhagemann P, De Jong W, Gebhardt C, Bonnel E, Waugh R: QTL for field resistance to late blight in potato are strongly correlated with maturity and vigour. *Mol Breed* 1999,5(5):387–398.

40. Kumar JR, Kumar BT: Quantitative trait loci (QTL) mapping for crop improvement. *Res J Biotech* 2009,4(2):67–79.

41. Leonard KJ, Levy Y, Smith DR: Proposed nomenclature for pathogenic races of *Exserohilum turcicum* on corn. *Plant Dis*1989,73(9):776–777.

42. Vélez H: *Alternaria alternata* mannitol metabolism in plant-pathogen interactions. *PhD thesis*. North Carolina State University, Department of Plant Pathology; 2005.

43. Hood ME, Shew HD: Applications of KOH-aniline blue fluorescence in the study of plant-fungal interactions. *Phytopathology*1996,86(7):704–708.

44. Qi M, Yang Y: Quantification of *Magnaporthe grisea* during infection of rice plants using real-time polymerase chain reaction and northern blot/phosphoimaging analyses. *Phytopathology* 2002,92(8):870.

45. Gaurilcikiene I, Deveikyte I, Petraitiene E: Epidemic progress of *Cercospora beticola* Sacc. in *Beta vulgaris* L. under different conditions and cultivar resistance. *Biologija* 2006, 4:54–59.

46. Pataky JK: Influence of host resistance and growth stage at the time of inoculation on Stewart›s wilt and Goss›s wilt development and sweet corn hybrid yield. *Plant Dis* 1989,73(4):339–345.

47. Blanco MH, Johnson MG, Colbert TR, Zuber MS: An inoculation technique for Stewart›s wilt disease of corn. *Plant Dis Rep*1977,61(5):413–416.

48. Chang C-M, Hooker AL, Lim SM: An inoculation technique for determining Stewart's bacterial leaf blight reaction in corn. *Plant Dis Rep* 1977,61(12):1077–1079.

49. Keller NP, Bergstrom GC: Development predisposition of maize to anthracnose stalk rot. *Plant Dis* 1988,72(11):977–980.

50. Muimba-Kankolongo A, Bergstrom GC: Transitory wound predisposition of maize to Anthracnose stalk rot. *Can J Plant Pathol*1990,12(1):1–10.

51. du Toit LJ, Pataky JK: Variation associated with silk channel inoculation for common smut of sweet corn. *Plant Dis* 1999,83(8):727–732.

52. Webb CA, Richter TE, Collins NC, Nicolas M, Trick HN, Pryor T, Hulbert SH: Genetic and molecular characterization of the maize *rp3*rust resistance locus. *Genetics* 2002,162(1):381–394.

53. Pataky JK, Campana MA: Reduction in common rust severity conferred by the *Rp1 D* gene in sweet corn hybrids infected by mixtures of *Rp1D* -virulent and avirulent *Puccinia sorghi* . *Plant Dis* 2007,91(11):1484–1488.

54. Doyle JJ, Doyle JL: A rapid DNA isolation procedure for small quantities of fresh leaf tissue. *Phytochem Bull* 1987, 19:11–15.

55. Qiu F, Wang H, Chen J, Zhuang J, Hei L, Cheng S, Wu J: A rapid DNA mini-prep method for large-scale rice mutant screening. *Rice Sci*2006,13(4):299–302.

56. Schuelke M: An economic method for the fluorescent labeling of PCR fragments. *Nat Biotechnol* 2000,18(2):233–234.

57. Wisser RJ, Murray SC, Kolkman JM, Ceballos H, Nelson RJ: Selection mapping of loci for quantitative disease resistance in a diverse maize

population. *Genetics* 2008,180(1):583–599.

58. Patterson HD, Williams ER: A new class of resolvable incomplete block designs. *Biometrika* 1976,63(1):83–92.

59. van Dam J, Levin I, Struik PC, Levy D: Identification of epistatic interaction affecting glycoalkaloid content in tubers of tetraploid potato (*Solanum tuberosum* L.). *Euphytica* 2003,134(3):353–360.

60. Gonzalez J, Reyes F, Salas C, Santiago M, Codriansky Y, Coliheuque N, Silva H: *Arabidopsis thaliana* : A model host plant to study plant-pathogen interaction using Chilean field isolates of *Botrytis cinerea* . *Biol Res* 2006,39(2):221–228.

61. Johnston PR, Sutherland PW, Joshee S: Visualising endophytic fungi within leaves by detection of (1->3)-beta-D-glucans in fungal cell walls. *Mycologist* 2006,20(4):159–162.

62. Soylu S: Accumulation of cell-wall bound phenolic compounds and phytoalexin in *Arabidopsis thaliana* leaves following inoculation with pathovars of *Pseudomonas syringae* . *Plant Sci* 2006,170(5):942–952.

63. Ficke A, Gadoury DM, Seem RC: Ontogenic resistance and plant disease management: A case study of grape powdery mildew. *Phytopathology* 2002,92(6):671–675.

64. Bennett M, Gallagher M, Fagg J, Bestwick C, Paul T, Beale M, Mansfield J: The hypersensitive reaction, membrane damage and accumulation of autofluorescent phenolics in lettuce cells challenged by *Bremia lactucae* . *Plant J* 1996,9(6):851–865.

65. Hooker AL: Inheritance of chlorotic-lesion resistance to *Helminthosporium turcicum* in seedling corn. *Phytopathology* 1963,53(6):660–662.

66. Hooker AL: A second major gene locus in corn for chlorotic lesion resistance to *Helminthosporium turcicum* . *Crop Sci* 1977,17(1):132–135.

67. Hooker AL: Resistance to *Helminthosporium turcicum* from *Tripsacum floridanum* incorporated into corn. *Maize Genet Coop Newsl* 1981, 55:87–88.

68. Ogliari JB, Guimaraes MA, Geraldi IO, Aranha Camargo LE: New resistance genes in the *Zea mays - Exserohilum turcicum* pathosystem. *Genet Mol Biol* 2005,28(3):435–439.

69. Ogliari JB, Guirnaraes MA, Aranha Carnargo LE: Chromosomal locations of the maize (*Zea mays* L.) *HtP* and *rt* genes that confer resistance to *Exserohilum turcicum* . *Genet Mol Biol* 2007,30(3):630–634.

70. Carson ML: A new gene in maize conferring the «chlorotic halo» reaction to infection by *Exserohilum turcicum* . *Plant Dis* 1995,79(7):717–720.

71. Gevers HO: A new major gene for resistance to *Helminthosporium turcicum* leaf blight of maize. *Plant Dis Rep* 1975,59(4):296–299.

72. Freymark PJ, Lee M, Woodman WL, Martinson CA: Quantitative and qualitative trait loci affecting host-plant response to*Exserohilum turcicum* in maize (*Zea mays* L.). *Theor Appl Genet* 1993,87(5):537–544.

73. Welz HG, Xia XC, Bassetti P, Melchinger AE, Luebberstedt T: QTLs for resistance to *Setosphaeria turcica* in an early maturing Dent × Flint maize population. *Theor Appl Genet* 1999,99(3–4):649–655.

74. Carson ML: Inheritance of latent period length in maize infected with *Exserohilum turcicum* . *Plant Dis* 1995,79(6):581–585.

75. Smith DR, Kinsey JG: Latent period-a possible selection tool for*Exserohilum turcicum* resistance in corn (*Zea mays* L.). *Maydica*1993, 38:205–208.

76. Welz HG, Schechert AW, Geiger HH: Dynamic gene action at QTLs for resistance to *Setosphaeria turcica* in maize. *Theor Appl Genet*1999,98(6–7):1036–1045.

77. Schechert AW, Welz HG, Geiger HH: QTL for resistance to *Setosphaeria turcica* in tropical African maize. *Crop Sci* 1999,39(2):514–523.

78. Brewster VA, Carson ML, Wicks ZW III: Mapping components of partial resistance to northern leaf blight of maize using reciprocal translocation. *Phytopathology* 1992,82(2):225–229.

79. Tucker SL, Talbot NJ: Surface attachment and pre-penetration stage development by plant pathogenic fungi. *Annu Rev Phytopathol*2001, 385–417.

80. Bubeck DM, Goodman MM, Beavis WD, Grant D: Quantitative trait loci controlling resistance to gray leaf spot in maize. *Crop Sci*1993,33(4):838–847.

81. Leonard KJ, Thompson DL: Effects of temperature and host maturity on lesion development of *Colletotrichum graminicola* on corn. *Phytopathology* 1976,66(5):635–639.

82. Sofi P, Rather AG: QTL analysis in rice improvement: concept, methodology and application. *Biotechnology* 2007,6(1):1–13.

83. Trognitz BR, Orrillo M, Portal L, Roman C, Ramon P, Perez S, Chacon G: Evaluation and analysis of reduction of late blight disease in a diploid potato progeny. *Plant Pathol* 2001,50(3):281–291.

84. Dinh SQ, Joyce DC, Irving DE, Wearing AH: Effects of combined methyl jasmonate and ethylene-inhibitor treatments against *Botrytis*

cinerea infecting Geraldton waxflower. *Acta Hort (ISHS)* 2007, 755:527–533.

85. Balint-Kurti PJ, Wisser R, Zwonitzer JC: Use of an advanced intercross line population for precise mapping of quantitative trait loci for gray leaf spot resistance in maize. *Crop Sci* 2008,48(5):1696–1704.

86. Mitchell-Olds T, James RV, Palmer MJ, Williams PH: Genetics of *Brassica rapa* (syn. *campestris*). 2. Multiple disease resistance to three fungal pathogens. *Peronospora parasitica* , *Albugo candida* and *Leptosphaeria maculans* . *Heredity* 1995,75(4):362–369.

87. Asai T, Tena G, Plotnikova J, Willmann MR, Chiu W-L, Gomez-Gomez L, Boller T, Ausubel FM, Sheen J: MAP kinase signalling cascade in *Arabidopsis* innate immunity. *Nature* 2002,415(6875):977–983.

88. Durrant WE, Dong X: Systemic acquired resistance. *Annu Rev Phytopathol* 2004, 42:185–209.

89. Yang Y, Shah J, Klessig DF: Signal perception and transduction in plant defense responses. *Gene Dev* 1997,11(13):1621–1639.

90. Mou Z, Fan W, Dong X: Inducers of plant systemic acquired resistance regulate NPR1 function through redox changes. *Cell* 2003,113(7):935–944.

91. Consonni C, Humphry ME, Hartmann HA, Livaja M, Durner J, Westphal L, Vogel J, Lipka V, Kemmerling B, Schulze-Lefert P, Somerville SC, Panstruga R: Conserved requirement for a plant host cell protein in powdery mildew pathogenesis. *Nat Genet* 2006,38(6):716–720.

92. Pataky JK, Bohn MO, Lutz JD, Richter PM: Selection for quantitative trait loci associated with resistance to Stewart)s wilt in sweet corn. *Phytopathology* 2008,98(4):469–474.

93. Ming R, Brewbaker JL, Moon HG, Musket TA, Holley RN, Pataky JK, McMullen MD: Identification of RFLP makers linked to a major gene, *sw1* , conferring resistance to Stewart)s wilt in maize. *Maydica* 1999, 44:319–323.

94. Freymark PJ, Lee M, Martinson CA, Woodman WL: Molecular-marker-facilitated investigation of host-plant response to *Exserohilum turcicum* in maize (*Zea mays* L.): components of resistance. *Theor Appl Genet* 1994,88(3–1):305–313.

95. Carson ML, Van Dyke CG: Effect of light and temperature on expression of partial resistance of maize to *Exserohilum turcicum* .*Plant Dis* 1994,78(5):519–522.

96. Sigulas KM, Hill RRJ, Ayers JE: Genetic analysis of *Exserohilum turcicum* lesion expansion on corn. *Phytopathology* 1988,78(2):149–153.

97. Brown AF, Juvik JA, Pataky JK: Quantitative trait loci in sweet corn associated with partial resistance to Stewart›s wilt, northern corn leaf blight, and common rust. *Phytopathology* 2001,91(3):293–300.

98. Balint-Kurti PJ, Zwonitzer JC, Wisser RJ, Carson ML, Oropeza-Rosas MA, Holland JB, Szalma SJ: Precise mapping of quantitative trait loci for resistance to southern leaf blight, caused by *Cochliobolus heterostrophus* race O, and flowering time using advanced intercross maize lines. *Genetics* 2007,176(1):645–657.

Chapter 2

PROTEOMICS OF PLANT PATHOGENIC FUNGI

Raquel González-Fernández,[1] Elena Prats,[2] and Jesús V. Jorrín-Novo[1]

[1]Agricultural and Plant Biochemistry and Proteomics Research Group, Department of Biochemistry and Molecular Biology, University of Córdoba, 14071 Córdoba, Spain

[2]CSIC, Institute of Sustainable Agriculture, 14080 Córdoba, Spain

ABSTRACT

Plant pathogenic fungi cause important yield losses in crops. In order to develop efficient and environmental friendly crop protection strategies, molecular studies of the fungal biological cycle, virulence factors, and interaction with its host are necessary. For that reason, several approaches have been performed using both classical genetic, cell biology, and biochemistry and the modern, holistic, and high-throughput, omic techniques. This work briefly overviews the tools available for studying Plant Pathogenic Fungi and is amply focused on MS-based Proteomics analysis, based on original papers published up to December 2009. At a methodological level, different steps in a proteomic workflow experiment are discussed. Separate sections are devoted to fungal descriptive (intracellular, subcellular, extracellular) and differential expression proteomics and interactomics. From the work published we can conclude that Proteomics, in combination with other techniques, constitutes a powerful tool for providing important information about pathogenicity and virulence factors, thus opening up new possibilities for crop disease diagnosis and crop protection.

INTRODUCTION: PLANT PARASITIC FUNGI

Fungi form a large and heterogeneous eukaryotic group of living organisms characterized by their lack of photosynthetic pigment and their chitinous cell wall. It has been estimated that the fungal kingdom contains more than 1.5 million species, but only around 100,000 have so far been described, with yeast, mold, and mushroom being the most familiar [1]. Although the majority of fungal species are saprophytes, a number of them are parasitics, in order to complete their biological cycle, animals or plants, with around 15,000 of them causing disease in plants, the majority belonging to the Ascomycetes and Basidiomycetes [2] (Table 1). Within a fungal plant pathogen species, for example, in Fusarium oxysporum, up to 120 different formae specialis can be found based on specificity to host species belonging to a wide range of plant families [3].

Table 1: Main plant pathogenic fungi causing disease in plants

Phylum	Genus	Anamorphic stage	Hosts	Disease	Example
Chytridio-mycota	Olpidium		cabbage	root diseases	O. brassicae
	Physoderma		corn	brown spot	P. maydis
			alfalfa	crown wart	P. (= Urophlyctis) alfalfae
	Synchytrium		potato	potato wart	S. endobioticum
Zygomy-cota	Rhizopus		fruits and vegetables	bread molds and soft rot	R. oligosporus
	Choanephora		squash	soft rot	C. cucurbitarum
	Mucor		fruits and vegetables	bread mold and storage rots	M. indicus
Ascomy-cota	Taphrina		peach plum oak	leaf curl leaf blister and so forth	T. deformans
	Galactomyces		citrus	sour rot	G. candidum
	Blumeria		cereals and grasses	powdery mildew	B. graminis [1]
	Erysiphe	Oidium	many herbaceous plants	powdery mildew	E. pisi
	Leveillula		tomato	powdery mildew	L. taurica
	Microsphaera		lilac	powdery mildew	M. penicullata
	Oidium		tomato	powdery mildew	O. neolycopersici
	Podosphaera		apple	powdery mildew	P. leucotricha
	Sphaerotheca		roses and peach	powdery mildew	S. pannosa
	Uncinula		grape	powdery mildew	U. necator

Nectria		trees	twig and stem cankers	N. galligena
Gibberella		corn and small grains	foot or stalk rot	F. gra-minearum [1]
	Fusarium	several plants	vascular wilts root rots stem rots seed infections	F. oxysporum[1]
Claviceps		grain crops	ergot	C. purpurea
Ceratocystis	Chalara	oak	oak wilt	C. fagacearum
		stone fruit and sweet potato	cankers and root rot	C. fimbriata
		pineapple	butt rot	C. paradoxa
Monosporascus		cucurbits	root rot and collapse	M. cannonbal-lus
Glomerella		apple	anthracnoses and bitter rot	G. cingulata
	Colletotri-chum	many plants	anthracnoses	C. lindemuthi-anum
Phyllachora		grasses	leaf spots	P. graminis
Ophiostoma	Sporo-thrix and Graphium	elm	Dutch elm disease	O. novo-ulm
Diaporthe			citrus melanose	D. citri
			eggplant fruit rot	D. vexans
			soybean pod and stem rot	D. phaseolo-rum
Gaeumannomy-ces		grain crops and grasses	take-all disease	G. graminis
Magnaporthe		rice	rice blast	M. grisea [1]
Cryphonectria		chestnut	blight disease	C. parasitica
Leucostoma		peach and other trees	canker diseases	L. persoonii
Hypoxylon		poplars	canker disease	H. mamma-tum
Rosellinia		fruit trees and vines	root diseases	R. necatrix
Xylaria		trees	tree cankers and wood decay	X. longipes
Eutypa		fruit trees and vines	canker	E. armeniacae

Mycosphaerella	Cercospora	Banana	Sigatoka disease	M. musicola and M. fijiensis
	Septoria	cereals and grasses	leaf spots	M. graminicola
		strawberry	leaf spot	M. fragariae
Elsinoë		citrus trees	citrus scab	E. fawcetti
		grape	anthracnose	E. ampelina
		raspberry	anthracnose	E. veneta
Capnodium		most plants	sooty molds	C. elaeophilum
Cochliobolus	Bipolaris	grain crops and grasses	leaf spots and root rots	C. carbonum and B. maydis
	Curvularia	grasses	leaf spots	C. lunata [1]
Pyrenophora	Drechslera	cereals and grasses	leaf spots	P. graminea
Setosphaera		cereals and grasses	leaf spots	S. turcica
Pleospora	Stemphylium	tomato	black mold rot	P. lycopersici and S. solani
Leptosphaeria		cabbage	black leg and foot rot	L. maculans [1]
Venturia		apple	apple scab	V. inaequalis
		pear	pear scab	V. pyrina
	Cladosporium	tomato	leaf mold	C. fulvum
		peach and almond	scab	C. carpophilum
Guignardia	Phyllosticta	grapes	black rot	G. bidwellii
Apiosporina		cherries and plums	black knot	A. morbosa
Hypoderma		pines	needle cast	H. desmazierii
Lophodermium		pines	needle cast	L. pinastri
Rhabdocline		pines	Douglas fir needle cast	R. weirii
Rhytisma		maple	tar spot of leaves	R. acerinum
Monilinia		stone fruit	brown rot disease	M. fruticola
Sclerotinia		vegetables	white mold	S. sclerotiorum [1]
Stromatinia		gladiolus	corm rot	S. gladioli
Pseudopeziza		alfalfa	leaf spot	P. trifolii

	Diplocarpon		quince and pear	black spot	D. maculatum
	Talaromyces	Penicillium	fruits	blue mold rot	P. digitatum
		Aspergillus	seeds	bread mold and seed decays	A. niger [1]
	Hypocrea	Verticillium	many plants	vascular wilts	V. dahliae [1]
	Lewia	Alternaria	many plants	leaf spots and blights	A. alternata
	Setosphaera	Exserohilum	grasses	leaf spots	E. longirostratum
	Botryosphaeria	Sphaeropsis	apple	black rot	S. pyriputrescens
	Botryotinia	Botrytis	many plants	gray mold rots	B. cinerea [1]
	Monilinia	Monilia	stone fruits	brown rot	M. fruticola
	Diplocarpon	Entomosporium	pear	leaf and fruit spot	E. mespili
	Greeneria	Melanconium	grape	bitter rot	M. fuligineum
Basidiomycota	Ustilago		corn	smut	U. maydis [1]
			oats	loose smuts	U. avenae
			barley	loose smuts	U. nuda
			wheat	loose smuts	U. tritici
	Tilletia		wheat	covered smut or bunt	T. caries
			wheat	Karnal bunt	T. indica
	Urocystis		onion	smut	U. cepulae
	Sporisorium		sorghum	covered kernel smut	S. sorghi
			sorghum	loose sorghum smut	S. cruentum
	Sphacelotheca		sorghum	head smut	S. reiliana
	Cronartium		pines	blister rust	C. ribicola
	Gymnosporangium		apple	cedar-apple rust	G. juniperivirginianae
	Hemileia		coffee	rust	H. vastatrix
	Melampsora		flax	rust	M. lini
	Phakopsora		soybeans	rust	P. pachyrrhizi
	Puccinia		cereals	rust	P. recondita
	Uromyces		beans	rust	U. appendiculatus [1]

Exobasidium		ornamentals	leaf flower and stem galls	E. japonicum
Athelia		many plants	Southern blight	A. rolfsii
	Sclerotium	onions	white rot	S. cepivorum
Thanatephorus	Rhizoctonia	many plants	root and stem rots damping-off and fruit rots	T. cucumeris and R.solani
Typhula		turf grasses	snow mold	T. incarnata
Armillaria		trees	root rots	A. mellea
Crinipellis		cacao	witches'-broom	C. perniciosus
Marasmius		turf grasses	fairy ring disease	M. oreades
Pleurotus		trees	white rot on logs tree stumps and living trees	P. ostreatus [1]
Pholiota		trees	brown wood rot	P. squarrosa
Chondrostereum		trees	silver leaf disease	C. purpureum
Corticium		turf grasses	red thread disease	C. fuciforme
Heterobasidion		trees	root and butt rot	H. annosum
Ganoderma		trees	root and basal stem rots	G. boninense
Inonotus		trees	heart rot	I. hispidus
Polyporus		trees	heart rot	P. glomeratus
Postia		trees	wood and root rot	P. fragilis

[1]These phytopathogenic fungi are named in the text.

According to the type of parasitism and infection strategy, fungi are classified as biotrophic (e.g., Blumeria graminis), necrotophic (e.g., Botrytis cinerea), or hemibiotrophic (e.g., Colletotrichum destructivum). While the former derives nutrients from dead cells, the latter takes nutrients from the plant but does not kill it [4]. Hemibiotrophes sequentially deploy a biotrophic and then a necrotrophic mode of nutrition. Necrotrophic species tend to attack a broad range of plant species; on the contrary, biotrophes usually exhibit a high degree of specialization for individual plant species. Most biotrophic fungi are obligatory parasites, surviving only limited saprophytic phases. Differently from necrotrophes, the cultivation of biotrophic fungi succeeds only in a few exceptions, for example, Podosphaera fusuca [5] or B. graminis (M. M. Corbitt, personal communication, adapted from [5]). Fungal diseases are, in nature, more the exception than the rule. Thus, only a limited number of fungal species are able to penetrate and invade host tissues, avoiding recognition and plant defence responses, in order to obtain nutrients from them, causing disease and sometimes host death. In agriculture, annual crop losses due to pre- and post harvest fungal diseases exceed 200 billion euros, and, in the United Stated alone, over $600 million are annually spent on fungicides [6].

Fungal pathogens have complicated life cycles, with both asexual and sexual reproduction, and stages involving the formation of different infective, vegetative, and reproductive structures [7]. The primary events in a disease cycle are establishment of infection, colonization (invasion), growth and reproduction, of the pathogen, dissemination of the pathogen, and survival of the pathogen in the absence of the host, that is, overwintering or oversummering (overseasoning) of the pathogen (Figure 1). However, the execution of each stage largely differs depending on the pathogen [8]. In polycyclic diseases there are several infection cycles within one, the so-called secondary cycles [3] (Figure 1).

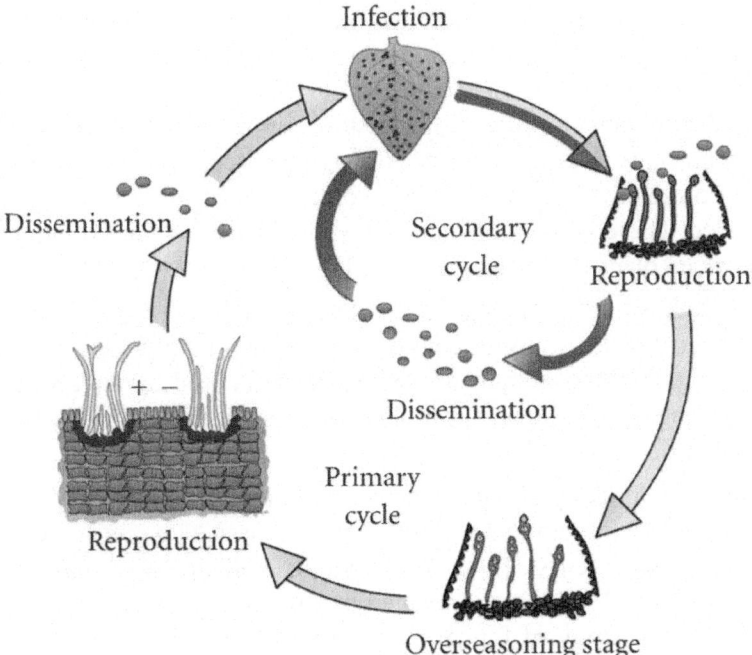

Figure 1: Diagram of monocyclic (yellow) and polycyclic (yellow and blue) fungi. In monocyclic diseases the fungus produces spores at the end of the season that serve as primary and only inoculum for the following year. The primary inoculum infects plants during the growth season and, at the end of the growth season, produces new spores in the infected tissues. These spores remain in the soil (overseasoning stage) and serve as the primary inoculum the following season. In polycyclic fungal patho-

gens, the primary inoculum often consists of the sexual (perfect) spore or, in fungi that lack the sexual stage, some other structures such as sclerotia, pseudosclerotia, or mycelium in infected tissue. This inoculum causes the primary infection and then large numbers of asexual spores (secondary inoculum) are produced at each infection site and these spores can themselves cause new (secondary) infections that produce more asexual spores for more infections.

The fungal plant interplay depends on mutual recognition, signalling, and the expression of pathogenicity and virulence factors, from the fungal side, and the existence of passive, preformed, or inducible defence mechanisms in the plant, resulting in compatible (susceptibility) or incompatible (nonhost, basal or host specific resistance) interactions. From a genetic point of view, and according to the gene-for-gene interaction hypothesis, proposed by Flor while studying flax rust [9], resistance results from the combination of a dominant avirulence (Avr) gene in the pathogen and a cognate resistance (R) gene in the host; the interaction of both gene products leads to the activation of host defence responses, such as the hypersensitive response, that arrests the growth of fungi. This hypothesis has been experimentally demonstrated for a number of pathosystems, mainly involving biotrophic fungi, with a number of avirulent genes identified [10]. A number of fungal mechanisms and molecules have been shown to contribute to fungal pathogenicity or virulence, understood as the capacity to cause damage in a host, in absolute or relative terms. Among them, cell wall degrading proteins, inhibitory proteins [11], and enzymes involved in the synthesis of toxins [12–15] are included. These virulence factors are typically involved in evolutionary arms races between plants and pathogens [16, 17].

Knowledge of the pathogenic cycle and that of virulence factors [18, 19] is crucial for designing effective crop protection strategies, including the development of resistant plant genotypes through classical plant breeding [20] or genetic engineering [21], fungicides [22], or the use of biological control strategies [23].

Studies of fungal pathogens and their interactions with plants have been performed using several approaches, from classical genetic, cell biology, and biochemistry [24–33], to the modern, holistic, and high-throughput omic techniques [34, 35] accompanied by proper bioinformatic tools [36]. In recent years, the study of fungal plant pathogens has been greatly promoted by the availability of their genomic sequences and resources for functional genomic analysis, including transcriptomics, proteomics, and metabolomics [37], which, in combination with targeted mutagenesis or transgenic studies, are unravelling molecular host-pathogen crosstalk, the complex mechanisms involving pathogenesis and host avoidance [38]. This review work makes an overview and summarizes the contribution of the most recent molecular techniques to the knowledge of phytopathogenic fungi biology and is mainly focused on the MS-based Proteomics approach.

FROM STRUCTURAL TO FUNCTIONAL GENOMICS

The importance of plant pathogenic fungi studies is underlined by the increasing number of fungal genome sequencing projects. Currently, over 40 fungal genomes have been sequenced, 16 of which are phytopathogenic (Table 2), with more than 300 sequencing projects being in progress (Genomes Online database,http://www.genomesonline.org/). Sequence information, while valuable and a necessary starting point, is insufficient alone to answer questions concerning gene function, regulatory networks, and the biochemical pathways activated during pathogenesis. Based on the accumulation of a wealth of fungal genomic sequences, the traditional pursuit of a gene starting with a phenotype (forward genetics) has given way to the opposite situation where the gene sequences are known but not their functions. Thus, the challenge is now to decipher the function of the thousands of genes identified by genome projects, and reverse genetics methodologies are key tools in this endeavour [39].

Table 2: Publicly available plant pathogenic fungal Genome sequences

Phytopathogen Speciesa	URL
Ascomycota	
Dothydeomycetes	
Mycosphaerella fijensis (Banana black leaf streak)	http://genome.jgi-psf.org/Mycfi1/Mycfi1.home.html
Mycosphaerella graminicola (Wheat leaf blotch)	http://genome.jgi-psf.org/Mycgr1/Mycgr1.home.html
Pyrenophora tritici-repentis (Wheat disease)	http://www.broad.mit.edu/annotation/genome/pyrenophora_tritici_repentis/Home-html
Stagonospora nodorum (Wheat glume blotch)	http://www.broad.mit.edu/annotation/fungi/stagonospora_nodorum
	http://www.acnfp.murdoch.edu.au/Mission.htm
Eurotiomycetes	
Aspergillus flavus	http://www.aspergillusflavus.org/
Leotiomycetes	
Botrytis cinerea (Grape/other host grey rot) BO5.10	http://www.broad.mit.edu/annotation/fungi/botrytis_cinerea

T4	http://urgi.versailles.inra.fr/proyects/Botrytis/index.php
Sclerotinia sclerotium (Multi-host rot diseases)	http://www.broad.mit.edu/annotation/fungi/sclerotinia_sclerotium
Saccharomycetes	
Ashbya gossypii (Cotton/citrus fruits disease)	http://agd.vital-it.ch/index.html
Sordariomycetes	
Fusarium graminearum (Wheat/barley head blight)	http://www.broad.mit.edu/annotation/genome/fusarium_group
Fusarium oxysporum (Multi-host wilt disease)	http://www.broad.mit.edu/annotation/genome/fusarium_group
Fusarium verticillioides (Maize seed rot)	http://www.broad.mit.edu/annotation/genome/fusarium_group
Magnaporthe grisea (Rice blast disease)	http://www.broad.mit.edu/annotation/genome/magnaporthe_grisea/MultiHome.htlm
Nectria haematococca (Pea wilt)	http://genome.jgi-psf.org/Necha2/ Necha2.home.html
Verticillium dahliae VdLs.17 (Multi-host wilt)	http://www.broad.mit.edu/annotation/genome/verticillium_dahliae/MultiHome.html
Basidiomycota	
Pucciniomycetes	
Puccinia graminis (Cereal rusts)	http://www.broad.mit.edu/annotation/genome/puccinia_graminis
Ustilaginomycetes	
Ustilago maydis (Corn smut disease)	http://www.broad.mit.edu/annotation/fungi/ustilago_maydis

[a]Species are grouped by phylum and class. In parenthesis below the species's name and associated with each species, the most common or most widely recognized diseases are listed.

The study of gene function in filamentous fungi has made great advances in recent years [69]. Some of the techniques used in high-throughput reverse genetics approaches are targeted gene disruption/replacement (knock-out) [70], gene silencing (knock-down) [71], insertional mutagenesis [72], or targeting induced local lesions in genomes (TILLING) [73] (for review and examples see [39]). Thus, a number of pathogenicity factors have been targeted [74–76], and among them, several signaling pathways such as the cAMP and a mitogen activated protein kinase (MAPK) pathways have been shown to be crucial to virulence in several phytopathogenic fungi [77–82]. Random insertional mutagenesis is an excellent approach for dissecting complex biological traits, such as pathogenicity, because it does not require any prior information or assumptions on gene function. Recently, transposable elements (TEs) have been used for insertional mutagenesis and large-scale transposon mutagenesis has been developed as a tool for the genome-wide identification of virulence determinants in F. oxysporum [83].

Otherwise, transcriptomics, the global analysis of gene expression at the mRNA level, is also an attractive method for analyzing the molecular basis of fungal-plant interactions and pathogenesis [84–87]. For understanding the transcriptional activation or repression of genes during the infection process tools such as Differential Display (DD) [88], cDNA-Amplified Fragment-Length Polymorphism (cDNA-AFLP) [89], Suppression Subtractive Hybridization (SSH) [90], Serial Analyse of Gene Expression (SAGE) [91], expressed sequence tags (ESTs) [92], or DNA microarrays [91] have been developed in addition to older techniques such as Northern blotting, and they are reviewed in [84, 93].

PROTEOMICS

Until the early 1990s most biological research was focused on the in vitro studies of individual components in which genes and proteins were investigated one at a time. This strategy shifted in the early and mid 1990s to in vivo and molecular large-scale research, starting with structural genomics and transcriptomics research projects, then moving to proteomics, and recently to metabolomics. All these together constitute the methodological bases of the Modern Systems Biology [37, 94]. Since no single approach can fully clear up the complexity of living organisms, each approach does contribute and must be validated, this being considered as part of a multidisciplinary integrative analysis at different levels, extending from the gene to the phenotype through proteins and metabolites.

Within the "-omics" techniques, Proteomics constitutes, nowadays, priority research for any organism and configures a fundamental discipline in the postgenomic era. It is true that, at 2010, the realities are below the expectations originally generated and that the results gained over the last 15 years have shown that the dynamism, variability, and behaviour of proteins are more complex than had ever been imagined, especially as refers to a number of protein species per gene as a result of alternative splicing, reading frame, and posttranslational modifications, trafficking, and interactions, and considering that protein complexes, rather than individual proteins, are the functional units of the biological machines. However, and differently from other biological systems, mainly yeast [95] and humans [96], the full potential of proteomics is far from being fully exploited in fungal pathogen research, as refers to the low number of fungal pathogen species under investigation at the proteomic level, the low proteome coverage in those species investigated, and the almost unique use of classical, first generation techniques, those based on 2-DE coupled to MS.

The term proteomics was coined by Marc Wilkins, back during the 1994 Siena Meeting, to simply refer to the "PROTein complement of a genOME" [97]. Fifteen years later proteomics has become more than just an appendix of genomics or an experimental approach but a complex scientific discipline dealing with the study of the cell proteome. In the broadest sense, the proteome can be defined as being the total set of protein species present in a biological unit (organule, cell, tissue, organ, individual, species, ecosystem) at any developmental stage and under specific environmental conditions. By using proteomics we aim to know how, where, when, and what for are the several hundred thousands of individual protein species produced in a living organism, how they interact with one another and with other molecules to construct the cellular building, and how they work with each other to fit in with programmed growth and development, and to interact with their biotic and abiotic environment. In the last ten years, excellent reviews and monographs on the fundamentals, concepts, applications, power, and limitations of proteomics have appeared [95, 98–104], some of them dealing with fungal pathogens [37, 105, 106]. It is not the objective of this review to comment or discuss every aspect but instead to show which one has been its contribution to the knowledge of fungal pathogens.

In Proteomics, several areas can be defined:

- Descriptive Proteomics, including Intracellular and Subcellular Proteomics;
- Differential Expression Proteomics,
- Post-translational Modifications;
- Interactomics; and
- Proteinomics (targeted or hypothesis-driven Proteomics).

In the case of fungi, a new area can also be defined as Secretomics (the secretome is defined as being the combination of native proteins and cell machinery involved in their secretion), since many fungi secrete a vast number of proteins to accommodate their saprotrophic life-style; this would be the case of proteins implicated in the adhesion to the plant surface [107], host-tissue penetration and invasion effectors, [11, 108, 109], and other virulence factors [110].

Proteomics is constantly being renewed to respond to the question of the role of the proteins expressed in a living organism, experiencing, in the last ten years, an explosion of new protocols, and platforms with continuous improvements made at all workflow stages, starting from the laboratory (tissue and cell fractionation, protein extraction, depletion, purification, separation, MS analysis) and ending at the computer (algorithms for protein identification and bioinformatics tools for data analysis, databases, and repositories). These techniques will be briefly introduced and discussed in the next section.

Despite the technological achievements in proteomics, only a tiny fraction of the cell proteome has been characterized so far, and only for a few biological systems (human, fruit fly, Arabidopsis, rice). Even for these organisms, the function of quite a number of proteins remains to be investigated [99]. Proteomics techniques have a number of limitations, such as sensitivity, resolution, and speed of data capture. They also face a number of challenges, such as deeper proteome coverage, proteomics of unsequenced "orphan" organisms, top-down Proteomics [100], protein quantification [98], PTMs [105], and Interactomics [55, 57]. Most of these limitations and challenges reflect the difficulty of working with the biological diversity of proteins and their range of physicochemical properties.

The relevance of proteomics in plant fungal pathogens research is very well illustrated by the pioneer work on the Cladosporium fulvum-tomato interaction carried out by the Pierre de Wit research group back in 1985 [111] that allowed the characterization of the first avirulence gene product (Avr9) after purification from tomato apoplastic fluids by preparative polyacrylamide gel electrophoresis followed by reverse-phase HPLC and EDMAN N-terminal sequencing [112]. Later on, a number of avirulence gene product effectors have been discovered, mainly by genomic approaches [113]. Curiously, this pioneer work followed the typical proteomics strategy even before MS-based powerful techniques were developed. Another good example is the tomato-F. oxysporum pathosystem, in which the first effector of root invading fungi was identified and sequenced, in this case by MS, the Six1, corresponding to a 12 kDa cysteine-rich protein [114]. Other further protein effectors have been characterized in different fungi [115].

Next, we describe the state of the art in the methodology of fungal plant pathogen proteomics and summarize the works published in this field up to December 2009, which so far are 30 (Table 3) out of a total of over 14000 original papers in Proteomics in the last ten years.

Table 3: Original proteomics papers and reviews published on plant pathogenic fungi

Fungus	Proteomic approach (reference)
Aspergillus ssp.	1-DE, MALDI-TOF-MS [40]
Aspergillus flavus	1-DE/2-DE, nanoLC-MS/MS [41]
	1-DE/2-DE, MALDI-TOF-MS [42]
Blumeria graminis f.sp. hordei	2-DE, MALDI-TOF/TOF-MS/MS [43]
	nanoLC-MS/MS [44]
Botrytis cinerea	2-DE, MALDI-TOF/TOF-MS/MS, nanoLC-MS/MS, ESI-IT-MS/MS [45]
	2-DE, MALDI-TOF/TOF-MS/MS, ESI-IT-MS/MS [46]
	2-DE, MALDI-TOF/TOF-MS/MS [47]
	1-DE, nanoLC MS/MS [48]
	1-DE, nanoLC MS/MS [49]
Curvularia lunata	2-DE, MALDI-TOF/TOF-MS/MS [50]
Fusarium graminearum	1-DE/2-DE, nanoLC-Q-TOF-MS/MS [51]
	1-DE, CID-LTQ-MS [52]
	2-DE, IT-MS/MS, iTRAQ-MS/MS [53]
	2-DE, ESI-MS/MS [54]
	Interactome [55]
Leptosphaeria maculans	1-DE, liquid-phase IEF, 2-DE [56]
Magnaporthe grisea	Interactome [57]
Phytophthora infestans	2-DE, MALDI-TOF-MS [58]
	2-DE, nanoLC-MS/MS [59]
Phytophthora palmivora	2-DE, MALDI-TOF-MS [58]
Phytophthora ramorum	HPLC-ESI-Q-TOF-nanoLC-MS/MS [60]
Pyrenophora tritici-repentis	2-DE, ESI-Q-TOF-MS/MS [61]
Pleorotus ostreatus	1-DE, MALDI-TOF-MS, ESI-Q-TOF-MS/MS/ de novo sequencing [62]
Rhizoctonia solani	2-DE, MALDI-TOF-MS [63]
Sclerotinia sclerotiorum	2-DE, ESI-Q-TOF-nanoLC-MS/MS [64]
Stagnospora nodorum	2 -DE, LC-MS/MS [65]
	2-D LC-MALDI-MS/MS [66]
Uromyces appendiculatus	MudPit-MS/MS [67]
Ustilago maydis	2-DE, MALDI-TOF-MS, nanoLC-Q-TOF-MS/MS [68]

Methodology

The workflow of a standard MS-based proteomics experiment includes all or most of the following steps: experimental design, sampling, tissue/cell or organelle preparation, protein extraction/fractionation/purification, labeling/ modification, separation, MS analysis, protein identification, statistical analysis of data, validation of identification, protein inference, quantification, and data analysis, management, and storage (Figure 2) [102, 116]. The most appropriate protocol to be used depends on and must be optimized for the biological system (i.e., fungal species, plant species, organ, tissue, cells), as well as the objectives of the research (descriptive, comparative, PTMs, interactions, targeted Proteomics) [102].

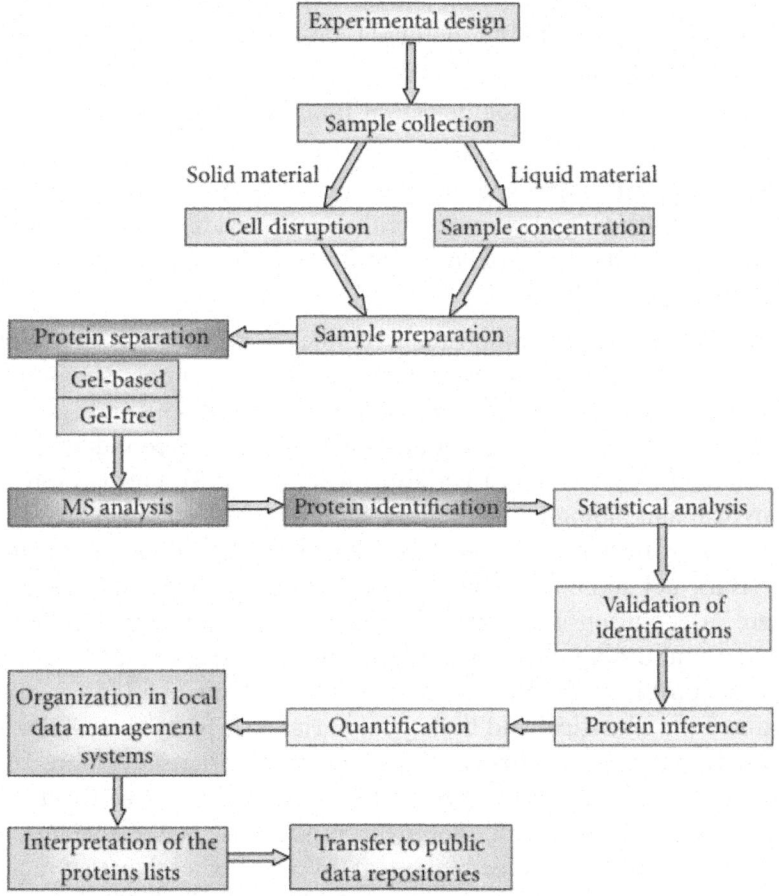

Figure 2: Schematic overview of the work flow in a fungal proteomatic approach (adapted from Deutsch et al., 2008 [116]).

Experimental Design

A good experimental design is crucial to the success of any proteomics experiment. Eriksson and Fenyö [117] have developed a simulation tool for evaluating the success of current designs and for predicting the performance of future, better-designed proteomics experiments. The simulation gives a holistic view of a general analytical experiment and attempts to identify the factors that affect the success rate. It has been used to predict the success of proteome analyses of Human tissue and body fluids that use various experimental design principles. Several parameters, are required to simulate the steps of a proteome analysis:

- the distribution of protein amounts in the sample analyzed,
- the loss of analyte material and the maximal limit of the amount loaded at each step of sample manipulation (e.g., separation, digestion, and chemical modification),
- the dynamic range, the detection limit, and the losses associated with MS analysis.

Depending on what experiment is being modelled, the detection limit used in a simulation can represent either protein identification only (lower identification limit) or protein identification with quantification (lower quantification limit) [102].

The establishment of an adequate number of replicates is essential to any differential expression proteomics experiment. This number should be set up while taking into account the dynamic nature of the proteome, and a good number will allow a correct interpretation of the results and the confident assignment of any protein to the group of variable ones [102, 118]. Furthermore, a study of analytical and biological variability should be carried out. Thus, analytical variability examines both the experiment procedures (protein extraction, IEF, SDS-PAGE, gel staining-destaining) and the accuracy of the hardware and software in acquiring and analysing images, and biological variability tests look at several different samples. For example, in the studies to characterize the protein profile of the fungal phytopathogen B. cinerea [45], it previously determined the analytical and biological variability. Sixty-four major spots in three 2-DE replicate gels were analyzed and the average CVs were 16.1% for the analytical variance and 37.5% for the biological variance for this fungus. The analytical variance was similar to that reported for bacteria, plants, mice, or human extract and the average biological CV was higher than that reported for other organisms grown under controlled conditions [45]. While characterizing the Pinus radiata needle proteome [119], we found differences in the standard error of mean spot quantity, depending on the number of replicates; the error ranged from 111 and 115 ng for two analytical and biological replicates to 58 and 59 ng for 10 analytical and 12 biological gel replicates.

Using more than six biological replicates did not significantly reduce the standard error; so this figure could be optimal for comparative proteomic experiments. Since normally this is not feasible, most papers in our literature review used only three biological replicates. Given the susceptibility of the data to variation, in 2-DE comparative proteomics it is necessary to be restrictive when deciding whether a spot showed variation. First, all the spots considered had to be consistent, that is, present or absent in all the biological replicates of the particular stage in question; second, when not qualitative (presence vs. absence), differences had to be statistically significant (ANOVA); finally, the variance with respect to a control had to be higher than the average biological coefficient of variance determined for a representative set of at least 150 spots.

Cell Disruption and Sample Preparation

The importance of the extraction protocol in a proteomic experiment can be summarized in the following statement: only if you can extract and solubilize a protein, you have a chance of detecting and identifying it. This sentence sums up the importance of the preparation of protein samples in a proteomic experiment [102]. This is more important in the case of plant tissue or fungal material. In the case of filamentous fungi, the protein extraction is also influenced by the presence of a cell wall which makes up the majority of the cell mass. This cell wall is exceptionally robust [120], and the effective extraction of proteins is a critical step and essential for reproducible results in fungal proteomic studies. Therefore, cell breakdown is an important element in sample preparation for fungal proteomics.

Several early studies were performed to overcome this challenge by providing an effective means of cell lysis for an adequate release of intracellular proteins. For example, mechanical lysis was used to release the cytoplasmic protein via glass beads [59, 121–125], using a cell mill [68], or by sonication [40, 126, 127], these being more efficient than either chemical or enzyme extraction methods [128]. Shimizu and Wariishi [129] utilized an alternative approach to avoid the difficulty of lysing the fungal cell wall by generating protoplasts of Tyromyces palustris, since 2-DE patterns from protoplast were better visualized than protein obtained from disrupting the fungal cell wall using SDS extraction. However, the most widely used method for cell disruption is pulverizing the fungus material in liquid nitrogen using a mortar and pestle [45–47, 64, 126, 130, 131].

The extraction protocol most amply utilized implicates the use of protein precipitation media containing organic solvents, such as trichloroacetic acid

(TCA)/acetone, followed by solubilization of the precipitate in an appropriate buffer. It allows an increase in the protein concentration and removal of contaminants (salts, lipids, polysaccharides, phenols, and nucleic acids) which can be a problem during IEF [132] and prevents protease activities. This method has often been applied to the preparation of extracts from plants [133–135] and fungi [123, 136]. TCA-treatment makes subsequent protein solubilization for IEF difficult, especially with hydrophobic proteins. These problems have been partly overcome by the use of new zwitterionic detergents [137–142] and thiourea [143], or by a brief treatment with sodium hydroxide [123], which led to an increase in the resolution and capacity of 2-DE gels. Other protein extraction methods have reported an improvement when using acidic extraction solution to reduce streaking of fungal samples caused by their cell wall [144], as well as the use of a phosphate buffer solubilization before the precipitation [45, 46]. In a study to develop an optimized protein extraction protocol for Rhizoctonia solani, Lakshman et al. [63] compared TCA/acetone precipitation and phosphate solubilization before TCA/acetone precipitation. Both protocols worked well for R. solani protein extraction, although selective enrichment of some proteins was noted with either method. Finally, the combined use of TCA precipitation and phenol extraction gave a better spot definition, because it reduced streaking and led to a higher number of detected spots [47, 145].

In B. cinerea, our group has optimized a protein precipitation protocol from mycelium based on a combination of TCA/acetone and phenol/methanol protein extraction methods described by Maldonado et al.[135] with some modifications (Gonzalez-Fernandez et al., unpublished data). Mycelium is lyophilized, pulverized with liquid nitrogen using a mortar and pestle, sonicated in 10% (w/v) TCA/acetone solution with a sonic probe, washed sucessively with methanol and acetone, and finally a protein extraction is released with phenol/methanol precipitation method. We have used a similar protocol from conidia of this fungus by sonicating directly the spores collected in 10% (w/v) TCA/acetone solution (Gonzalez-Fernandez et al., unpublished results). In this sense, specific protocols for protein extraction from spores were optimized in Aspergillus ssp. using acidic conditions, step organic gradient, and variable sonication treatment (sonic probe and water bath) [40]. In this study, the use of a sonic probe was the best method to break the robust cell wall of conidia and obtain more proteins.

Special protocols are required in the case of secreted proteins, facing problems such as the very low protein concentration, sometimes below the detection limit of colorimetric methods for determining protein concentration such as Bradford, Lowry, or BCA, and the presence of polysaccharides, mucilaginous material, salts, and secreted metabolites (low-molecular organic acids, fatty acids, phenols, quinines, and other aromatic compounds). Moreover, the presence of these extracellular compounds may impair standard methods for protein quantification and can result in a strong overestimation of total protein [146]. This determination can also be affected by the high concentration of reagents from the solubilisation buffer (such as urea, thiourea or DTT) that may interfere in the spectrophotometric measurement, producing an overestimation of the total amount of protein. Fragner et al., showed that, depending on the method, the differences varied in the order of two magnitudes, indicating that only the Bradford assay does not lead to an overestimation of the proteins [147].

Francisco and colleagues provided pioneering contributions to this field, establishing a sample preparation protocol for fungal secretome [148], including steps of lyophilization or ultrafiltration plus dialysis, precipitation (TCA/ethanol or chloroform/methanol), deglycosylation, and solubilization for SDS-PAGE or 2-DE.

Comparison of different standard methods for protein precipitation has demonstrated their limited applicability to analyzing the whole fungal secretome [149–152]. Usually, the fungal liquid culture is clarified by filtering and centrifuging, then dialyzed and concentrated up by lyophilizing. Recently, a new optimized protocol has been developed to obtain extracellular proteins from several higher basidiomycetes (Coprinopsis cinerea, Pleorotus ostreatus, Phanerochaete chrysosporium, Polyporus brumalis, and Schizophyllum commune) [147]. In this work, several precipitation methods, (TCA/acetone, phenol/methanol, other precipitation methods and an optimized protocol by high-speed centrifugation/TCA precipitation/Tris-acetone washes) were compared from liquid cultures of these fungi. The best method was the use of high-speed centrifugation, since it removed a considerable gelatinous material from the liquid culture supernatants. Then fungal proteins were effectively enriched by TCA precipitation and coprecipitated metabolites hampering 2-DE were removed through the application of Tris/acetone washes [147].

Vincent et al. [56], using the plant pathogenic fungus Leptosphaeria maculans and symbiont Laccaria bicolorgrown in culture, have established a proteomic protocol for the analysis of the secretome. They evaluate different protocols, including ultrafiltration followed by TCA/acetone precipitation or phenol/ammonium acetate-phase partition, successive TCA/acetone precipitations without ultrafiltration, phenol/ether extraction without ultrafiltration, lyophilization, and prefractionation of secretome samples using liquid-phase IEF. Liquid-phase IEF followed by dialysis and lyophilization as a prefractionation prior to IPG-IEF produced the best results, with up to 2000 spots well resolved on 2-DE. This protocol can be applicable to a reduced number of samples and be very useful for descriptive proteomics but not for comparative proteomics experiments because of the excessive manipulation of the sample. Thus, in our work with B. cinerea, we aim to compare the proteome of wild-type and a high number of mutant strains (close to 100) affected in pathogenicity, and therefore we have optimized a protocol including the following steps: liquid media filtering, dialyzing, lyophilizing, and TCA/ acetone-phenol/methanol protein precipitation. It is true that the number of resolved spots is much lower, but still enough for our purposes (Gonzalez-Fernandez et al., unpublished data).

A new field of study has been opened up with the analysis of infection structures such as appressorium and haustoria. In this case, specific protocols for the isolation of such structures are required [85] and the main problem is the large amount of material required for proteomic analysis if compared to transcriptomics. Godfrey et al. have developed a procedure for isolating haustoria from the barley powdery mildew fungus based on filtration and the use of differential and gradient centrifugation [44].

Up to now we have made reference to experiments with in vitro grown fungi; studies in planta are much more complicated due to the presence of both proteomes than of the plant and the pathogen. Bindschedler et al.[153] undertook a systematic shotgun proteomics analysis of the obligate biotroph B. graminis f. sp. hordei at different stages of development in the host: ungerminated spores, epiphytic sporulating hyphae, and haustoria, this being, as far as we know, the only large-scale comparative study of proteomes of phytopathogenic fungi during in planta colonization in addition to those analyses of whole extracts from host infected tissue [154,155] or intercellular washing fluids [156].

In short, since no single protein extraction protocol can capture the full proteome, the chosen protocol should be optimized for the research objective. The ideal method should be highly reproducible and should extract the greatest number of protein species, while at the same time reduce the level of contaminants and minimize artifactual protein degradation and modifications [148, 157, 158]. Fortunately, each protocol takes us to a specific fraction of the proteome, thus being complementary [135]. Another issue to consider is the extreme complexity of the proteome and the wide dynamic range in protein abundance, which exceeds the capability of all currently available analytical platforms. Sample prefractionation is a good approach to reducing the complexity of the proteome sample and decreasing the dynamic range [159], with EQUALIZER being the last developed technology [160].

Protein Separation, Mass Spectrometry and Protein Identification

In fungal plant pathogen research electrophoresis, including denaturing 1-DE SDS-PAGE and 2-DE, with IEF as first dimension, and SDS-PAGE as the second, is almost the only protein separation technique employed, with both crude total protein extracts and protein fractions obtained from various prefractionation procedures [161]. Despite its simplicity, 1-DE can still be quite a valid technique providing relevant information, especially in the case of comparative proteomics with large numbers of samples to be compared. Thus, using this technique, it is possible to distinguish between genotypes of different wild-type strains of B. cinerea and identify proteins involved in the pathogenicity mechanisms (Figure 3) (Gonzalez-Fernandez et al., unpublished results). With appropriate software, 1-DE is a simple and reliable technique for finger-printing crude extracts and it is especially useful in the case of hydrophobic and low-molecular-weight proteins [162]. The combination of 1-DE, band cutting, trypsin digestion, and LC separation of the resulting peptides remains the proteomic technique capable of providing the greatest protein coverage [163, 164]. Therefore, the 1-DE is a good approach to obtaining preliminary results before the study by 2-DE [41, 42, 50, 165]. For example, the use of 1-DE in combination with MS/MS analysis has allowed the detection of both known [166] or new proteins [62] of interest in fungal pathogenicity.

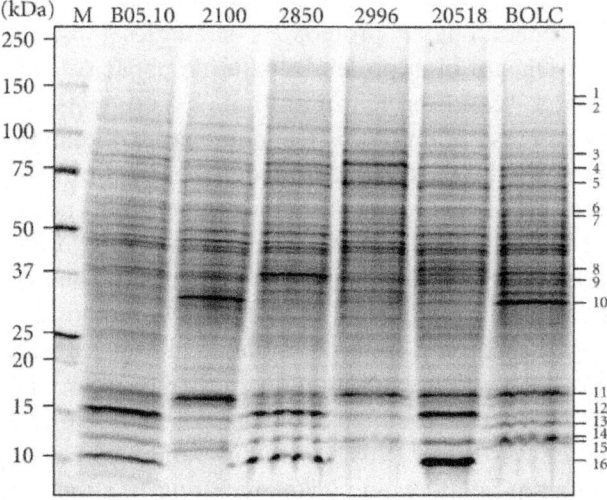

Figure 3: One-DE of 15 g of mycelium protein extract of six different strains of B. cinerea (B05.10, CECT 2100, CECT 2850, CECT 2996, CECT 20518, BOLC (isolated from infected lentil plants)). This approach allowed us to observe differences in the

protein band patterns among strains. The bands were cut out and the protein identification was made using MALDI-TOF/TOF MS/MS, and PMF search and a combined search (+MS/MS) were performed in nrNCBI database of proteins using MASCOT. Some of these proteins identified have been reported to be involved in pathogenicity in B. cinereaor in other phytopathogenic fungi, such as malate dehydrogenase (10), woronin body major protein (11), peptidyl-prolyl cis-trans isomerase (14) and PIC5 protein (15), or implicated in fungal growth and differentiation, such as nucleoside diphosphate kinase (12). The abundance of these proteins was different among isolated (Gonzalez-Fernandez et al., unpublished results).

In fungal proteomics, 2-DE separation techniques [132, 167, 168] are widely used (Table 3), these having the advantage of separating the proteins at the protein species level with a high resolution of up to 10,000 spots [169]. Briefly, the 2-DE consists of a tandem pair of electrophoretic separations: in the first dimension, proteins are resolved according to their isoelectric points (pIs) normally using IEF, and in the second dimension, proteins are separated according to their approximate molecular weight using SDS-PAGE. These proteins can then be detected by a variety of staining techniques:

- organic dyes, such as colloidal Coomassie blue staining,
- zinc-imidazole staining,
- silver staining, and
- fluorescence-based detection.

Excellent reviews describe and discuss the features and protocols of electrophoretic separations in proteomics strategies [132, 170]. The main advantages of 2-DE are high protein separation capacity and the possibility of making large-scale protein-profiling experiments. However, the reproducibility and resolution of this technique are still remaining challenges. Moreover, this method was reported to under-represent proteins with extreme physicochemical properties (size, isoelectric point, transmembrane domains) and those of a low abundance [171]. These limitations to analytical protein profiling have led to the more recent development of techniques based on LC separation of proteins or peptides, including two-dimensional liquid-phase chromatography (based on a high-perfomance chromatofocusing in the first dimension followed by high-resolution reversed-phase chromatography in the second one) [172], and one-dimensional 1-DE-nanoscale capillary LC-MS/MS, namely, GeLC-MS/MS (this technique combines a size-based protein separation with an in-gel digestion of the resulting fractions). Recorbert et al. [173] explored the efficiency of GeLC-MS/MS to identify proteins from the mycelium of Glomus intraradices developed on root organ cultures, reporting on the identification of 92 different proteins. Overall, this GeLC-MS/MS strategy paves the way towards analysing on a large-scale fungal response environmental cues on the basis of quantitative shotgun protein profiling experiments. Despite the existence of quite a number of different methods developed for protein extraction and separation, it is clear that, all in all, it is not enough to allow for the analysis of entire proteomes (organelle, cell, tissue, or organ). Some methodologies have proven to be more powerful and decisive than others, with regard to the number of proteins identified. This is the case of Multidimensional Protein Identification Technology (MudPIT), an LC-based strategy, which allows the detection of a much larger number of proteins compared to gel-based methods, its drawback being the lack of quantitative data [174]. MudPIT was first applied to the fungal proteome of the S. cerevisiae and yielded the largest proteome analysis to date, in which a total of 1484 proteins were detected and identified [175]. The categorization of these hits demonstrated the ability of this technology to detect and identify proteins rarely seen in proteome analysis, including low abundance proteins like transcription factors and protein kinases [175]. It has been reported that a set of proteins can only be detected by a specific technology [176, 177], which is in agreement with the idea that a combination of different methodologies is still needed to characterize entire proteomes [131]. MudPIT has been used to compare the proteome from germinating and ungerminated asexual uredospores of the biotrophic fungal pathogen Uromyces appendiculatus, which is the causal agent of rust disease in beans [67].

Mass spectrometry is the basic technique for global proteomic analysis due to its accuracy, resolution, and sensitivity, small amounts of sample (femtomole to attomole concentrations), and having the capacity for a high throughput. It allows not only to profile a proteome from a qualitative and quantitative point of view but also, and more important, to identify protein species and characterize postranslational modifications. Proteins are identified from mass spectra of intact proteins (top-down proteomics) or peptide fragments obtained after enzymatic (mostly tryptic peptides) or chemical treatment (bottom-up proteomics). Protein species are identified by comparison of the experimental spectra and the theoretical one obtained in silico from protein, genomic, ESTs sequence, or MS spectra databases. For that purpose, different instrumentation, algorithms, databases, and repositories are available [178, 179].

Different strategies and algorithms can be used for protein identification (i) including peptide mass fingerprinting (PMF, only valid if the protein sequence is present in the database of interest, generally used if the genome of the organism of study is fully sequenced); (ii) tandem mass spectrometry (where peptide sequences are identified by correlating acquired fragment ion spectra with theoretical spectra predicted for each peptide contained in a protein sequence database, or by correlating acquired fragment ion spectra with libraries of experimental MS/MS spectra identified in previous experiments); (iii) de novo sequencing, where peptide sequences are explicitly read out directly from fragment ion spectra; and (iv) hybrid approaches, such as those based on the extraction of short sequence tags of 3–5 residues in length, followed by "error-tolerant" database searching [179]. In a genomic-based proteomics strategy the percentage of proteins identified from MS data is dependent on the availability of genomic DNA or EST sequences. The construction of protein repositories with signature peptides and derived MS spectra will open up new possibilities for protein identification. The availability of ESTs from unsequenced "orphan" organisms as is the case of most plant fugal pathogens of interest will increase the percentage of identified proteins [93]. There are relatively few and a slow accumulation of EST data derived from a number of plant fungal pathogens and related species in public databases (e.g., dbEST at the National Centre for Biotechnology Information (NCBI),http://www.ncbi.nlm.nih.gov/dbEST/; and COGEME, Phytopathogenic Fungi and Oomycete EST Database [180, 181] at Exeter University, UK, http://cogeme.ex.ac.uk/). In Table 4, the number of EST entries for a number of fungi at the NCBI and Dana Faber databases is listed.

Table 4: Number of ESTs entries for some fungi of interest up to December 2009

Fungus	Dana Faber/NBCIa
Aspergillus flavus	20372/22452
Aspergillus niger	No entries/47082
Blumeria graminis hordei	No entries/17142
Botrytis cinerea	No entries/28531
Claviceps purpurea	No entries/8789
Curvularia lunata	No entries/1488
Fusarium graminearum	No entries/58011
Fusarium oxysporum	No entries/17478
Fusarium verticillioides	86908/87134
Leptosphaeria maculans	No entries/20034
Magnaporthe grisea	87403/110613
Puccinia graminis	No entries/209
Phytophthora infestans	90287/164143
Phytophthora palmivora	No entries/14824
Pyrenophora tritici-repentis	No entries/1
Nectria haematococca mpVI	No entries/33122

Sclerotinia sclerotiorum	No entries/2578
Stagonospora nodorum	No entries/16447
Ustilago maydis	No entries/39717

[a]Data taken from The Gene Index Proyect at the Dana Farber Cancer Institute

Second-Generation Techniques for Quantitative Proteomics

By using proteomics it is not only aimed to identify protein species (main objective of descriptive proteomics) but also quantify them, at least in relative terms, by comparing two biological units (genotypes, cells, organules) under different spatiotemporal parameters and environmental conditions. Absolute, rather than relative, protein quantitation remains one of the main challenges in proteomics [102]. There are different methods to dissect the proteome in a quantitative manner:

- methods based on 2-DE with poststaining, such as colloidal Comassie blue staining [182] and fluorescence staining [183], or prelabeling such as two-dimensional fluorescence difference gel electrophoresis (2-DIGE) [184, 185]; and

- gel-free methods based on in vitro or in vivo protein targetting with a stable isotope, such as isotope-coded affinity tags (ICATs) [186,187], [18]O labeling [188], or stable isotope labelling in cell culture (SILAC) [189], or isobaric tags, such as isobaric tag for relative and absolute quantitation (iTRAQ) [190].

Novel, label-free approaches, such as spectral counting, are being developed [191, 192]. The use of second generation proteomic techniques based on protein labelling and those label-free ones are far from being fully exploited in fungal pathogen research.

In 2-DIGE, proteins in two samples are labeled in vitro through cysteine or lysine residues using two different fluorescent cyanine dyes differing in their excitation and emission wavelengths, but with an identical relative mass. Labeled samples are then mixed and subjected to 2-DE on the same gel. After consecutive excitation with both wavelengths, the images are overlaid and normalized, whereby only differences between the two samples are visualized [193]. 2-DIGE enables to perform high-throughput, differential protein expression analysis to compare directly, on a single gel, the differences in protein expression levels between different complex protein samples. The main advantage of 2-DIGE on 2-DE is its unrivalled performance, attributable to a unique experiment, in which each protein spot on the gel is represented by its own internal standard [105].

The classical proteomic quantification electrophoretic methods utilizing dyes, fluorophores, or radioactivity have provided very good sensitivity, linearity, and dynamic ranges, but they suffer from two important shortcomings: first, they require high-resolution protein separation typically provided by 2-DE gels, which limits their applicability to abundant and soluble proteins; and second, they do not reveal the identity of the underlying protein [194]. Both of these problems are overcome by modern LC-MS/MS techniques [95,195–197]. However, the MS-based techniques are not inherently quantitative because proteolytic peptides exhibit a wide range of physicochemical properties (size, charge, hydrophobicity) that lead to large differences in mass spectrometric response. Therefore, in MS-based gel-free quantification it is necessary to use isotopic labeling. Observed peak ratios for isotopic analogs are highly accurate, because there are no chemical differences between the species, and they are analyzed in the same experiment. Mass spectrometry can recognize the mass difference between the labeled and unlabeled forms of a peptide and the quantification is achieved by comparing their respective signal intensities [194].

A number of isotopic labeling techniques have recently been proposed that share the requirement of the chemical modification of the peptides or proteins. One of these strategies is the ICAT method for relative quantitation of protein abundance [186]. In this approach, an isotopically labeled affinity reagent is attached to particular amino acids in all proteins in the population. After digestion of the protein to peptides, as a necessary step in all mainstream proteomic protocols, the labeled peptides are affinity-purified using the newly incorporated affinity tag, thereby achieving a simplification of the peptide mixture at the same time as incorporating the isotopic label. This method addresses many of the above limitations and leads to a larger number of identifications of cysteine-containing peptides. However, the method is performed by cross-linking peptides to beads via their cysteine groups and photo-releasing them afterwards, which may compromise low-level analysis. The iTRAQ is used to identify and quantify proteins from different sources in one experiment. The method is based on the covalent labeling of the N-terminus and side-chain amines of peptides from protein digestions with tags of varying mass. The fragmentation of the attached tag generates a low molecular mass reporter ion that can be used to relatively quantify the peptides and the proteins from which they originated [198]. In SILAC (stable isotope labeling by amino acids in cell culture), labeled, essential amino acids are added to amino acid-deficient cell culture media and are therefore incorporated into all proteins as they are synthesized, "encoded into the proteome" [189]. No chemical labeling or affinity purification steps are performed, and the method is compatible with virtually all cell culture conditions. Finally, label-free protein quantification methods are promising alternatives. It is based on precursor signal intensity, which is, in most cases, applied to data acquired on high mass precision spectrometers. The mass spectral peak intensities of the peptide ion correlate well with protein abundances in complex samples [199–201]. Another label-free method is spectral counting, which simply counts the number of MS/MS spectra identified for a given peptide and then integrates the results for all measured peptides of proteins that are quantified [202]. An advantage of this technique is that relative abundances of different proteins can in principle be measured. These new quantification techniques have become powerful tools to overcome the inherent problems of the 2-DE including identification of proteins of a low abundance, high hydrophobicity, extreme pl or high MW.

These second-generation MS technologies for Quantitative Proteomics have not begun to be applied to Fungal Proteomics yet. The only one nongel-based quantitative proteomics example is the use of iTRAQ to study the profile protein expression differences on F. graminearum which allowed the identification of numerous candidate pathogenicity proteins[53].

Although this review is focused on MS-based proteomics, we want to make a brief mention of protein microarrays because they are powerful tools for individual studies as well as systematic characterization of proteins and their biochemical activities and regulation [203]. The arrays can be used to screen nearly the entire proteome in an unbiased fashion and have an enormous utility for a variety of applications. These include protein-protein interactions, identification of novel lipid- and nucleic acid-binding proteins, and finding targets of small molecules, protein kinases, and other modification enzymes.

In short, all these technologies have a great potential in protein separation and remain a challenge for future research works in Fungal Proteomics.

Data Analysis and Statistical Validation

Proteomics tools generate an important amount of data, because a single proteomics experiment reveals the expression information for hundreds or thousands of proteins. Therefore, data analysis and bioinformatics are essential for this type of research and in many cases take more time than the actual experiment and require special skills and tools (for review see[116, 118]). All 2-DE software permits a fast and reliable gel comparison, and multiple gel analyses, including filtering of 2-DE images, automatic spot detection, normalization of the volume of each protein spot, and differential and statistical analyses [204–206]. A great resource for finding software tools for proteomics can be found in the website https://proteomecommons.org/.

Before the protein identification, the remaining challenge is to determine whether the putative identification is, in fact, correct. Statistical tools help us to validate information. Although postsearch statistical validation still does not enjoy universal application, its importance has been recognized by most researchers and codified in the editorial policies of some leading journal [207]. In this decade, an important number of commercial software involving even more powerful algorithms and statistical tools than those of the previous generations have been designed to help researchers deal with the sheer quantity of data produced [194, 195, 208].

Statistical data analyses can be classified as univariated or multivariated [209]. The univariated methods, such as the Student's -Test, are used to detect significant changes in the expression of individual proteins. They are the simplest to interpret conceptually and the most common ones used. The multivariated methods, such as principle component analysis (PCA), look for patterns in expression changes and utilize all the data simultaneously. Early expression studies compared one sample with another, generally by the calculation of a ratio, and the analyses were restricted to looking for changes above a threshold determined by the system experimental noise. This method of analysis limits the sensitivity of the system, as biologically relevant changes smaller than the threshold cannot be detected. Using a threshold is a rather simplistic approach and does not take into account the variability of each protein, running the risk of selecting variable proteins due to sample selection. The use of a fold change has a potential role in preliminary experiments but the limitation of this method must always be considered when interpreting the data [209].

Hypothesis tests, for example, Student's -Test, assess whether the differences between groups are an effect of chance arising from a sampling effect or reflect a real statistical significant difference between groups [209]. Hypothesis tests are usually stated in terms of both a condition that is in doubt (the null hypothesis) and a condition that is believed to exist (the alternative hypothesis). The tests calculate a -score, which is the probability of obtaining these results assuming that the null hypothesis is correct. Hypothesis tests can be divided into two groups: parametric and nonparametric. Parametric tests assume that the distribution of the variables being assessed belong to known probability distributions. For example, the Student's -Test assumes that the variable comes from a normal distribution. Nonparametric tests, also called distribution-free methods, do not rely on estimation of parameters such as the mean. An example of nonparametric tests is the Mann-Whitney -Test, which ranks all values from low to high and then compares the mean ranking the two groups. Otherwise, Student's -Test or Mann-Whitney -Test compares two groups and ANOVA or Krustal-Wallis compares more than two groups. These tests analyze individual spots instead of the complete set, omitting information about correlated variables. Biron et al. recommend assessing the normality for each protein species and then selecting either a parametric or nonparametric test [208].

In expression studies, many thousands of statistical tests are conducted, one for each protein species. A substantial number of false positives may accumulate which is termed the multiple testing problem and is a general property of a confidence-based statistical test when applied many times [209]. One approach to addressing the multiple testing problem is to control the family wise error rate (FWER), which control the probability of one or more false rejections among all the tests conducted. The simplest and most conservative approach is the Bonferroni correction, which adjusts the threshold of significance by dividing the percomparison error rate (PCER) by the number of comparisons being completed [210]. This has led to the application of methodologies to control the false discovery rate (FDR), where the focus is on achieving an acceptable ratio of true- and false-positives. The FDR is a proportion of changes identified as significant that are false [211]. An extension to the FDR calculates a -value for each tested feature and is the expected proportion of false positives incurred when making a call that this feature has a significant change in the expression [212]. For each -value, a -value will be reported on an overall estimation for the proportion of species changing in the study.

Multiple testing correction methods, such as the Bonferroni correction and testing for the false discovery rate (FDR) [213], fit the Student -test or ANOVA values for each protein spot to keep the overall error rate as low as possible. Multivariate data analysis methods, such as PCA, are now used to pinpoint spots that differ between samples. These multivariate methods focus not only on differences in individual spots but also on the covariance structure between proteins [214]. However, the results of these methods are sensitive to data scaling, and they may fail to produce valid multivariate models due to the large number of spots in the gels that do not contribute to the discrimination process [215]. One of the limitations of PCA analysis is that it does not allow to miss values, a problem that can be avoided by imputing them when possible (if enough replicates are available) [216].

Databases and Repositories

The huge amount of data generated are being deposited and organized in several databases available to the scientific community: the UniProt knowledgebase reported by Schneider et al. [217] and other Proteome Databases mentioned in Table 5. After 20 years of Proteomics research, it is possible to look back at previous research and publications, identifying errors from the experimental design, the analysis, and the interpretation of the data [179]. In addition, data validation is done in a purely descriptive or speculative manner, as well as it is common to find low-confidence protein identification in the literature, especially in the case of unsequenced organisms and inappropriate statistical analyses of results have often been performed. It is interesting to see how many manuscripts contain the term "proteome" when probably only a tiny fraction of the total proteome has been analysed. About this problem, HUPO's Proteomic Standard Initiative has developed guidance modules [218] that have been translated into Minimal Information about a Proteomic Experiment (MIAPE) documents. The MIAPE documents recommend proteomic techniques that should be considered and followed when conducting a proteomic experiment. Proteomics journals should be, and in fact are, extremely strict when recommending that investigators follow the MIAPE standards for publishing a proteomic experiment. On the other hand, many journals recommend or require the original data generated in a proteomic experiment to be submitted to public repositories [207, 219]. A shift in the protein identification paradigm is currently underway, moving from sequencing and database searching to spectrum searching in spectral libraries. This underscores the importance of repositories for Proteomics [220–222]. The main public peptide and protein identification repositories are GPMDB (Global Proteome Machine database) [223], PeptideAtlas [224], and Proteomics Identifications database (PRIDE) [225]. Other emerging and smaller systems include Genome Annotating Proteomic Pipeline database (GAPP database) [226], Tranche (Falkner, J. A., Andrews, P. C., HUPO Conference 2006, Long Beach, USA, Poster presentation), PepSeeker [227], Max-Planck Unified database (MAPU) [228], the Open Proteomic Database (OPD) [229], and the Yeast Resource Center Public Data Repository (YRC PDR) [230].

Table 5: Useful online resources and Fungal Genome and Proteome Databases

Name/description	URL
Genome Databases	
National Center for Biotechnology Information (NCBI).	http://www.ncbi.nlm.nih.gov/
NIH genetic sequences database.	http://www.ncbi.nlm.nih.gov/Genbank/
Fungal Genomes Central, information and resources pertaining to fungi and fungal sequencing projects.	http://www.ncbi.nlm.nih.gov/projects/genome/guide/fungi/

The Gene Index proyect (GI). The Computational Biology and Functional Genomics Laboratory, and the Dana-Faber Institute and Public School of Public Health.	http://compbio.dfci.harvard.edu/tgi/fungi.html
Fungal Genome Initiative of The Broad Institute (FGI).	http://www.broadinstitute.org/science/projects/fungal-genome-initiative/fungal-genome-initiative
Genoscope, Sequencing National Centre.	http://www.genoscope.cns.fr/spip/Fungi-sequenced-at-Genoscope.html
Joint Genome Institute (JGI).	http://www.jgi.doe.gov/
The Genome Center at Washington University (WU-GSC).	http://genome.wustl.edu/genomes/list/plant_fungi
The Sanger Institute fungal sequencing.	http://www.sanger.ac.uk/Projects/Fungi/
Genome projects.	http://genomesonline.org/
The MIPS F. Graminearum Genome Database.	http://mips.gsf.de/projects/fungi/Fgraminearum/
The MIPS U. Maydis Database.	http://mips.gsf.de/genre/proj/ustilago
The MIPS Neurospora crassa Genome Database.	http://mips.helmholtz-muenchen.de/genre/proj/ncrassa/
COGEME, Phytopathogenic Fungi and Oomycete EST Database (v1.6), constructed and maintained by Darren Soanes (University of Exeter, UK).	http://cogeme.ex.ac.uk/
SGD, Saccaromyces Genome Database, scientific database of the molecular biology and genetics of the yeast Saccharomyces cerevisiae.	http://www.yeastgenome.org/
e-Fungi, warehouse which integrates sequence data (genomic data, EST data, Gene Ontology annotation, KEGG pathways and results of the following analyses performed on the genomic data) from multiple fungal sequences in a way that facilitates the systematic comparative study of those genomes (School of Computer Science and the Faculty of Life Sciences at the University of Manchester and the Departments of Computer Science and Biological Sciences at the University of Exeter).	http://beaconw.cs.manchester.ac.uk/efungi/execute/welcome

CADRE, Central Aspergillus Database Repository, resource for viewing assemblies and annotated genes arising from various Aspergillus sequencing and annotation projects.	http://www.cadre-genomes.org.uk/
FungalGenome, website with several links and references for the currently available fungal genomes sequences or proposed fungal genomes.	http://fungalgenomes.org/wiki/Fungal_Genomes_Links
Proteome Databases	
The Expasy (Expert Protein Analysis System) proteomics server of Swiss Institute of Bioinformatics (SIB). Analysis of protein sequences, structures and 2-D-PAGE.	http://ca.expaxy.org/
MIPS, Munich Information Center for Protein Sequences.	http://mips.gsf.de/http://ca.expaxy.org/
The PRIDE, Proteinomics IDEntifications Database. EMBL-EBI (European Bioinformatic Institute).	http://www.ebi.ac.uk/pride/
Integr8, Integrated information about deciphered genomes and their corresponding proteomes. EMBL-EBI.	http://www.ebi.ac.uk/integr8/EBI-Integr8-HomePage.do
SNAPPVIEW (Structure, iNterfaces and Aligments for Protein-Protein Interactions).	http://www.compbio.dundee.ac.uk/SNAPPI/download.jsp
Phospho3. Database of three-dimensional structures of phosphorylation sites.	http://cbm.bio.uniroma2.it/phospho3d/
Proteome Analyst PA-GOSUB 2.5. Sequences, predicted GO molecular functions and subcellular localisations.	http://www.cs.ualberta.ca/~bioinfo/PA/GOSUB/
RCSB, The Research Collaboratory for Structural Bioinformatics. Protein Database (PDB).	http://www.rcbs.org/pdb/home/home.do
PDB-Site. Comprehensive structural and functional information on PTMs, catalytic active sites, ligand binding (protein-protein, protein-DNA, protein-RNA interacions) in the Protein Data Bank (PDB).	http://wwwmgs.bionet.nsc.ru/mgs/gnw/pbbsite/
WoLF PSORT, Protein Subcellular Localization Prediction.	http://wolfpsort.org/

NMPdb, Nuclear Matrix Associated Proteins.	http://cubic.bioc.columbia.edu/db/NMPdb/
TargetP, predicts the subcellular location of eukaryotic proteins, based on the predicted presence of the N-terminal presequences.	http://www.cbs.dtu.dk/services/TargetP/
MitoP2, Mitochondrial Database. This database provides a comprehensive list of mitochondrial proteins of yeast, mouse, Arabidopsis thaliana, neurospora and human.	http://www.mitop.de:8080/mitop2/
The SecretomeP, Prediction of protein secretion and information on various PTMs and localisational aspect of the protein.	http://www.cbs.dtu.dk/services/SecretomeP/
MASCOT, a powerful search engine that uses MS data to identify proteins from primary sequence databases.	http://www.matrixscience.com/
VEMS, Virtual Expert Mass Spectrometrist. Program for integrated proteome analysis.	http://www.yass.sdu.dk/
The NetPhos server produces neural network predictions for serine, threonine and tyrosine phosphorylation sites in eukaryotic proteins.	http://www.cbs.dtu.dk/services/NetPhos/
ProPrInt, Protein-Protein Interaction Predictor. Compilation of web-resources in the field of Protein-Protein Interaction (PPI).	http://www.imtech.res.in/raghava/proprint/resources.htm
ProteomeCommons, public proteomics database for annotations and other information linked to the Tranche data repository and to other resources. It provides public access to free, open-source proteomics tools and data.	https://proteomecommons.org
Fungal Proteome Specialized Databases	
MPID, Protein-protein interaction Database of M. grisea.	http://bioinformatics.cau.edu.cn/cgi-bin/zzd-cgi/ppi/mpid.pl
FPPI, Protein-protein interaction database of F. graminearum.	http://csb.shu.edu.cn/fppi

Proteomics of Plant Pathogenic Fungi

Several proteomic studies have been carried out in order to understand fungal pathogenicity or plant-fungus interactions (for reviews see [266–268]), although the plant-fungus association has been the one most studied by Proteomics approaches (Table 6), which is outside the scope of this review. On the other hand, some fungal species have attracted an increasing interest in the biotechnological industry, in food science, or in agronomy as biocontrol agents (Table 7), which is also beyond the objectives of this work. At this point, this review describes studies published up to December 2009 in plant pathogenic fungi in descriptive proteomics (intracellular proteomics, subproteomics, and secretomics), differential expression proteomics, as well as some basic knowledge about the Interactomics in fungi (Table 3).

Table 6: Original proteomics papers published in plant-pathogenic fungi interactions

Pathogen-Host	Description of study (References)
Alternaria bassicicola-Arabidopsis	Study of change in the Arabidopsis secretome in response to salicylic acid and identifying of several proteins involved in pathogen response such as GDSL LIPASE1 (GLIP1) [231].
Aphanomyces eutiches-Medicago truncatula	Identification of several proteins which play a major role during root adaptation to various stress conditions [232], and study of parasitic plant-pathogen interactions formed between legumes and this oomycete [233, 234].
Black point disease-Barley	Identification of a novel late embryogenesis abundant (LEA) protein and a barley grain peroxidase 1 (BP1) that were specifically more abundant in healthy grain and black pointed grain, respectively [235].
B. graminis hordei-Barley	Systematic shotgun proteomics analysis at different stages of development of powdery mildew in the host to gain further understanding of the biology during infection of this fungus [153].
Cladosporium fulvum-Tomato	Identified 3 novel fungal secretory proteins [236].
Cronartium ribicola-Pinus strobus	Study of molecular basis of white pine blister rust resistance [237].
Diploidia scrobiculata-Pinus nigra Sphaeropsis sapinea-Pinus nigra	Study about defense protein responses in phloem of Austrian pine inoculated with D. scrobiculata and S. sapinea[238].
Erysiphe pisi-Pea	Identification of proteins implicated in powdery mildew resistance [239].

F. graminearum-Barley	Identification of proteins associated with resistance to Fusarium head blight in barley [240].
F. graminearum-Wheat	Identification of proteins associated with resistance to Fusarium head blight in wheat [241] and which have a role in interaction between F. graminearum and T. aestivum [155].
F. graminearum-Wheat	Identification of proteins associated with resistance to scab in wheat spikes [238].
F. moniliforme-Arabidopsis	Study of changes in the extracellular matrix of A. thaliana cell suspension cultures with fungal pathogen elicitors of F. moniliforme [242].
F. oxysporum-Sugar beet	Study of resistance to F. oxysporum disease [243].
F. oxysporum-Tomato	Identification of 21 tomato and 7 fungal proteins in the xylem sap of tomato plants infected by F. oxysporum [156].
F. verticillioides-Maize	Identification of protein change patterns in germinating maize embryos in response to infection with F. verticillioides[244].
Fusarium-Arabidopsis	Identification of differentially expressed proteins in response to treatments with pathogen-derived elicitors to identify pivotal genes' role in pathogen defence systems [245].
Fusarium-Maize	Study of the role of the extracellular matrix in signal modulation during pathogen-induced defence responses [246].
Gossypium hirsutum-Cotton	Identified pathogen-induced cotton proteins implicated in post-invasion defence reponses (PR-proteins related to oxidative burst), nitrogen metabolism, amino acid synthesis and isoprenoid synthesis [247].
Hyaloperonospora parasitica-Arabidopsis B. cinerea-Arabidopsis	Study of pathogenic resistance of Arabidopsis wild-type and CaHIR1-overexpressing transgenic plants inoculated with these fungi, among other pathogens [248].
Leptosphaeria maculans-Brassica	Identification of Brassica proteins involved in resistance to this fungus [249].
L. maculans-Brassica carinata L. maculans-Brassica napus	Study of changes in the leaf protein profiles of Brassica napus (highly susceptible) and Brassica carinata (highly resistant) in order to understand the biochemical basis for the observed resistance to L. maculans [250].

M. grisea-Rice	Change protein analysis during blast fungus infection of rice leaves with different levels of nitrogen nutrient [251]. Analysis of differentially expressed proteins induced by blast fungus in suspension-cultured cells [252] in leaves [253] and during appressorium formation [254]. Proteomic approach of differentially expressed proteins in rice plant leaves at 12 h and 24 h after treatment with the glycoprotein elicitor CSB I, purified from ZC(13), a race of the rice blast fungus M. grisea [255].
Marsonina brunnea f. sp. Multigermtub-Populus euramericana	Identification of proteins related to black spot disease resistance in poplar leaves [256].
Moliniophtora perniciosa-Cocoa	Optimization of protein extraction for cocoa leaves and meristemes infected by this fungus that causes witches' broom disease [257].
Nectria haematococca-Pea	Study of extracellular proteins in pea roots inoculated with N. haematococca [258].
Penincillium exposum-Piccia membranefaciens-Peach fruit	Peach fruit inoculated with P. exposum and treated with SA and P. membranefaciens [259].
Phellinus sulphurasces-Pseudotsuga menzeii	Comparative proteomic study to explore the molecular mechanisms that underlie the defense response of Douglas-fir to laminated root rot disease caused by P. sulphurascens [260].
Peronospora viciae-Pea	Catalogued host (pea) leaf proteins, which showed alternation in their abundance levels during a compatible interaction with P. viciae [261].
Plasmodiophora brassicae-Brassica napus	Study of changes in the root protein profile of canola with clubroot disease [262].
Puccinia triticina-Wheat	Change analysis in the proteomes of both host and pathogen during development of wheat leaf rust disease [154].
Rhizoctonia solani-Rice	Identification of proteins and DNA markers in rice associated with response to infection by R. solani [263].
Rust-Phaseolus vulgaris	Study of basal and R-gene-mediated plant defense in bean leaves against this pathogen [264].
S. sclerotiorum-Brassica napus	Study of changes in the leaf proteome of B. napus accompanying infection by S. sclerotiorum [265].

Table 7: Original proteomics papers published on fungi for biotechnological or agricultural applications

Fungus	Interest (reference)
Boletus edilus	Study of salinity stress of this ectomycorrhizal fungus for its importance in reforestation in saline areas [269].
Coprinopsis cinerea	Optimization of a protocol for 2-DE of extracellular proteins from these wood-degrading fungi [147].
Pleorotus ostreatus	
Phanerochaete chryso-sporium	
Polyporus brumalis	
Schizophyllum com-mune	
Glomus intraradices	Study of arbuscular mycorrhiza symbiosis [173].
Metarhizium aniso-pliae	Study of bioinsecticidal activity of this fungus to develop novel compounds or produce genetically modified plants resistant to insect pests [270].
Monascus pilosus	Study of the influence of nitrogen limitation for industrial production of many poliketide secondary metabolites [271].
Phanerochaete chryso-sporium	Several studies of ligninolytic processes for wood biodelignification in cellulose pulp industries [130, 131, 149, 150, 272].
Pleurotus sapidus	Study of secretome for wood biodelignification for peanuts industry applications [152].
Trichoderma atro-viride	Several studies in these fungus for their biocontrol properties [126, 273–275]
T. harzianum	
T. reesei	Study of cell-wall-degrading enzymes in A. bisporigera (an ectomycorrhizal basidiomycetous fungus) comparing with a MS/MS-based shotgun proteomics of the secretome of T. reesei [276].
Amanita bisporigera	

Descriptive, Subcellular, and Differential Expression Proteomics

Within this section, papers devoted to establishing reference proteome maps of fungal cells and structures and subcellular fractions, and to study changes in the protein profile between species, races, populations, mutants, growth and developmental stages, as well as growth conditions, are discussed, paying special attention to proteins related to pathogenicity and virulence.

Most of the reported work mainly uses mycelia from in vitro grown fungi, and 2-DE coupled to MS as proteomic strategy. Thus, a partial proteome map has been reported for the ascomycete B. cinerea Pers. Fr. (teleomorf Botryotinia fuckeliana (de Bary) Whetzel), a phytopathogenic necrotroph pathogen causing significant yield losses in a number of crops, by Fernández-Acero et al. They have reported the detection of 400 spots in Coomassie-stained 2-DE gels, covering the 5.4–7.7 pH and 14–85 kDa ranges. Out of 60 spots subjected to MS analysis, twenty-two proteins were identified by MALDI-TOF or ESI IT MS/MS, with some of them corresponding to forms of malate dehydrogenase, glyceraldehyde-3-phosphate dehydrogenase, and a cyclophilin, proteins that have been related to virulence [45]. In a second study, comparative proteomic analysis of two B. cinerea strains differing in virulence and toxin production revealed the existence of qualitative and quatitative differences in the 2-DE protein profile. Some of them were the same proteins mentioned above and they appeared overexpressed or exclusively in the most virulent strain [46]. A third and more exhaustive work tried to establish a proteomic map of B. cinerea during cellulose degradation [47]. Using 2-DE and MALDI-TOF/TOF MS/MS, 306 proteins were identified, mostly representing unannoted proteins. The authors conclude that since cellulose is one of the major components of the plant cell wall, many of the identified proteins may have a crucial role in the pathogenicity process, be involved in the infection cycle, and be potential antifungal targets.

A close relative to B. cinerea is the soil-borne Sclerotinia sclerotiorum. Yajima and Kav [64] performed the first comprehensive proteome-level study in this important phytopathogenic fungus, in order to gain a better understanding of its life cycle and its ability to infect susceptible plants. For the high-throughput identification of secreted as well as mycelial proteins, they employed 2-DE and MS/MS. Eighteen secreted and 95 mycelial proteins were identified. Many of the annotated secreted proteins were cell wall-degrading enzymes that had been previously identified as pathogenicity or virulence factors of S. Sclerotiorum. Furthermore, this study has allowed the annotation of a number of proteins that were unnamed, predicted, or hypothetical proteins with undetermined functions in the available databases.

Xu et al. [50] analyzed the proteome profile of six different isolates of Curvularia lunata, a maize phytopatogenic fungus, by means of 1-DE and 2-DE, in an attempt to correlate the band or spot pattern with virulence. According to the 1-DE band pattern, isolates were clustered into three groups consisting of different virulent types. By 2-DE 423 spots were resolved with 29 of them being isolate-specific, and 39 showed quantitative differences. Twenty proteins were identified by MALDI-TOF-TOF, most of them associated with virulence differentiation, metabolisms, stress response, and signal transduction. One of them was identified as Brn1 protein which has been reported to be related to melanin biosynthesis and the virulence differentiation in fungi.

The fungal pathogen F. graminearum (teleomorph Gibberella zeae) is the causal agent of Fusarium head blight in wheat, barley, and oats and Gibberella ear rot in maize in temperate climates worldwide. It synthesizes trichothecene mycotoxins during plant host attack to facilitate spread within the host. In order to study proteins and pathways that are important for successful host invasion, Taylor et al. [53] conducted experiments in which F. graminearum cells were grown in aseptic liquid culture conditions conducive to trichothecene and butenolide production in the absence of host plant tissue. Protein samples were extracted from three biological replicates of a time course study and subjected to iTRAQ (isobaric tags for relative and absolute quantification) analysis. Statistical analysis of a filtered dataset of 435 proteins revealed 130 F. graminearum proteins that exhibited significant changes in expression, 72 of which were upaccumulated relative to their level at the initial phase of the time course. There was good agreement between upaccumulated proteins identified by 2-DE-MS/MS and iTRAQ. RT-PCR and northern hybridization confirmed that genes encoding proteins that were upregulated based on iTRAQ were also transcriptionally active under mycotoxin-producing conditions. Numerous candidate pathogenicity proteins were identified using this technique, including many predicted secreted proteins. Curiously, enzymes catalyzing reactions in the mevalonate pathway leading to trichothecene precursors were either not identified or only identified in one replicate, indicating that proteomics approaches cannot always probe biological characteristics. Two-DE with MS has been used to compare the proteome of virus-free and virus-(FgV-DK21-) infected F. graminearum cultures [54]. The virus perturbs fungal developmental processes such as sporulation, morphology, and pigmentation and attenuates its virulence. A total of 148 spots showing differences in abundance were identified. Among these spots, 33 spots were subjected to ESI-MS/MS, with 23 identified. Seven proteins including sporulation-specific gene SPS2, triose phosphate isomerase, nucleoside diphosphate kinase, and woronin body major protein precursor were upaccumulated while 16, including enolase, saccharopine dehydrogenase, flavohemoglobin, mannitol dehydrogenase, and malate dehydrogenase, were downaccumulated. Variations in protein abundance were investigated at the mRNA level by real-time RT-PCR analysis, which confirmed the proteomic data for 9 out of the representative 11 selected proteins.

There are a few proteomics studies on fungal spores published. The results in the characterization of Penicillium spores by MALDI-TOF MS with different matrices demonstrated its ability for the classification of fungal spores [277]. Recently, Sulc et al. [40] have reported protein profiling of intact Aspergillus ssp. spores, including some plant pathogenic species, by MALDI-TOF MS, and they built up a mass spectral database with twenty-four Aspergillus strains. Thus, these mass finger-printing generated by MS can be used for typing and characterizing different fungal strains and finding new biomarkers in host-pathogen interactions. Another study on B. graminis f.sp. hordei (Bgh) spores concluded with the first proteome of Bgh, using a combination of 2-DE and MS analyses and matched to NCBInr EST on Bgh-translated genome databases [43]. The identity of 123 distinct fungal gene products was determined, most of them with a predicted function in carbohydrate, lipid, or protein metabolism indicating that the conidiospore is geared for the breakdown of storage compounds and protein metabolites during germination correlating with previously reported transcriptomic data [278, 279]. These results allowed a functionally annotated reference proteome for Bgh conidia.

Holzmuller et al. [118] have reported a technique to isolate the fungal haustorium (specialised structures, existing in intimate contact with the host cell, are required by the pathogen to acquire nutrients from the host cell) from infected plants, using the barley powdery mildew as an experimental system. The technique is of relevance in the study of the molecular bases of biotrophy considering that biotrophic fungi, including downy mildews (Oomycota), powdery mildews (Ascomycota), and rust fungi (Basidiomycota), are some of the most destructive pathogens on many plants. Extracted proteins were separated and analyzed by LC-MS/MS. The searches were made against a custom Bgh EST sequence database and the NCBInr fungal protein database, using the MS/MS data, and 204 haustorium proteins were identified. The majority of the proteins appeared to have roles in protein metabolic pathways and biological energy production. Surprisingly, pyruvate decarboxylase (PDC), involved in alcoholic fermentation and commonly abundant in fungi and plants, was absent in both their Bgh proteome data set and in their EST sequence database. Significantly, BLAST searches of the recently available Bgh genome sequence data also failed to identify a sequence encoding this enzyme, strongly indicating that Bgh does not have a gene for PDC [44].

In order to overcome the low proteome coverage of most of the proteomic platforms available, this being related to the physicochemical and biological complexity and high dynamism range of proteins, different strategies directed at subfractionating the whole proteome have been developed, most of them involving cell fractionation. The analysis of the subcellular proteomes [280] not only allows a deeper proteome coverage but also provides relevant informaction on the biology of the different organules, protein location, and trafficking. The number of intracellular subproteomic studies carried out with fungal plant pathogens is minimum. Next we introduce a couple of papers appearing in the literature. The number of them devoted to the cell wall and extracellular fraction is much higher, and because of that a specific section is devoted to them. Hernández-Macedo et al. [131] have reported differences in the patterns of cellular and membrane proteins obtained from iron-sufficient and iron-deficient mycelia from P. chrysosporium and L. edodes by using SDS-PAGE and 2-DE. Mitochondria have also received attention. Grinyer et al. [121] were the first to publish a mitochondrial subproteome, describing a successful sample preparation protocol and mitochondrial proteome map for T. harzianum. Based on protein databases of N. crassa, A. nidulans, A. oryzae, S. cerevisiae, andSchizosaccharomyces pombe, they identified 25 unique mitochondrial proteins involved in the tricarboxylic acid cycle, chaperones, binding-proteins and transport proteins, as well as mitochondrial integral membrane proteins. More recently, the same researchers separated and identified 13 of the 14 subunits of the T. reesei 20S proteasome [281], providing the first filamentous fungal proteasome proteomics and paving the way for future differential display studies addressing intracellular degradation of endogenous and foreign proteins in filamentous fungi.

Relevant information on biological systems and processes comes from comparative studies in which genotypes, including mutants, developmental stages, or environmental conditions supply the knowledge inferred from the observed differences. Fungal pathogenicity requires the coordinated regulation of multiple genes (and their protein products) involved in host recognition, spore germination, hyphal penetration, appressorium formation, toxin production, and secretion. To study the infection cycle and to identify virulence factors, proteomics provides us with a powerful tool for analyzing changes in protein expression between races and stages. However, most of these studies are made in planta after the plant inoculation, which is outside the scope of this review. In the case of plant fungal pathogens at least four papers have reported changes in the proteome at different developmental stages or strains. The dimorphic phytopathogenic fungusUstilago maydis has been established as a valuable model system to study fungal dimorphism and pathogenicity. In its haploid stage, the fungus is unicellular and multiplies vegetatively by budding and undergoing a dimorphic transition infective filamentous growth. This process is coordinately regulated by the bW/bE transcription factor. Böhmer et al. [68] reported the first proteome reference map of U. maydis cells, in which proteins were identified combining 2-DE with MALDI-TOF MS and ESI-MS/MS analyses. The authors observed 13 proteins spots accumulated in greater abundance in the bW/bE-induced filamentous form than in the budding state. The majority of the identified proteins might have putative roles in energy and general metabolism. Comparison of Rac1- and -b-regulated protein sets supports the hypothesis that filament formation during pathogenic development occurs via stimulation of a Rac1-containing signalling module. The proteins identified in this study might prove to be potential targets for antibiotic substances specifically targeted at dimorphic fungal pathogens. Detailed information on the proteins can be found in an interactive map accessible at the MIPS Ustilago Maydis Database (MUMDB;http://mips.gsf.de/genre/proj/ustilago/Maps/2D/). The reference map generated from U. maydis had a coverage of 4% of all annotated genes, indicating the low proteome coverage encompassed by standard proteomic techniques.

By using 2-DE and MALDI-TOF MS, specific proteins of asexual life stages from Phytophthora palmivora, a pathogen of cocoa and other economically important tropical crops, were analyzed [58]. From 400 (cyst and germinated cyst) to 800 (sporangial) could be resolved. Approximately 1% of proteins appeared to be specific for each of the mycelial, sporangial, zoospore, cyst, and germinated cyst stages of the life cycle. Moreover, they made the protein profiles of parallel samples of P. palmivora and P. infestans and demonstrated that precisely 30% of proteins comigrated suggesting that proteomics could be used to proteptype Phytophthora spp. In this work, only three identified proteins were reported, corresponding to actin isoforms. More recently, Ebstrup et al. [59] performed a proteomic study of proteins from cysts, germinated cysts, and appressoria on P. infestansgrown in vitro, identifying significant changes in the amount of several proteins. These identified proteins were most likely important for disease establishment and some of the proteins could therefore be putative targets for disease control. For example, downregulation of the crinkling- and necrosis-inducing (CRN2) protein in appressoria compared to germinated cysts and the discovery of upregulation of a putative elongation factor (EF-3) are of great interest. On the one hand, CRN2 protein might have an important function in the interaction with the host-plant before and after penetration into the leaf, this being a putative target for disease control. Since plants presumably do not contain EF-3, it could represent a putative antioomycete as well as a putative antifungal target. Furthermore, several representatives of housekeeping systems were upaccumulated, and these changes are most likely involved in the runup to the establishment of the infection of the host plant. The biotrophic fungal pathogen U. appendiculatus is the causal agent of rust disease of beans. Cooper et al. [67] surveyed the proteome from germinating and ungerminated asexual uredospores of this pathogen, using MudPIT MS/MS. The proteins identified revealed that uredospores require high energy and structural proteins during germination, indicating a metabolic transition from dormancy to germination.

The role of signal transduction in the pathogenecity of Stagonospora nodorum is well established and the inactivation of heterotrimeric G protein signaling caused developmental defects and reduced pathogenicity [282]. In a follow-up study, the S. nodorum wild-type and Galpha-defective mutant (gnal) proteomes were compared via 2-DE coupled to LC-MS/MS. By matching the protein mass spectra to the translate S. nodorumgenome, the study identified several Gnal-regulated proteins, including a positively regulated short-chain dehydrogenase (Sch1) [65].

Cao et al. [61] released an evaluation of pathogenic ability of Pyrenophora tritici-repentis and the possible adaptation to a saprophytic habit of an avirulent race. This fungus causes tan spot, an important foliar disease of wheat, and produces multiple host-specific toxins, including Ptr ToxB, which is also found in avirulent isolates of the fungus. In order to improve the understanding of the role of this homolog and evaluate the general pathogenic ability of P. tritici-repentis, the authors compared both full mycelial and secreted proteomes of avirulent and virulent isolates of the pathogen, by 2-DE and ESI-q-TOF MS/MS. The proteomic analysis revealed a number of the proteins found to be upregulated in a virulent race, which has been implicated in microbial virulence in other pathosystems, such as the secreted enzymes a-mannosidase and exo-b-1,3-glucanase, heat-shock and bip proteins, and various metabolic enzymes, which suggests a reduced general pathogenic ability in avirulent race of P. tritici-repentis, irrespective of toxin production.

Extracellular and Cell Wall Proteins: The Secretome

Most eukaryotic plant pathogens initially invade the space between host cell walls (the apoplastic space), and much of the initial host defence and pathogen counter defence happens in the apoplast and commonly involves secreted pathogen and host-derived proteins and metabolites [283]. While some pathogens remain exclusively in the apoplast, such as Cladosporium fulvum, others, including mildews, rusts smuts, Phytophthora, and Magnaporthe species, breach host cell walls but remain external to and separated from the host cytoplasm by host and pathogen cell membranes. Some host wall-breaching pathogens, like rusts, mildews, and oomycetes, form specialised expanded hyphal protuberances called haustoria whereas others, like maize smut and the rice blast fungi, use unexpanded but probably specialised intrahost cell wall hyphae [284]. The role of these structures was initially thought to be primarily nutrient acquisition, but recently their additional role in secretion of effectors, some of which are translocated to the host cytoplasm, has become more apparent. These issues have been recently reviewed by Ellis et al. [113].

The secretome has been defined as being the combination of native secreted proteins and the cell machinery involved in their secretion [285]. A defining characteristic of plant pathogenic fungi is the secretion of a large number of degradative enzymes and other proteins, which have diverse functions in nutrient acquisition, substrate colonization, and ecological interactions [286–288]. Several extracellular fungal enzymes, such as polygalacturonase, pectate lyase, xylanase, and lipase, have been shown or postulated to be required for virulence in at least one host-pathogen interaction [289–295]. Proteomics is the right approach to study the interaction between plants and microbes mediated by excreted molecules, the role of the cell wall and the interface, and to identify fungal protein effectors facilitating either infection (virulence factors, enzymes of the toxin biosynthesis pathways) or trigger defence responses (avirulence factors). In the light of this, it has been said that, unlike animals, "fungi digest their food and [then] eat it" [296], illustrating the large number of extracellular hydrolytic enzymes necessary to digest a plethora of potential substrates. Therefore, many of these proteins are of special interest in the study of plant pathogens [46, 64]. This might also be owing to the fact that secretome sample preparation is much faster and simpler than extraction and preparation of intracellular proteins. Next, a number of papers covering this topic are presented, including those dealing with the secretome of Trichoderma spp, a study directed at identifying proteins related to its biofungicidal activity.

Pioneering work on this field comes before the arrival of proteomics during the 1990s, with typical studies focused on the identification, purification, and characterization of single secreted proteins, under the influence of the biotechnology industry for the production of enzymes for commercial and industrial use [297]. The first complete proteomic study of secreted proteins was released on the filamentous fungus A. flavus [41, 42]. The interest of this study was the ability of both A. flavus and A. parasiticus to degrade the flavonoids that plants produce as typical secondary metabolites against invading microorganisms. The secreted proteins were analyzed by 2-DE and MALDI-TOF mass spectrometry, with 15 rutin-induced proteins and 7 noninduced proteins identified, among them enzymes of routine catabolism pathway and glycosidases.

In F. graminearum, a devastating pathogen of wheat, maize, and other cereals, Phalip et al. [51] investigated the exoproteome of this fungus grown on glucose and on plant cell wall (Humulus lupulus, L.). The culture medium was found to contain a larger amount of proteins and these were more diverse when the fungus grew on the cell wall. Using both 1-DE and 2-DE coupled to LC-MS/MS analysis and protein identification based on similarity searches, 84 unique proteins were identified in the cell wall-grown fungal exoproteome and 45% were implicated in plant cell wall degradation. These cell wall-degradating enzymes were predominantly matches to putative carbohydrate active enzymes implicated in cellulose, hemicelluloses, and pectin, catabolism. As expected,

F. graminearum grown on glucose produced relatively few cell wall-degrading enzymes. These results indicated that fungal metabolism becomes oriented towards the synthesis and secretion of a whole arsenal of enzymes able to digest almost the complete plant cell wall.

The secretome has also been analyzed in S. sclerotium as commented above [64]. In this study, 52 secreted proteins were identified and many of the annotated secreted proteins were cell wall-degrading enzymes that had been identified previously as pathogenic or virulence factors of S. sclerotium. However, one of them, α-L-arabinofuranosidase, which is involved in the initiation or progression of plant diseases, was not detected by previous EST studies, clearly demonstrating the merit of performing proteomic research.

Two studies have been published reporting the B. cinerea (B05.10) secreted proteins analysis [48, 49]. First, secretions were collected from fungus grown on a solid substrate of cellophane membrane while mock infecting media supplemented with the extract of full red tomato, ripened strawberry, or Arabidopsis leaf extract. Overall, 89 B. cinerea proteins were identified by high-throughput LC-MS/MS from all growth conditions. Sixty of these proteins were predicted to contain a SignalP motif indicating the extracellular location of the proteins. The proteins identified were transport proteins, proteins well-characterized for carbohydrate metabolism, peptidases, oxidation/reduction, and pathogenicity factors that could provide important insights into how B. cinerea might use secreted proteins for plant infection and colonization [48]. In the second work, the impact of degree of esterification of pectin on secreted enzyme of B. cinerea was studied, because changes during the ripening process of fruits appear to play an important role in the activation of the dormant infection. All the major components of the fruit cell wall (pectin, cellulose, hemicellulose) undergo these changes. By 1-DE and LC-MS/MS, 126 proteins were identified and 87 proteins were predicted secreted by SignalP, some of them being pectinases. The results showed that the growth of B. cinerea and the secretion of proteins were similar in cultures containing differently esterified pectins, and therefore it is likely that the activation of this fungi from dormant state is not solely dependent on changes in the degree of esterification of the pectin component of the plant cell wall [49]. Therefore, future studies of the B. cinerea secretome in infections of ripe and unripe fruits will provide important information for describing the mechanisms that the fungus employs to access nutrients and decompose tissues.

Using the plant pathogenic fungus L. maculans and symbiont Laccaria bicolor grown in culture, Vincent et al.[56] established a proteomic protocol for extraction, concentration, and resolution of the fungal secretome. These authors used both broad and narrow acidic and basic pH range in IEF. The quality of protein extracts was assessed by both 1-DE and 2-DE and MS identification. Compared with the previously published protocols for which only dozens of 2-DE spots were recovered from fungal secretome samples, in this study, up to approximately 2000 2-DE spots were resolved. This high resolution was confirmed with the identification of proteins along several pH gradients as well as the presence of major secretome markers such as endopolygalacturonases, beta-glucanosyltransferases, pectate lyases, and endoglucanases. Thus, shotgun proteomic experiments evidenced the enrichment of secreted protein within the liquid medium.

One of the earliest works was released on Trichoderma reesei mycelium cell wall, one of the most powerful producers of extracellular proteins, this study being justified in order to find out the protein secretory pathways and the effect of the fungal genus, strain, and media condition on the excretion through the cell wall [298]. A total of 220 cell envelope-associated proteins were successfully extracted and separated by 2-DE fromTrichoderma reesei mycelia actively secreting proteins and from mycelia in which the secretion of proteins is low. Out of the 52 2-DE spots subjected to ESI-TOF MS, 20 were identified, with HEX1, the major protein in Woronin body, a structure unique to filamentous fungi, being the most abundant one. Suárez et al. [299] studied the secretome of T. harzianum grown using either chitin (a key cell wall component) or cell wall of other fungi (R. solani, B. cinerea, or Pythium ultimum) as a nutrient source. For each different substrate, they found significant differences in 2-DE maps of extracellular proteins. However, despite these differences, the most abundant protein under all conditions was a novel aspartic protease (P6281), which showed a strong homology with polyporopepsin from Irpex lacteus. This led to speculation that this protein plays a fundamental role in the parasitic activity of Trichoderma spp. Marra et al. [273] have studied interactions between T. atroviride, two different fungal phytopathogens (B. cinerea and R. solani), and plants (bean). Two-DE was used to analyze separately collected proteomes from each single, two- or three-partner interaction. Then, differential proteins were subjected to MALDI-TOF MS and in silico analysis to search homologies with known proteins. Thus, a large number of protein factors associated with the multiplayer interactions examined were identified, including protein kinases, cyclophilines, chitine synthase, and ABC transporters. Recently, another similar study was released between T. harzianum and R. solani by analysing the secretome to identify the target proteins that are directly related to biocontrol mechanism [274]. Seven cell-wall degrading enzymes, chitinase, cellulase, xylanase, β-1,3-glucanase, β-1,6-glucanase, mannanase, and protease, were revealed by activity assay, in-gel activity stain, 2-DE, and LC-MS/MS analysis, these being increased in response to R. solani.

A cell wall proteome has been proposed for the oomycete Phytophthora ramorum, the causal agent of sudden oak death, in order to study its pathogenic factors [60]. This study showed an inventory of cell wall-associated proteins based on MS sequence analysis. Seventeen secreted proteins were identified by homology searches. The functional classification revealed several cell wall-associated proteins, thus suggesting that cell wall proteins may also be important for fungal pathogenicity.

The filamentous fungus Neurospora crassa is a model laboratory organism but in nature is commonly found growing on dead plant material, particularly grasses. Using functional genomics resources available for N. crassa, which include a near-full genome deletion strain set and whole genome microarrays, Tian et al. [300] undertook a system-wide analysis of plant cell wall and cellulose degradation, identifying approximately 770 genes that showed expression differences when N. crassa was cultured on ground Miscanthus stems as a sole carbon source. An overlap set of 114 genes was identified from expression analysis of N. crassa grown on pure cellulose. Functional annotation of upregulated genes showed enrichment for proteins predicted to be involved in plant cell wall degradation, but also many genes encoding proteins of an unknown function. As a complement to expression data, the secretome associated with N. crassa growth on Miscanthus and cellulose was determined using a shotgun MudPIT proteomic strategy. Over 50 proteins were identified, including 10 of the 23 predicted N. crassa cellulases. Strains containing deletions in genes encoding 16 proteins detected in both the microarray and mass spectrometry experiments were analyzed for phenotypic changes during growth on crystalline cellulose and for cellulase activity. While growth of some of the deletion strains on cellulose was severely diminished, other deletion strains produced higher levels of extracellular proteins that showed increased cellulase activity. These results show that proteomics in combination with other powerful tools available in model systems such as N. crassa allow for a comprehensive system level understanding of fungal biology.

Interactomics

The biological organization in living cells can be regarded as being part of a complex network [301–303]. Traditional approaches studied a single gene or unique protein and therefore did not provide a complete knowledge of the biological processes. Proteins release their functional roles through their interactions with one another in vivo. Thus, developing a protein-protein interaction (PPI) network can lead to a more comprehensive understanding of the cell processes [304]. Interactomics is a discipline at the intersection of bioinformatics and biology that deals with studying both the interactions and the consequences of those interactions between and among proteins and other molecules within a cell [305]. The network of all such interactions is called the interactome. Interactomics thus aims to compare these interaction networks (i.e., interactomes) between and within species in order to find how the traits of such networks are either preserved or varied. Interactomics is an example of top-down systems biology, which takes an overhead, as well as overall, view of a biosystem or organism.

In recent years, high-throughput methods have been implemented to identify PPIs [306–310] and these have recently been reviewed in [311, 312]. Using these experimental methods, such as yeast two-hybrid screens, PPI networks for a series of model organisms were determined and allowed us to better understand the function of proteins at the level of system biology. Two-hybrid screening (also known as yeast two hybrid system or Y2H) is a powerful tool for identifying PPI. The premise behind the test is the activation of dowstream reporter gene(s) by the binding of a transcription factor onto an upstream activating sequence (UAS). For the purposes of two-hybrid screening, the transcription factor is split into two separate fragments, called the binding domain (BD) and activating domain (AD). The BD is the domain responsible for binding to the UAS and the AD is the domain responsible for activation of transcription [313].

In parallel with the large-scale experimental determination of PPI, many PPI prediction methods were also developed. These methods are based on diverse attributes, concepts, or data types, such as interolog [314], gene expression profiles [315], gene ontology (GO) annotations [316], domain interactions [317], coevolution [318], and structural information [319]. Some machine learning methods, such as support vector machines (SVMs) have also been used to predict PPIs [320, 321]. Among the above-mentioned computational methods the interolog approach has been widely implemented [322] and has proved to be reliable for predicting PPI from model organisms [323]. The core idea of the interolog approach is that many PPIs are conserved in different organisms [324]. Accumulated PPI data from model organisms as well as advances in detecting orthologous proteins in different organisms [280] have continuously made the interolog method an increasingly powerful tool for constructing PPI maps for entire proteomes.

Using the interolog method, He et al. [57] constructed the first PPI network for M. grisea. Thus, 11674 PPIs among 3017 M. grisea proteins were deduced from the experimental PPI data in different organisms, although the predicted PPI network covered approximately only one-fourth of the fungal proteome and may still contain many false-positives. Moreover, they built two subnets called pathogenicity and secreted proteins networks, which may be helpful in constructing an interactome between the rice blast fungus and rice (MPID website, http://bioinformatics.cau.edu.cn/zzd_lab/MPID.html).

A F. graminearum protein-protein interaction database providing comprehensive information on protein-protein interactions based on both interologs from several protein-protein interaction of seven species and domain-domain interactions experimentally determined based on protein is available athttp://csb.shu.edu.cn/fppi [55]. It contains 223 166 interactions among 7406 proteins for F. graminearum, covering 52% of the whole F. graminearum proteome.

CONCLUDING REMARKS

In the current scientific scenario, Proteomics should be understood to be part of a multidisciplinary approach. A combination of high-throughput "Omics" (Genomics, Transcriptomics, Proteomics, and Metabolomics) and classical biochemistry and cell biology techniques should be used for data validation and to deepen the knowledge of living organisms. Proteomic techniques are used to characterize a specific protein or a structural or functional group of proteins. This is what we can call "Hypothesis-driven Proteomics", "Targeted Proteomics", or "Proteinomics". This type of study will provide relevant information on protein structure and function, isoforms, organs, cells, and subcellular location and trafficking, processing, signal peptides, PTMs, expression kinetics, and correlation with RNA and metabolites. At the same time, it is a method for validating data obtained using one specific approach [102].

Despite the continuous development and improvement of powerful proteomic techniques, protocols, equipments, and bioinformatic tools, just a minimal fraction of the cell proteome, and for only a few organisms, has been characterized so far. This is mainly due to the enormous diversity and complexity of proteomes, and to technical limitations in quantification, sensitivity, resolution, speed of data capture, and analysis.

In the field of Fungal MS-based Proteomics, great progress has been made in past years. This is because of the increasing number of fungal genomes available and the developments in sample preparation, high-resolution protein separation techniques, MS, MS software for effective protein identification and characterization, and bioinformatics technology. The tremendous diversity and genome flexibility in fungi, however, will make this task a difficult one. Thus, a key step in sequence analysis is the annotation. The existing programs for automated gene prediction are not perfect and need to be improved or trained better. Follow-up manual annotation is also necessary to improve the accuracy of automated annotation, but this is time-consuming and labor-intensive.

Ultimately, a comprehensive genome database similar to YPD (http://www.yeastgenome.org/) will be desirable for fungal pathogens.

To date, most proteomic studies in plant pathogenic fungi have been limited to 1- and 2-DE analysis. However, various powerful proteomic methods have been developed for genomewide analysis of protein expression, protein localization, and protein-protein interaction in fungi. Whole-genome protein arrays and systematic yeast two-hybrid assays have been used to characterize the yeast proteome and interactome. Integration of large-scale genomics and proteomics data enables the elucidation of global networks and system biology studies in yeast. It is necessary for similar advanced proteomic resources to be soon available for some fungal plant pathogens. Moreover, second-generation MS technologies for Quantitative Proteomics such as 2-DIGE, stable isotope labeling, (ICAT, iTRAQ and SILAC), or label-free methods (peak integration, spectral counting) have not yet begun or are beginning to be applied to Fungal Proteomics research.

Otherwise, the major challenge is the analysis and significance of PTMs because proteins have properties arising from their folded structure and so generic methods are difficult to design and apply.

In conclusion, since plant pathogenic fungi cause important losses in a number of crops, it is necessary to make high-throughput studies on these organisms to identify pathogenicity factors. Although genomics-based investigation of host-pathogen interactions can provide valuable information on the changes in gene expression, the investigation into changes in protein abundance is also important, in order to identify those proteins that are essential during such interactions. This is because there is often a poor correlation between transcript and protein abundance [325]. Proteomics analysis is an excellent tool that can give us a great deal of information about fungal pathogenicity by high-throughput studies. This approach has allowed the identification of new fungal virulence factors, characterizing signal transduction or biochemical pathways, studying the fungal life cycle and their life-style. We can use this information to provide new targets for disease crop diagnosis focused on fungicide design. Otherwise, the secretome analysis is especially important because fungi secrete an arsenal of extracellular enzymes to break down the plant cell wall for pathogen penetration and nutrient consumption. In this sense, Proteomics allows us to identify numerous differential proteins involved in multiple-player cross-talk normally occurring in nature between plants and pathogens, the so-called "interaction proteomes". Finally, MS-based Proteomics can help us to characterize fungal strains and find new biomarkers in host-pathogen interactions. In short, Fungal Proteomics is in the first step. Therefore, we still have a long way to go in the "Omics" of Plant Pathogenic Fungi compared to studies made in Humans, Bacteria, Yeast, or Plants. The important investment made by both the public and private sector in recent years augurs good prospects in fungal proteomics research in the future.

ABBREVIATIONS

1-DE:	One-dimentional electrophoresis
2-DE:	Two-dimentional electrophoresis
ESI:	Electrospray ionization
HPLC:	High performance liquid chromatograohy
IEF:	Isoelectrofocusing
LC:	Liquid chromatography
MALDI:	Matrix-assisted laser desorption/ionization
MS:	Mass spectrometry
SDS-PAGE:	Sodium dodecyl sulphate polyacrylamide gel electrophoresis
TOF:	Time of flight.

ACKNOWLEDGMENTS

This work was supported by the Spanish "Ministerio de Ciencia e Innovación" (Project BOTBANK EUI2008-03686), the "Junta de Andalucía", and the "Universidad de Córdoba" (AGR 164: Agricultural and Plant Biochemistry and Proteomics Research Group).

REFERENCES

1.   D. L. Hawksworth, "The fungal dimension of biodiversity: magnitude, significance, and conservation,"Mycological Research, vol. 95, no. 6, pp. 641–655, 1991. ·

2.   D. L. Hawksworth, P. M. Kirk, B. C. Sutton, and D. N. Pegler, Ainsworth and Bisby›s Dictionary of the Fungi, International Mycological Institute, Surrey, UK, 2001.

3.   G. N. Agrios, Plant Pathology, Elsevier Academic Press, San Diego, Calif, USA, 2005.

4.   A. M. Green, U. G. Mueller, and R. M. Adams, "Extensive exchange of fungal cultivars between sympatric species of fungus-growing ants," Molecular Ecology, vol. 11, no. 2, pp. 191–195, 2002.

5.   A. Pérez-García, E. Mingorance, M. E. Rivera, et al., "Long-term

preservation of Podosphaera fusca using silica gel," Journal of Phytopathology, vol. 154, no. 3, pp. 190–192, 2006.

6.   D. K. Arora, P. D. Bridge, and D. Bhatnagar, Fungal Biotechnology in Agricultural, Food and Environmental Applications, Marcel Dekker, New York, NY, USA, 2004.

7.   D. A. Glawe, "The powdery mildews: a review of the world›s most familiar (yet poorly known) plant pathogens," Annual Review of Phytopathology, vol. 46, pp. 27–51, 2008.

8.   S. Isaac, Fungal-Plant Interactions, Chapman & Hall, London, UK, 1992.

9.   H. H. Flor, "Inheritance of pathogenicity in Melampsora lini," Phytopathology, vol. 32, pp. 653–669, 1942.

10.   P. J. De Wit, R. Mehrabi, H. A. Van den Burg, and I. Stergiopoulos, "Fungal effector proteins: past, present and future," Molecular Plant Pathology, vol. 10, no. 6, pp. 735–747, 2009.

11.   H. P. van Esse, J. W. Van›t Klooster, M. D. Bolton, et al., "The Cladosporium fulvum virulence protein Avr2 inhibits host proteases required for basal defense," Plant Cell, vol. 20, no. 7, pp. 1948–1963, 2008. · ·

12.   T. L. Friesen, J. D. Faris, P. S. Solomon, and R. P. Oliver, "Host-specific toxins: effectors of necrotrophic pathogenicity," Cellular Microbiology, vol. 10, no. 7, pp. 1421–1428, 2008.

13.   X. Gao and M. V. Kolomiets, "Host-derived lipids and oxylipins are crucial signals in modulating mycotoxin production by fungi," Toxin Reviews, vol. 28, no. 2-3, pp. 79–88, 2009.

14.   K. S. Kim, J. Y. Min, and M. B. Dickman, "Oxalic acid is an elicitor of plant programmed cell death during Sclerotinia sclerotiorum disease development," Molecular Plant-Microbe Interactions, vol. 21, no. 5, pp. 605–612, 2008.

15.   C. B. Lawrence, T. K. Mitchell, K. D. Craven, Y. Cho, R. A. Cramer, and K. H. Kim, "At death›s door: alternaria pathogenicity mechanisms," Plant Pathology Journal, vol. 24, no. 2, pp. 101–111, 2008.

16.   H. C. van der Does and M. Rep, "Virulence genes and the evolution of host specificity in plant-pathogenic fungi," Molecular Plant-Microbe Interactions, vol. 20, no. 10, pp. 1175–1182, 2007.

17.   Z. Zhang, T. L. Friesen, K. J. Simons, S. S. Xu, and J. D. Faris, "Development, identification, and validation of markers for marker-assisted selection against the Stagonospora nodorum toxin sensitivity genes Tsn1 and Snn2 in wheat," Molecular Breeding, vol. 23, no. 1, pp.

35–49, 2009.

18. B. Pariaud, V. Ravigné, F. Halkett, H. Goyeau, J. Carlier, and C. Lannou, "Aggressiveness and its role in the adaptation of plant pathogens," Plant Pathology, vol. 58, no. 3, pp. 409–424, 2009.

19. D. Parker, M. Beckmann, H. Zubair, et al., "Metabolomic analysis reveals a common pattern of metabolic re-programming during invasion of three host plant species by Magnaporthe grisea," Plant Journal, vol. 59, no. 5, pp. 723–737, 2009.

20. S. McCouch, "Diversifying selection in plant breeding," PLoS Biology, vol. 2, no. 10, article e347, 2004. · ·

21. Y. Yang, H. Zhang, G. Li, W. Li, X. Wang, and F. Song, "Ectopic expression of MgSM1, a Cerato-platanin family protein from Magnaporthe grisea, confers broad-spectrum disease resistance inArabidopsis," Plant Biotechnology Journal, vol. 7, no. 8, pp. 763–777, 2009.

22. B. S. Kim and B. K. Hwang, "Microbial fungicides in the control of plant diseases," Journal of Phytopathology, vol. 155, no. 11-12, pp. 641–653, 2007. · ·

23. G. Berg, "Plant-microbe interactions promoting plant growth and health: perspectives for controlled use of microorganisms in agriculture," Applied Microbiology and Biotechnology, vol. 84, no. 1, pp. 11–18, 2009.

24. G. Aguileta, M. E. Hood, G. Refrégier, and T. Giraud, "Chapter 3 genome evolution in plant pathogenic and symbiotic fungi," Advances in Botanical Research, vol. 49, pp. 151–193, 2009.

25. G. Aguileta, G. Refrégier, R. Yockteng, E. Fournier, and T. Giraud, "Rapidly evolving genes in pathogens: methods for detecting positive selection and examples among fungi, bacteria, viruses and protists," Infection, Genetics and Evolution, vol. 9, no. 4, pp. 656–670, 2009.

26. M. D. C. Alves and E. A. Pozza, "Scanning electron microscopy applied to seed-borne fungi examination," Microscopy Research and Technique, vol. 72, no. 7, pp. 482–488, 2009.

27. V. Farkaš, "Structure and biosynthesis of fungal cell walls: methodological approaches," Folia Microbiologica, vol. 48, no. 4, pp. 469–478, 2003.

28. A. R. Hardham and H. J. Mitchell, "Use of molecular cytology to study the structure and biology of phytopathogenic and mycorrhizal fungi," Fungal Genetics and Biology, vol. 24, no. 1-2, pp. 252–284, 1998.

29. R. J. Howard, "Cytology of fungal pathogens and plant-host interactions," Current Opinion in Microbiology, vol. 4, no. 4, pp. 365–373, 2001.

30. S. Koh and S. Somerville, "Show and tell: cell biology of pathogen invasion," Current Opinion in Plant Biology, vol. 9, no. 4, pp. 406–413, 2006.

31. P. Pérez and J. C. Ribas, "Cell wall analysis," Methods, vol. 33, no. 3, pp. 245–251, 2004.

32. J. A. Poland, P. J. Balint-Kurti, R. J. Wisser, R. C. Pratt, and R. J. Nelson, "Shades of gray: the world of quantitative disease resistance," Trends in Plant Science, vol. 14, no. 1, pp. 21–29, 2009.

33. J. Xu, "Fundamentals of fungal molecular population genetic analyses," Current Issues in Molecular Biology, vol. 8, no. 2, pp. 75–89, 2006. ·

34. M. Choquer, E. Fournier, C. Kunz, et al., "Botrytis cinerea virulence factors: new insights into a necrotrophic and polyphageous pathogen," FEMS Microbiology Letters, vol. 277, no. 1, pp. 1–10, 2007. · ·

35. M. J. Egan and N. J. Talbot, "Genomes, free radicals and plant cell invasion: recent developments in plant pathogenic fungi," Current Opinion in Plant Biology, vol. 11, no. 4, pp. 367–372, 2008.

36. L. A. Hadwiger, "Localization predictions for gene products involved in non-host resistance responses in a model plant/fungal pathogen interaction," Plant Science, vol. 177, no. 4, pp. 257–265, 2009.

37. K.-C. Tan, S. V. Ipcho, R. D. Trengove, R. P. Oliver, and P. S. Solomon, "Assessing the impact of transcriptomics, proteomics and metabolomics on fungal phytopathology," Molecular Plant Pathology, vol. 10, no. 5, pp. 703–715, 2009.

38. S. Walter, P. Nicholson, and F. M. Doohan, "Action and reaction of host and pathogen during Fusariumhead blight disease," New Phytologist, vol. 185, no. 1, pp. 54–66, 2010.

39. V. Bhadauria, S. Banniza, Y. Wei, and Y.-L. Peng, "Reverse genetics for functional genomics of phytopathogenic fungi and oomycetes," Comparative and Functional Genomics, vol. 2009, Article ID 380719, 11 pages, 2009.

40. M. Sulc, K. Peslova, M. Zabka, M. Hajduch, and V. Havlicek, "Biomarkers of Aspergillus spores: strain typing and protein identification," International Journal of Mass Spectrometry, vol. 280, no. 1–3, pp. 162–168, 2009.

41. M. L. Medina, P. A. Haynes, L. Breci, and W. A. Francisco, "Analysis of secreted proteins fromAspergillus flavus," Proteomics, vol. 5, no. 12, pp. 3153–3161, 2005.

42. M. L. Medina, U. A. Kiernan, and W. A. Francisco, "Proteomic analysis

of rutin-induced secreted proteins from Aspergillus flavus," Fungal Genetics and Biology, vol. 41, no. 3, pp. 327–335, 2004.

43. S. Noir, T. Colby, A. Harzen, J. Schmidt, and R. Panstruga, "A proteomic analysis of powdery mildew (Blumeria graminis f.sp. hordei) conidiospores," Molecular Plant Pathology, vol. 10, no. 2, pp. 223–236, 2009.

44. D. Godfrey, Z. Zhang, G. Saalbach, and H. Thordal-Christensen, "A proteomics study of barley powdery mildew haustoria," Proteomics, vol. 9, no. 12, pp. 3222–3232, 2009.

45. F. J. Fernández-Acero, I. Jorge, E. Calvo, et al., "Two-dimensional electrophoresis protein profile of the phytopathogenic fungus Botrytis cinerea," Proteomics, vol. 6, supplement 1, pp. S88–S96, 2006. ·

46. F. J. Fernández-Acero, I. Jorge, E. Calvo, et al., "Proteomic analysis of phytopathogenic fungus Botrytis cinerea as a potential tool for identifying pathogenicity factors, therapeutic targets and for basic research," Archives of Microbiology, vol. 187, no. 3, pp. 207–215, 2007.

47. F. J. Fernández-Acero, T. Colby, A. Harzen, J. M. Cantoral, and J. Schmidt, "Proteomic analysis of the phytopathogenic fungus Botrytis cinerea during cellulose degradation," Proteomics, vol. 9, no. 10, pp. 2892–2902, 2009.

48. P. Shah, J. A. Atwood, R. Orlando, H. E. Mubarek, G. K. Podila, and M. R. Davis, "Comparative proteomic analysis of Botrytis cinerea secretome," Journal of Proteome Research, vol. 8, no. 3, pp. 1123–1130, 2009.

49. P. Shah, G. Gutierrez-Sanchez, R. Orlando, and C. Bergmann, "A proteomic study of pectin-degrading enzymes secreted by Botrytis cinerea grown in liquid culture," Proteomics, vol. 9, no. 11, pp. 3126–3135, 2009.

50. S. Xu, J. I. E. Chen, L. Liu, X. Wang, X. Huang, and Y. Zhai, "Proteomics associated with virulence differentiation of Curvularia iunata in maize in China," Journal of Integrative Plant Biology, vol. 49, no. 4, pp. 487–496, 2007.

51. V. Phalip, F. Delalande, C. Carapito, et al., "Diversity of the exoproteome of Fusarium graminearumgrown on plant cell wall," Current Genetics, vol. 48, no. 6, pp. 366–379, 2005.

52. J. M. Paper, J. S. Scott-Craig, N. D. Adhikari, C. A. Cuomo, and J. D. Walton, "Comparative proteomics of extracellular proteins in vitro and in planta from the pathogenic fungus Fusarium graminearum," Proteomics, vol. 7, no. 17, pp. 3171–3183, 2007.

53. R. D. Taylor, A. Saparno, B. Blackwell, et al., "Proteomic analyses of Fusarium graminearum grown under mycotoxin-inducing conditions," Proteomics, vol. 8, no. 11, pp. 2256–2265, 2008.

54. S.-J. Kwon, S.-Y. Cho, K.-M. Lee, J. Yu, M. Son, and K.-H. Kim, "Proteomic analysis of fungal host factors differentially expressed by Fusarium graminearum infected with Fusarium graminearum virus-DK21," Virus Research, vol. 144, no. 1-2, pp. 96–106, 2009. · ·

55. X.-M. Zhao, X.-W. Zhang, W.-H. Tang, and L. Chen, "FPPI: Fusarium graminearum protein-protein interaction database," Journal of Proteome Research, vol. 8, no. 10, pp. 4714–4721, 2009.

56. D. Vincent, M.-H. Balesdent, J. Gibon, et al., "Hunting down fungal secretomes using liquid-phase IEF prior to high resolution 2-DE," Electrophoresis, vol. 30, no. 23, pp. 4118–4136, 2009.

57. F. He, Y. Zhang, H. Chen, Z. Zhang, and Y. L. Peng, "The prediction of protein-protein interaction networks in rice blast fungus," BMC Genomics, vol. 9, article 519, 2008.

58. S. J. Shepherd, P. van West, and N. A. R. Gow, "Proteomic analysis of asexual development ofPhytophthora palmivora," Mycological Research, vol. 107, no. 4, pp. 395–400, 2003.

59. T. Ebstrup, G. Saalbach, and H. Egsgaard, "A proteomics study of in vitro cyst germination and appressoria formation in Phytophthora infestans," Proteomics, vol. 5, no. 11, pp. 2839–2848, 2005.

60. H. J. G. Meijer, P. J. I. van de Vondervoort, Q. Y. Yin, et al., "Identification of cell wall-associated proteins from Phytophthora ramorum," Molecular Plant-Microbe Interactions, vol. 19, no. 12, pp. 1348–1358, 2006.

61. T. Cao, Y. M. Kim, N. N. V. Kav, and S. E. Strelkov, "A proteomic evaluation of Pyrenophora tritici-repentis, causal agent of tan spot of wheat, reveals major differences between virulent and avirulent isolates," Proteomics, vol. 9, no. 5, pp. 1177–1196, 2009. · ·

62. M. Matis, M. Žakelj-Mavrič, and J. Peter-Katalinić, "Mass spectrometry and database search in the analysis of proteins from the fungus Pleurotus ostreatus," Proteomics, vol. 5, no. 1, pp. 67–75, 2005.

63. D. K. Lakshman, S. S. Natarajan, S. Lakshman, W. M. Garrett, and A. K. Dhar, "Optimized protein extraction methods for proteomic analysis of Rhizoctonia solani," Mycologia, vol. 100, no. 6, pp. 867–875, 2008.

64. W. Yajima and N. N. V. Kav, "The proteome of the phytopathogenic fungus Sclerotinia sclerotiorum,"Proteomics, vol. 6, no. 22, pp. 5995–6007, 2006.

65. K. A. R. C. Tan, J. L. Heazlewood, A. H. Millar, G. Thomson, R. P. Oliver, and P. S. Solomon, "A signaling-regulated, short-chain dehydrogenase of Stagonospora nodorum regulates asexual development," Eukaryotic Cell, vol. 7, no. 11, pp. 1916–1929, 2008.

66. S. Bringans, J. K. Hane, T. Casey, et al., "Deep proteogenomics; high throughput gene validation by multidimensional liquid chromatography and mass spectrometry of proteins from the fungal wheat pathogen Stagonospora nodorum," BMC Bioinformatics, vol. 10, article 301, 2009.

67. B. Cooper, A. Neelam, K. B. Campbell, et al., "Protein accumulation in the germinating Uromyces appendiculatus uredospore," Molecular Plant-Microbe Interactions, vol. 20, no. 7, pp. 857–866, 2007. · ·

68. M. Böhmer, T. Colby, C. Böhmer, A. Bräutigam, J. Schmidt, and M. Bölker, "Proteomic analysis of dimorphic transition in the phytopathogenic fungus Ustilago maydis," Proteomics, vol. 7, no. 5, pp. 675–685, 2007.

69. R. J. Weld, K. M. Plummer, M. A. Carpenter, and H. J. Ridgway, "Approaches to functional genomics in filamentous fungi," Cell Research, vol. 16, no. 1, pp. 31–44, 2006.

70. J. Schumacher, I. F. de Larrinoa, and B. Tudzynski, "Calcineurin-responsive zinc finger transcription factor CRZ1 of Botrytis cinerea is required for growth, development, and full virulence on bean plants,"Eukaryotic Cell, vol. 7, no. 4, pp. 584–601, 2008.

71. N. Ajiro, Y. Miyamoto, A. Masunaka, et al., "Role of the host-selective ACT-toxin synthesis gene ACTTS2 encoding an enoyl-reductase in pathogenicity of the tangerine pathotype of alternaria alternata," Phytopathology, vol. 100, no. 2, pp. 120–126, 2010.

72. M.-H. Chi, S.-Y. Park, S. Kim, and Y.-H. Lee, "A novel pathogenicity gene is required in the rice blast fungus to suppress the basal defenses of the host," PLoS Pathogens, vol. 5, no. 4, Article ID e1000401, 2009.

73. K. H. Lamour, L. Finley, O. Hurtado-Gonzales, D. Gobena, M. Tierney, and H. J. Meijer, "Targeted gene mutation in Phytophthora spp.," Molecular Plant-Microbe Interactions, vol. 19, no. 12, pp. 1359–1367, 2006.

74. A. Di Pietro, M. P. Madrid, Z. Caracuel, J. Delgado-Jarana, and M. I. G. Roncero, "Fusarium oxysporum: exploring the molecular arsenal of a vascular wilt fungus," Molecular Plant Pathology, vol. 4, no. 5, pp. 315–325, 2003.

75. S. R. Herron, J. A. Benen, R. D. Scavetta, J. Visser, and F. Jurnak, "Structure and function of pectic enzymes: virulence factors of plant pathogens," Proceedings of the National Academy of Sciences of the

United States of America, vol. 97, no. 16, pp. 8762–8769, 2000.

76. S. Meng, T. Torto-Alalibo, M. C. Chibucos, B. M. Tyler, and R. A. Dean, "Common processes in pathogenesis by fungal and oomycete plant pathogens, described with Gene Ontology terms," BMC Microbiology, vol. 9, supplement 1, p. S7, 2009.

77. J. Delgado-Jarana, A. L. Martínez-Rocha, R. Roldán-Rodriguez, M. I. Roncero, and A. D. Pietro, "Fusarium oxysporum G-protein β subunit Fgb1 regulates hyphal growth, development, and virulence through multiple signalling pathways," Fungal Genetics and Biology, vol. 42, no. 1, pp. 61–72, 2005.

78. N. Lee, C. A. D›Souza, and J. W. Kronstad, "Of smuts, blasts, mildews, and blights: cAMP signaling in phytopathogenic fungi," Annual Review of Phytopathology, vol. 41, pp. 399–427, 2003. · ·

79. N. Rispail and A. Di Pietro, "Fusarium oxysporum Ste12 controls invasive growth and virulence downstream of the Fmk1 MAPK cascade," Molecular Plant-Microbe Interactions, vol. 22, no. 7, pp. 830–839, 2009.

80. J. R. Xu and J. E. Hamer, "MAP kinase and cAMP signaling regulate infection structure formation and pathogenic growth in the rice blast fungus Magnaporthe grisea," Genes and Development, vol. 10, no. 21, pp. 2696–2706, 1996. ·

81. L. Zheng, M. Campbell, J. Murphy, S. Lam, and J. R. Xu, "The BMP1 gene is essential for pathogenicity in the gray mold fungus Botrytis cinerea," Molecular Plant-Microbe Interactions, vol. 13, no. 7, pp. 724–732, 2000. ·

82. N. Rispail, D. M. Soanes, C. Ant, et al., "Comparative genomics of MAP kinase and calcium-calcineurin signalling components in plant and human pathogenic fungi," Fungal Genetics and Biology, vol. 46, no. 4, pp. 287–298, 2009.

83. M. S. López-Berges, A. Di Pietro, M. J. Daboussi, et al., "Identification of virulence genes in Fusarium oxysporum f. sp. lycopersici by large-scale transposon tagging," Molecular Plant Pathology, vol. 10, no. 1, pp. 95–107, 2009.

84. V. Bhadauria, L. Popescu, W. S. Zhao, and Y. L. Peng, "Fungal transcriptomics," Microbiological Research, vol. 162, no. 4, pp. 285–298, 2007.

85. Y. Oh, N. Donofrio, H. Pan, et al., "Transcriptome analysis reveals new insight into appressorium formation and function in the rice blast fungus Magnaporthe oryzae," Genome Biology, vol. 9, no. 5, R85, 2008.

86. H. Takahara, A. Dolf, E. Endl, and R. O›Connell, "Flow cytometric purification of Colletotrichum higginsianum biotrophic hyphae from Arabidopsis leaves for stage-specific transcriptome analysis,"Plant Journal, vol. 59, no. 4, pp. 672–683, 2009.

87. R. P. Wise, M. J. Moscou, A. J. Bogdanove, and S. A. Whitham, "Transcript profiling in host-pathogen interactions," Annual Review of Phytopathology, vol. 45, pp. 329–369, 2008.

88. B. Venkatesh, U. Hettwer, B. Koopmann, and P. Karlovsky, "Conversion of cDNA differential display results (DDRT-PCR) into quantitative transcription profiles," BMC Genomics, vol. 6, article 51, 2005. · ·

89. X. Wang, C. Tang, G. Zhang, et al., "cDNA-AFLP analysis reveals differential gene expression in compatible interaction of wheat challenged with Puccinia striiformis f. sp. tritici," BMC Genomics, vol. 10, article 289, 2009.

90. C. Fekete, R. W. Fung, Z. Szabó, et al., "Up-regulated transcripts in a compatible powdery mildew-grapevine interaction," Plant Physiology and Biochemistry, vol. 47, no. 8, pp. 732–738, 2009.

91. R. C. Venu, Y. Jia, M. Gowda, et al., "RL-SAGE and microarray analysis of the rice transcriptome after Rhizoctonia solani infection," Molecular Genetics and Genomics, vol. 278, no. 4, pp. 421–431, 2007.

92. C. Yin, X. Chen, X. Wang, Q. Han, Z. Kang, and S. Hulbert, "Generation and analysis of expression sequence tags from haustoria of the wheat stripe rust fungus Puccinia striiformis f. sp. Tritici," BMC Genomics, vol. 10, article 626, 2009.

93. W. Skinner, J. Keon, and J. Hargreaves, "Gene information for fungal plant pathogens from expressed sequences," Current Opinion in Microbiology, vol. 4, no. 4, pp. 381–386, 2001.

94. F. J. Bruggeman and H. V. Westerhoff, "The nature of systems biology," Trends in Microbiology, vol. 15, no. 1, pp. 45–50, 2007.

95. P. Picotti, B. Bodenmiller, L. N. Mueller, B. Domon, and R. Aebersold, "Full dynamic range proteome analysis of S. cerevisiae by targeted proteomics," Cell, vol. 138, no. 4, pp. 795–806, 2009.

96. N. L. Anderson, N. G. Anderson, T. W. Pearson, et al., "A human proteome detection and quantitation project," Molecular and Cellular Proteomics, vol. 8, no. 5, pp. 883–886, 2009.

97. M. R. Wilkins, J. C. Sanchez, A. A. Gooley, et al., "Progress with proteome projects: why all proteins expressed by a genome should be identified and how to do it," Biotechnology and Genetic Engineering Reviews, vol. 13,

pp. 19–50, 1996. ·

98. J. Cox and M. Mann, "Is proteomics the new genomics?" Cell, vol. 130, no. 3, pp. 395–398, 2007.

99. B. F. Cravatt, G. M. Simon, and J. R. Yates III, "The biological impact of mass-spectrometry-based proteomics," Nature, vol. 450, no. 7172, pp. 991–1000, 2007.

100. X. Han, M. Jin, K. Breuker, and F. W. McLafferty, "Extending top-down mass spectrometry to proteins with masses great than 200 kilodaltons," Science, vol. 314, no. 5796, pp. 109–112, 2006. · ·

101. X. Han, A. Aslanian, and J. R. Yates III, "Mass spectrometry for proteomics," Current Opinion in Chemical Biology, vol. 12, no. 5, pp. 483–490, 2008.

102. J. V. Jorrín-Novo, A. M. Maldonado, S. Echevarría-Zomeño, et al., "Plant proteomics update (2007-2008): second-generation proteomic techniques, an appropriate experimental design, and data analysis to fulfill MIAPE standards, increase plant proteome coverage and expand biological knowledge," Journal of Proteomics, vol. 72, no. 3, pp. 285–314, 2009.

103. M. Mann, "Comparative analysis to guide quality improvements in proteomics," Nature Methods, vol. 6, no. 10, pp. 717–719, 2009.

104. A. Schmidt, M. Claassen, and R. Aebersold, "Directed mass spectrometry: towards hypothesis-driven proteomics," Current Opinion in Chemical Biology, vol. 13, no. 5-6, pp. 510–517, 2009. · ·

105. V. Bhadauria, W. S. Zhao, L. X. Wang, et al., "Advances in fungal proteomics," Microbiological Research, vol. 162, no. 3, pp. 193–200, 2007.

106. Y. Kim, M. P. Nandakumar, and M. R. Marten, "Proteomics of filamentous fungi," Trends in Biotechnology, vol. 25, no. 9, pp. 395–400, 2007.

107. L. J. Newey, C. E. Caten, and J. R. Green, "Rapid adhesion of Stagonospora nodorum spores to a hydrophobic surface requires pre-formed cell surface glycoproteins," Mycological Research, vol. 111, no. 11, pp. 1255–1267, 2007.

108. P. N. Dodds, M. Rafiqi, P. H. P. Gan, A. R. Hardham, D. A. Jones, and J. G. Ellis, "Effectors of biotrophic fungi and oomycetes: pathogenicity factors and triggers of host resistance," New Phytologist, vol. 183, no. 4, pp. 993–1000, 2009.

109. P. M. Houterman, B. J. Cornelissen, and M. Rep, "Suppression of plant resistance gene-based immunity by a fungal effector," PLoS Pathogens,

vol. 4, no. 5, Article ID e1000061, 2008.

110. J. Song, J. Win, M. Tian, et al., "Apoplastic effectors secreted by two unrelated eukaryotic plant pathogens target the tomato defense protease Rcr3," Proceedings of the National Academy of Sciences of the United States of America, vol. 106, no. 5, pp. 1654–1659, 2009.

111. P. J. G. M. De Wit, M. B. Buurlage, and K. E. Hammond, "The occurrence of host-, pathogen- and interaction-specific proteins in the apoplast of Cladosporium fulvum (syn. Fulvia fulva) infected tomato leaves," Physiological and Molecular Plant Pathology, vol. 29, no. 2, pp. 159–172, 1986.

112. I. M. J. Schottens-Toma and P. J. G. M. De Wit, "Purification and primary structure of a necrosis-inducing peptide from the apoplastic fluids of tomato infected with Cladosporium fulvum (syn. Fulvia fulva)," Physiological and Molecular Plant Pathology, vol. 33, no. 1, pp. 59–67, 1988.

113. J. G. Ellis, M. Rafiqi, P. Gan, A. Chakrabarti, and P. N. Dodds, "Recent progress in discovery and functional analysis of effector proteins of fungal and oomycete plant pathogens," Current Opinion in Plant Biology, vol. 12, no. 4, pp. 399–405, 2009.

114. M. Rep, H. C. van der Does, M. Meijer, et al., "A small, cysteine-rich protein secreted by Fusarium oxysporum during colonization of xylem vessels is required for I-3-mediated resistance in tomato,"Molecular Microbiology, vol. 53, no. 5, pp. 1373–1383, 2004. · ·

115. M. Rep, "Small proteins of plant-pathogenic fungi secreted during host colonization," FEMS Microbiology Letters, vol. 253, no. 1, pp. 19–27, 2005.

116. E. W. Deutsch, H. Lam, and R. Aebersold, "Data analysis and bioinformatics tools for tandem mass spectrometry in proteomics," Physiological Genomics, vol. 33, no. 1, pp. 18–25, 2008.

117. J. Eriksson and D. Fenyö, "Improving the success rate of proteome analysis by modeling protein-abundance distributions and experimental designs," Nature Biotechnology, vol. 25, no. 6, pp. 651–655, 2007.

118. P. Holzmuller, P. Grébaut, J. P. Brizard, et al., "Pathogeno-proteomics," Annals of the New York Academy of Sciences, vol. 1149, pp. 66–70, 2008.

119. L. Valledor, M. A. Castillejo, C. Lenz, R. Rodríguez, M. J. Cañal, and J. Jorrín, "Proteomic analysis of Pinus radiata needles: 2-DE map and protein identification by LC/MS/MS and substitution-tolerant database searching," Journal of Proteome Research, vol. 7, no. 7, pp. 2616–2631,

2008.

120. J. Ruiz-Herrera, Fungal Cell Wall: Structure, Synthesis and Assembly, CRC Press, Boca Raton, Fla, USA, 1992.

121. J. Grinyer, M. McKay, B. Herbert, and H. Nevalainen, "Fungal proteomics: mapping the mitochodrial proteins of a Trichoderma harzianum strain applied for biological control," Current Genetics, vol. 45, no. 3, pp. 170–175, 2004.

122. P. Melin, J. Schnürer, and E. G. Wagner, "Proteome analysis of Aspergillus nidulans reveals proteins associated with the response to the antibiotic concanamycin A, produced by Streptomyces species,"Molecular Genetics and Genomics, vol. 267, no. 6, pp. 695–702, 2002.

123. M. P. Nandakumar, J. Shen, B. Raman, and M. R. Marten, "Solubilization of trichloroacetic acid (TCA) precipitated microbial proteins via NaOH for two-dimensional electrophoresis," Journal of Proteome Research, vol. 2, no. 1, pp. 89–93, 2003.

124. K. Ström, J. Schnürer, and P. Melin, "Co-cultivation of antifungal Lactobacillus plantarum MiLAB 393 and Aspergillus nidulans, evaluation of effects on fungal growth and protein expression," FEMS Microbiology Letters, vol. 246, no. 1, pp. 119–124, 2005. · ·

125. J. M. Bruneau, T. Magnin, E. Tagat, et al., "Proteome analysis of Aspergillus fumigatus identifies glycosylphosphatidylinositol-anchored proteins associated to the cell wall biosynthesis," Electrophoresis, vol. 22, no. 13, pp. 2812–2823, 2001.

126. J. Grinyer, S. Hunt, M. McKay, B. E. N. R. Herbert, and H. Nevalainen, "Proteomic response of the biological control fungus Trichoderma atroviride to growth on the cell walls of Rhizoctonia solani,"Current Genetics, vol. 47, no. 6, pp. 381–388, 2005.

127. J. Grinyer, M. McKay, H. Nevalainen, and B. E. N. R. Herbert, "Fungal proteomics: initial mapping of biological control strain Trichoderma harzianum," Current Genetics, vol. 45, no. 3, pp. 163–169, 2004. · ·

128. M. P. Nandakumar and M. R. Marten, "Comparison of lysis methods and preparation protocols for one- and two-dimensional electrophoresis of Aspergillus oryzae intracellular proteins," Electrophoresis, vol. 23, no. 14, pp. 2216–2222, 2002.

129. M. Shimizu and H. Wariishi, "Development of a sample preparation method for fungal proteomics,"FEMS Microbiology Letters, vol. 247, no. 1, pp. 17–22, 2005. · ·

130. M. Shimizu, N. Yuda, T. Nakamura, H. Tanaka, and H. Wariishi,

"Metabolic regulation at the tricarboxylic acid and glyoxylate cycles of the lignin-degrading basidiomycete Phanerochaete chrysosporium against exogenous addition of vanillin," Proteomics, vol. 5, no. 15, pp. 3919–3931, 2005. · ·

131. M. L. Hernández-Macedo, A. Ferraz, J. Rodrguez, L. M. M. Ottoboni, and M. P. De Mello, "Iron-regulated proteins in Phanerochaete chrysosporium and Lentinula edodes: differential analysis by sodium dodecyl sulfate polyacrylamide gel electrophoresis and two-dimensional polyacrylamide gel electrophoresis profiles," Electrophoresis, vol. 23, no. 4, pp. 655–661, 2002.

132. A. Görg, W. Weiss, and M. J. Dunn, "Current two-dimensional electrophoresis technology for proteomics," Proteomics, vol. 4, no. 12, pp. 3665–3685, 2004.

133. S. C. Carpentier, E. Witters, K. Laukens, P. Deckers, R. Swennen, and B. Panis, "Preparation of protein extracts from recalcitrant plant tissues: an evaluation of different methods for two-dimensional gel electrophoresis analysis," Proteomics, vol. 5, no. 10, pp. 2497–2507, 2005.

134. C. Damerval, D. de Vienne, M. Zivy, and H. Thiellement, "Technical improvements in two-dimensional electrophoresis increase the level of genetic variation detected in wheat-seedling proteins,"Electrophoresis, vol. 7, pp. 52–54, 1986.

135. A. N. A. M. Maldonado, S. Echevarría-Zomeño, S. Jean-Baptiste, M. Hernández, and J. V. Jorrín-Novo, "Evaluation of three different protocols of protein extraction for Arabidopsis thaliana leaf proteome analysis by two-dimensional electrophoresis," Journal of Proteomics, vol. 71, no. 4, pp. 461–472, 2008. · ·

136. R. Wildgruber, G. Reil, O. Drews, H. Parlar, and A. Görg, "Web-based two-dimensional database ofSaccharomyces cerevisiae proteins using immobilized pH gradients from pH 6 to pH 12 and matrix-assisted laser desorption/ionization-time of flight mass spectrometry," Proteomics, vol. 2, no. 6, pp. 727–732, 2002. ·

137. H. Everberg, N. Gustavasson, and F. Tjerned, "Enrichment of membrane proteins by partitioning in detergent/polymer aqueous two-phase systems," Methods in Molecular Biology, vol. 424, pp. 403–412, 2008. ·

138. S. Luche, V. Santoni, and T. Rabilloud, "Evaluation of nonionic and zwitterionic detergents as membrane protein solubilizers in two-dimensional electrophoresis," Proteomics, vol. 3, no. 3, pp. 249–253, 2003.

139. M. P. Molloy, B. R. Herbert, K. L. Williams, and A. A. Gooley,

"Extraction of Escherichia coli proteins with organic solvents prior to two-dimensional electrophoresis," Electrophoresis, vol. 20, no. 4-5, pp. 701–704, 1999. ·

140. T. Rabilloud, "Solubilization of proteins for electrophoretic analyses," Electrophoresis, vol. 17, no. 5, pp. 813–829, 1996.

141. T. Rabilloud, C. Adessi, A. Giraudel, and J. Lunardi, "Improvement of the solubilization of proteins in two-dimensional electrophoresis with immobilized pH gradients," Electrophoresis, vol. 18, no. 3-4, pp. 307–316, 1997.

142. O. Kniemeyer, F. Lessing, O. Scheibner, C. Hertweck, and A. A. Brakhage, "Optimisation of a 2-D gel electrophoresis protocol for the human-pathogenic fungus Aspergillus fumigatus," Current Genetics, vol. 49, no. 3, pp. 178–189, 2006.

143. T. Rabilloud, "Use of thiourea to increase the solubility of membrane proteins in two-dimensional electrophoresis," Electrophoresis, vol. 19, no. 5, pp. 758–760, 1998.

144. B. E. N. R. Herbert, J. Grinyer, J. T. McCarthy, et al., "Improved 2-DE of microorganisms after acidic extraction," Electrophoresis, vol. 27, no. 8, pp. 1630–1640, 2006.

145. O. Guais, G. Borderies, C. Pichereaux, et al., "Proteomics analysis of "Rovabiot Excel", a secreted protein cocktail from the filamentous fungus Penicillium funiculosum grown under industrial process fermentation," Journal of Industrial Microbiology & Biotechnology, pp. 1659–1668, 2008.

146. S. H. Kao, H. I. N. K. Wong, C. Y. Chiang, and H. A. N. M. Chen, "Evaluating the compatibility of three colorimetric protein assays for two-dimensional electrophoresis experiments," Proteomics, vol. 8, no. 11, pp. 2178–2184, 2008.

147. D. Fragner, M. Zomorrodi, U. Kües, and A. Majcherczyk, "Optimized protocol for the 2-DE of extracellular proteins from higher basidiomycetes inhabiting lignocellulose," Electrophoresis, vol. 30, no. 14, pp. 2431–2441, 2009.

148. M. L. Medina and W. A. Francisco, "Isolation and enrichment of secreted proteins from filamentous fungi," in 2D PAGE: Sample Preparation and Fractionation, pp. 275–285, Springer, New York, NY, USA, 2008.

149. A. Abbas, H. Koc, F. Liu, and M. Tien, "Fungal degradation of wood: initial proteomic analysis of extracellular proteins of Phanerochaete chrysosporium grown on oak substrate," Current Genetics, vol. 47, no. 1, pp. 49–56, 2005.

150. H. Ravalason, G. Jan, D. Mollé, et al., "Secretome analysis of Phanerochaete chrysosporium strain CIRM-BRFM41 grown on softwood," Applied Microbiology and Biotechnology, vol. 80, no. 4, pp. 719–733, 2008.

151. A. V. Wymelenberg, P. Minges, G. Sabat, et al., "Computational analysis of the Phanerochaete chrysosporium v2.0 genome database and mass spectrometry identification of peptides in ligninolytic cultures reveal complex mixtures of secreted proteins," Fungal Genetics and Biology, vol. 43, no. 5, pp. 343–356, 2006.

152. H. Zorn, T. Peters, M. Nimtz, and R. G. Berger, "The secretome of Pleurotus sapidus," Proteomics, vol. 5, no. 18, pp. 4832–4838, 2005.

153. L. V. Bindschedler, T. A. Burgis, D. J. S. Mills, J. T. C. Ho, R. Cramer, and P. D. Spanu, "In planta proteomics and proteogenomics of the biotrophic Barley fungal pathogen Blumeria graminis f. sp.hordei," Molecular and Cellular Proteomics, vol. 8, no. 10, pp. 2368–2381, 2009.

154. C. Rampitsch, N. V. Bykova, B. McCallum, E. V. A. Beimcik, and W. Ens, "Analysis of the wheat and Puccinia triticina (leaf rust) proteomes during a susceptible host-pathogen interaction," Proteomics, vol. 6, no. 6, pp. 1897–1907, 2006.

155. W. Zhou, F. Eudes, and A. Laroche, "Identification of differentially regulated proteins in response to a compatible interaction between the pathogen Fusarium graminearum and its host, Triticum aestivum,"Proteomics, vol. 6, no. 16, pp. 4599–4609, 2006.

156. P. M. Houterman, D. Speijer, H. L. Dekker, C. G. de Koster, B. J. C. Cornelissen, and M. Rep, "The mixed xylem sap proteome of Fusarium oxysporum-infected tomato plants," Molecular Plant Pathology, vol. 8, no. 2, pp. 215–221, 2007.

157. A. Harder, "Sample preparation procedure for cellular fungi," Methods in Molecular Biology, vol. 425, pp. 265–273, 2008. ·

158. A. Pitarch, C. Nombela, and C. Gil, "Cell wall fractionation for yeast and fungal proteomics," Methods in Molecular Biology, vol. 425, pp. 217–239, 2008. ·

159. X. Jiang, M. Ye, and H. Zou, "Technologies and methods for sample pretreatment in efficient proteome and peptidome analysis," Proteomics, vol. 8, no. 4, pp. 686–705, 2008.

160. E. Boschetti and P. G. Righetti, "The art of observing rare protein species in proteomes with peptide ligand libraries," Proteomics, vol. 9, no. 6, pp. 1492–1510, 2009.

161. D. Robertson, G. P. Mitchell, J. S. Gilroy, C. Gerrish, G. P. Bolwell, and A. R. Slabas, "Differential extraction and protein sequencing reveals major differences in patterns of primary cell wall proteins from plants," Journal of Biological Chemistry, vol. 272, no. 25, pp. 15841–15848, 1997. · ·

162. F. Supek, P. Peharec, M. Krsnik-Rasol, and T. Šmuc, "Enhanced analytical power of SDS-PAGE using machine learning algorithms," Proteomics, vol. 8, no. 1, pp. 28–31, 2008.

163. L. M. F. de Godoy, J. V. Olsen, G. A. de Souza, G. Li, P. Mortensen, and M. Mann, "Status of complete proteome analysis by mass spectrometry: SILAC labeled yeast as a model system," Genome Biology, vol. 7, no. 6, article R50, 2006.

164. F. Tribl, C. Lohaus, T. Dombert, et al., "Towards multidimensional liquid chromatography separation of proteins using fluorescence and isotope-coded protein labelling for quantitative proteomics,"Proteomics, vol. 8, no. 6, pp. 1204–1211, 2008.

165. B. G. Fryksdale, P. T. Jedrzejewski, D. L. Wong, A. L. Gaertner, and B. S. Miller, "Impact of deglycosylation methods on two-dimensional gel electrophoresis and matrix assisted laser desorption/ionization-time of flight-mass spectrometry for proteomic analysis," Electrophoresis, vol. 23, no. 14, pp. 2184–2193, 2002. ·

166. A. V. Wymelenberg, G. Sabat, M. Mozuch, P. J. Kersten, D. A. N. Cullen, and R. A. Blanchette, "Structure, organization, and transcriptional regulation of a family of copper radical oxidase genes in the lignin-degrading basidiomycete Phanerochaete chrysosporium," Applied and Environmental Microbiology, vol. 72, no. 7, pp. 4871–4877, 2006.

167. P. Jungblut and B. Thiede, "Protein identification from 2-DE gels by MALDI mass spectrometry," Mass Spectrometry Reviews, vol. 16, no. 3, pp. 145–162, 1997. ·

168. P. H. O›Farrell, "High resolution two dimensional electrophoresis of proteins," Journal of Biological Chemistry, vol. 250, no. 10, pp. 4007–4021, 1975. ·

169. J. Klose and U. Kobalz, "Two-dimensional electrophoresis of proteins: an updated protocol and implications for a functional analysis of the genome," Electrophoresis, vol. 16, no. 6, pp. 1034–1059, 1995. ·

170. T. Rabilloud, A. R. Vaezzadeh, N. Potier, C. Lelong, E. Leize-Wagner, and M. Chevallet, "Power and limitations of electrophoretic separations in proteomics strategies," Mass Spectrometry Reviews, vol. 28, no. 5, pp. 816–843, 2009.

171. P. A. Haynes and T. H. Roberts, "Subcellular shotgun proteomics in

plants: looking beyond the usual suspects," Proteomics, vol. 7, no. 16, pp. 2963–2975, 2007. · ·

172. A. Pirondini, G. Visioli, A. Malcevschi, and N. Marmiroli, "A 2-D liquid-phase chromatography for proteomic analysis in plant tissues," Journal of Chromatography B, vol. 833, no. 1, pp. 91–100, 2006. · ·

173. G. Recorbet, H. Rogniaux, V. Gianinazzi-Pearson, and E. DumasGaudot, "Fungal proteins in the extra-radical phase of arbuscular mycorrhiza: a shotgun proteomic picture," New Phytologist, vol. 181, no. 2, pp. 248–260, 2009.

174. M. Ye, X. Jiang, S. Feng, R. Tian, and H. Zou, "Advances in chromatographic techniques and methods in shotgun proteome analysis," Trends in Analytical Chemistry, vol. 26, no. 1, pp. 80–84, 2007.

175. M. P. Washburn, D. Wolters, and J. R. Yates III, "Large-scale analysis of the yeast proteome by multidimensional protein identification technology," Nature Biotechnology, vol. 19, no. 3, pp. 242–247, 2001.

176. T. McDonald, S. Sheng, B. Stanley, et al., "Expanding the subproteome of the inner mitochondria using protein separation technologies: one- and two-dimensional liquid chromatography and two-dimensional gel electrophoresis," Molecular and Cellular Proteomics, vol. 5, no. 12, pp. 2392–2411, 2006.

177. W. W. Wu, G. Wang, S. J. Baek, and R. F. Shen, "Comparative study of three proteomic quantitative methods, DIGE, cICAT, and iTRAQ, using 2D gel- or LC-MALDI TOF/TOF," Journal of Proteome Research, vol. 5, no. 3, pp. 651–658, 2006.

178. B. Domon and R. Aebersold, "Mass spectrometry and protein analysis," Science, vol. 312, no. 5771, pp. 212–217, 2006.

179. A. I. Nesvizhskii, O. Vitek, and R. Aebersold, "Analysis and validation of proteomic data generated by tandem mass spectrometry," Nature Methods, vol. 4, no. 10, pp. 787–797, 2007.

180. D. M. Soanes, W. Skinner, J. Keon, J. Hargreaves, and N. J. Talbot, "Genomics of phytopathogenic fungi and the development of bioinformatic resources," Molecular Plant-Microbe Interactions, vol. 15, no. 5, pp. 421–427, 2002. ·

181. D. M. Soanes and N. J. Talbot, "Comparative genomic analysis of phytopathogenic fungi using expressed sequence tag (EST) collections," Molecular Plant Pathology, vol. 7, no. 1, pp. 61–70, 2006. · ·

182. J. Fiévet, C. Dillmann, G. Lagniel, et al., "Assessing factors for reliable quantitative proteomics based on two-dimensional gel electrophoresis,"

Proteomics, vol. 4, no. 7, pp. 1939–1949, 2004.

183. G. B. Smejkal, M. H. Robinson, and A. Lazarev, "Comparison of fluorescent stains: relative photostability and differential staining of proteins in two-dimensional gels," Electrophoresis, vol. 25, no. 15, pp. 2511–2519, 2004.

184. Y. I. Hu, G. Wang, G. Y. J. Chen, X. I. N. Fu, and S. Q. Yao, "Proteome analysis of Saccharomyces cerevisiae under metal stress by two-dimensional differential gel electrophoresis," Electrophoresis, vol. 24, no. 9, pp. 1458–1470, 2003.

185. J. U. N. X. Yan, A. T. Devenish, R. Wait, T. I. M. Stone, S. Lewis, and S. U. E. Fowler, "Fluorescence two-dimensional difference gel electrophoresis and mass spectrometry based proteomic analysis ofEscherichia coli," Proteomics, vol. 2, no. 12, pp. 1682–1698, 2002.

186. S. P. Gygi, B. Rist, S. A. Gerber, F. Turecek, M. H. Gelb, and R. Aebersold, "Quantitative analysis of complex protein mixtures using isotope-coded affinity tags," Nature Biotechnology, vol. 17, no. 10, pp. 994–999, 1999.

187. H. Zhou, J. A. Ranish, J. D. Watts, and R. Aebersold, "Quantitative proteome analysis by solid-phase isotope tagging and mass spectrometry," Nature Biotechnology, vol. 20, no. 5, pp. 512–515, 2002.

188. X. Yao, A. Freas, J. Ramirez, P. A. Demirev, and C. Fenselau, "Proteolytic 18O labeling for comparative proteomics: model studies with two serotypes of adenovirus," Analytical Chemistry, vol. 73, no. 13, pp. 2836–2842, 2001.

189. P. A. Everley, J. Krijgsveld, B. R. Zetter, and S. P. Gygi, "Quantitative cancer proteomics: stable isotope labeling with amino acids in cell culture (SILAC) as a tool for prostate cancer research," Molecular and Cellular Proteomics, vol. 3, no. 7, pp. 729–735, 2004.

190. L. R. Zieske, "A perspective on the use of iTRAQTM reagent technology for protein complex and profiling studies," Journal of Experimental Botany, vol. 57, no. 7, pp. 1501–1508, 2006. · ·

191. W. M. Old, K. Meyer-Arendt, L. Aveline-Wolf, et al., "Comparison of label-free methods for quantifying human proteins by shotgun proteomics," Molecular and Cellular Proteomics, vol. 4, no. 10, pp. 1487–1502, 2005.

192. B. Zhang, N. C. VerBerkmoes, M. A. Langston, E. Uberbacher, R. L. Hettich, and N. F. Samatova, "Detecting differential and correlated protein expression in label-free shotgun proteomics," Journal of Proteome Research, vol. 5, no. 11, pp. 2909–2918, 2006. · ·

193. M. Ünlü, M. E. Morgan, and J. S. Minden, "Difference gel electrophoresis: a single gel method for detecting changes in protein extracts," Electrophoresis, vol. 18, no. 11, pp. 2071–2077, 1997. ·

194. M. Bantscheff, M. Schirle, G. Sweetman, J. Rick, and B. Kuster, "Quantitative mass spectrometry in proteomics: a critical review," Analytical and Bioanalytical Chemistry, vol. 389, no. 4, pp. 1017–1031, 2007.

195. L. N. Mueller, M. I. Y. Brusniak, D. R. Mani, and R. Aebersold, "An assessment of software solutions for the analysis of mass spectrometry based quantitative proteomics data," Journal of Proteome Research, vol. 7, no. 1, pp. 51–61, 2008.

196. V. J. Patel, K. Thalassinos, S. E. Slade, et al., "A comparison of labeling and label-free mass spectrometry-based proteomics approaches," Journal of Proteome Research, vol. 8, no. 7, pp. 3752–3759, 2009.

197. S. Xie, C. Moya, B. Bilgin, A. Jayaraman, and S. P. Walton, "Emerging affinity-based techniques in proteomics," Expert Review of Proteomics, vol. 6, no. 5, pp. 573–583, 2009.

198. I. A. N. P. Shadforth, T. P. J. Dunkley, K. S. Lilley, and C. Bessant, "i-Tracker: for quantitative proteomics using iTRAQTM," BMC Genomics, vol. 6, article 145, 2005.

199. P. V. Bondarenko, D. Chelius, and T. A. Shaler, "Identification and relative quantitation of protein mixtures by enzymatic digestion followed by capillary reversed-phase liquid chromatography-tandem mass spectrometry," Analytical Chemistry, vol. 74, no. 18, pp. 4741–4749, 2002.

200. D. Chelius and P. V. Bondarenko, "Quantitative profiling of proteins in complex mixtures using liquid chromatography and mass spectrometry," Journal of Proteome Research, vol. 1, no. 4, pp. 317–323, 2002. · ·

201. W. Wang, H. Zhou, H. U. A. Lin, et al., "Quantification of proteins and metabolites by mass spectrometry without isotopic labeling or spiked standards," Analytical Chemistry, vol. 75, no. 18, pp. 4818–4826, 2003.

202. H. Liu, R. G. Sadygov, and J. R. Yates III, "A model for random sampling and estimation of relative protein abundance in shotgun proteomics," Analytical Chemistry, vol. 76, no. 14, pp. 4193–4201, 2004. · ·

203. J. Fasolo and M. Snyder, "Protein microarrays," Methods in Molecular Biology, vol. 548, pp. 209–222, 2009.

204. J. Barrett, P. M. Brophy, and J. V. Hamilton, "Analysing proteomic data," International Journal for Parasitology, vol. 35, no. 5, pp. 543–553, 2005.

205. E. Marengo, E. Robotti, F. Antonucci, D. Cecconi, N. Campostrini, and P. G. Righetti, "Numerical approaches for quantitative analysis of two-dimensional maps: a review of commercial software and home-made systems," Proteomics, vol. 5, no. 3, pp. 654–666, 2005.

206. Å. M. Wheelock and S. Goto, "Effects of post-electrophoretic analysis on variance in gel-based proteomics," Expert Review of Proteomics, vol. 3, no. 1, pp. 129–142, 2006.

207. E. W. Deutsch, H. Lam, and R. Aebersold, "PeptideAtlas: a resource for target selection for emerging targeted proteomics workflows," EMBO Reports, vol. 9, no. 5, pp. 429–434, 2008.

208. D. G. Biron, C. Brun, T. Lefevre, et al., "The pitfalls of proteomics experiments without the correct use of bioinformatics tools," Proteomics, vol. 6, no. 20, pp. 5577–5596, 2006.

209. N. A. Karp and K. S. Lilley, "Design and analysis issues in quantitative proteomics studies," Proteomics, vol. 2, no. 1, pp. 42–50, 2007.

210. J. M. Bland and D. G. Altman, "Multiple significance tests: the Bonferroni method," British Medical Journal, vol. 310, no. 6973, article 170, 1995. ·

211. Y. Benjamini and Y. Hochberg, "Controlling the false discovery rate: a practical and powerful approach to multiple testing," Journal of the Royal Statistical Society. Series B, vol. 57, pp. 289–300, 1995.

212. J. D. Storey and R. Tibshirani, "Statistical significance for genomewide studies," Proceedings of the National Academy of Sciences of the United States of America, vol. 100, no. 16, pp. 9440–9445, 2003. · ·

213. K. Strimmer, "fdrtool: a versatile R package for estimating local and tail area-based false discovery rates,"Bioinformatics, vol. 24, no. 12, pp. 1461–1462, 2008.

214. S. Jacobsen, H. Grove, K. N. Jensen, et al., "Multivariate analysis of 2-DE protein patterns—practical approaches," Electrophoresis, vol. 28, no. 8, pp. 1289–1299, 2007.

215. K. N. Jensen, F. Jessen, and B. O. M. Jørgensen, "Multivariate data analysis of two-dimensional gel electrophoresis protein patterns from few samples," Journal of Proteome Research, vol. 7, no. 3, pp. 1288–1296, 2008.

216. R. Pedreschi, M. L. A. T. M. Hertog, S. C. Carpentier, et al., "Treatment of missing values for multivariate statistical analysis of gel-based proteomics data," Proteomics, vol. 8, no. 7, pp. 1371–1383, 2008.

217. M. Schneider, L. Lane, E. Boutet, et al., "The UniProtKB/Swiss-Prot knowledgebase and its Plant Proteome Annotation Program," Journal of

Proteomics, vol. 72, no. 3, pp. 567–573, 2009.

218. C. F. Taylor, N. W. Paton, K. S. Lilley, et al., "The minimum information about a proteomics experiment (MIAPE)," Nature Biotechnology, vol. 25, no. 8, pp. 887–893, 2007.

219. J. A. Mead, I. A. N. P. Shadforth, and C. Bessant, "Public proteomic MS repositories and pipelines: available tools and biological applications," Proteomics, vol. 7, no. 16, pp. 2769–2786, 2007.

220. M. Fälth, M. M. Savitski, M. L. Nielsen, F. Kjeldsen, P. E. R. E. Andren, and R. A. Zubarev, "SwedCAD, a database of annotated high-mass accuracy MS/MS spectra of tryptic peptides," Journal of Proteome Research, vol. 6, no. 10, pp. 4063–4067, 2007.

221. J. A. N. Hummel, M. Niemann, S. Wienkoop, et al., "ProMEX: a mass spectral reference database for proteins and protein phosphorylation sites," BMC Bioinformatics, vol. 8, article 216, 2007.

222. H. Lam, E. W. Deutsch, J. S. Eddes, J. K. Eng, S. E. Stein, and R. Aebersold, "Building consensus spectral libraries for peptide identification in proteomics," Nature Methods, vol. 5, no. 10, pp. 873–875, 2008. · ·

223. R. Craig, J. P. Cortens, and R. C. Beavis, "Open source system for analyzing, validating, and storing protein identification data," Journal of Proteome Research, vol. 3, no. 6, pp. 1234–1242, 2004.

224. F. Desiere, E. W. Deutsch, N. L. King, et al., "The peptide atlas project," Nucleic Acids Research, vol. 34, pp. D655–658, 2006. ·

225. L. Martens, H. Hermjakob, P. Jones, et al., "PRIDE: the proteomics identifications database," Proteomics, vol. 5, no. 13, pp. 3537–3545, 2005.

226. I. A. N. Shadforth, W. Xu, D. Crowther, and C. Bessant, "GAPP: a fully automated software for the confident identification of human peptides from tandem mass spectra," Journal of Proteome Research, vol. 5, no. 10, pp. 2849–2852, 2006.

227. T. McLaughlin, J. A. Siepen, J. Selley, et al., "PepSeeker: a database of proteome peptide identifications for investigating fragmentation patterns," Nucleic Acids Research, vol. 34, pp. D649–D654, 2006. ·

228. Y. Zhang, Y. Zhang, J. U. N. Adachi, et al., "MAPU: Max-Planck Unified database of organellar, cellular, tissue and body fluid proteomes," Nucleic Acids Research, vol. 35, database issue, pp. D771–D779, 2007. · ·

229. J. T. Prince, M. W. Carlson, R. Wang, P. Lu, and E. M. Marcotte, "The need for a public proteomics repository," Nature Biotechnology, vol. 22, no. 4, pp. 471–472, 2004.

230. M. Riffle, L. Malmström, and T. N. Davis, "The yeast resource center public data repository," Nucleic Acids Research, vol. 33, pp. D378–D382, 2005.

231. I. L. S. Oh, A. E. R. A. N. Park, M. I. N. S. Bae, et al., "Secretome analysis reveals an Arabidopsis lipase involved in defense against Altemaria brassicicola," Plant Cell, vol. 17, no. 10, pp. 2832–2847, 2005.

232. F. Colditz, O. Nyamsuren, K. Niehaus, H. Eubel, H. P. Braun, and F. Krajinski, "Proteomic approach: identification of Medicago truncatula proteins induced in roots after infection with the pathogenic oomycete Aphanomyces euteiches," Plant Molecular Biology, vol. 55, no. 1, pp. 109–120, 2004.

233. F. Colditz, H. P. Braun, C. Jacquet, K. Niehaus, and F. Krajinski, "Proteomic profiling unravels insights into the molecular background underlying increased Aphanomyces euteiches-tolerance of Medicago truncatula," Plant Molecular Biology, vol. 59, no. 3, pp. 387–406, 2005.

234. F. Colditz, K. Niehaus, and F. Krajinski, "Silencing of PR-10-like proteins in Medicago truncatula results in an antagonistic induction of other PR proteins and in an increased tolerance upon infection with the oomycete Aphanomyces euteiches," Planta, vol. 226, no. 1, pp. 57–71, 2007.

235. T. J. March, J. A. Able, C. J. Schultz, and A. J. Able, "A novel late embryogenesis abundant protein and peroxidase associated with black point in barley grains," Proteomics, vol. 7, no. 20, pp. 3800–3808, 2007.
· ·

236. M. D. Bolton, H. P. van Esse, J. H. Vossen, et al., "The novel Cladosporium fulvum lysin motif effector Ecp6 is a virulence factor with orthologues in other fungal species," Molecular Microbiology, vol. 69, no. 1, pp. 119–136, 2008.

237. J. A. Smith, R. A. Blanchette, T. A. Burnes, et al., "Proteomic comparison of needles from blister rust-resistant and susceptible Pinus strobus seedlings reveals upregulation of putative disease resistance proteins," Molecular Plant-Microbe Interactions, vol. 19, no. 2, pp. 150–160, 2006.

238. D. Wang, A. Eyles, D. Mandich, and P. Bonello, "Systemic aspects of host-pathogen interactions in Austrian pine (Pinus nigra): a proteomics approach," Physiological and Molecular Plant Pathology, vol. 68, no. 4–6, pp. 149–157, 2006.

239. M. Curto, E. Camafeita, J. A. Lopez, A. N. A. M. Maldonado, D. Rubiales, and J. V. Jorrín, "A proteomic approach to study pea (Pisum sativum) responses to powdery mildew (Erysiphe pisi)," Proteomics, vol. 6, supplement 1, pp. S163–S174, 2006. ·

240. J. Geddes, F. Eudes, A. Laroche, and L. B. Selinger, "Differential expression of proteins in response to the interaction between the pathogen Fusarium graminearum and its host, Hordeum vulgare," Proteomics, vol. 8, no. 3, pp. 545–554, 2008.

241. W. Zhou, F. L. Kolb, and D. E. Riechers, "Identification of proteins induced or upregulated by Fusarium head blight infection in the spikes of hexaploid wheat (Triticum aestivum)," Genome, vol. 48, no. 5, pp. 770–780, 2005.

242. B. K. Ndimba, S. Chivasa, J. M. Hamilton, W. J. Simon, and A. R. Slabas, "Proteomic analysis of changes in the extracellular matrix of Arabidopsis cell suspension cultures induced by fungal elicitors,"Proteomics, vol. 3, no. 6, pp. 1047–1059, 2003.

243. R. L. Larson, A. M. Y. L. Hill, and A. Nuñez, "Characterization of protein changes associated with sugar beet (Beta vulgaris) resistance and susceptibility to Fusarium oxysporum," Journal of Agricultural and Food Chemistry, vol. 55, no. 19, pp. 7905–7915, 2007.

244. S. Campo, M. Carrascal, M. Coca, J. Abián, and B. San Segundo, "The defense response of germinating maize embryos against fungal infection: a proteomics approach," Proteomics, vol. 4, no. 2, pp. 383–396, 2004.

245. S. Chivasa, J. M. Hamilton, R. S. Pringle, et al., "Proteomic analysis of differentially expressed proteins in fungal elicitor-treated Arabidopsis cell cultures," Journal of Experimental Botany, vol. 57, no. 7, pp. 1553–1562, 2006.

246. S. Chivasa, W. J. Simon, X. L. Yu, N. Yalpani, and A. R. Slabas, "Pathogen elicitor-induced changes in the maize extracellular matrix proteome," Proteomics, vol. 5, no. 18, pp. 4894–4904, 2005.

247. J. V. F. Coumans, A. Poljak, M. J. Raftery, D. Backhouse, and L. Pereg-Gerk, "Analysis of cotton (Gossypium hirsutum) root proteomes during a compatible interaction with the black root rot fungusThielaviopsis basicola," Proteomics, vol. 9, no. 2, pp. 335–349, 2009.

248. H. O. W. O. N. Jung, C. W. O. O. Lim, S. C. Lee, H. W. O. O. Choi, C. H. Hwang, and B. K. Hwang, "Distinct roles of the pepper hypersensitive induced reaction protein gene CaHIR1 in disease and osmotic stress, as determined by comparative transcriptome and proteome analyses," Planta, vol. 227, no. 2, pp. 409–425, 2008.

249. N. Sharma, N. Hotte, M. H. Rahman, M. Mohammadi, M. K. Deyholos, and N. N. V. Kav, "Towards identifying Brassica proteins involved in mediating resistance to Leptosphaeria maculans: a proteomics-based approach," Proteomics, vol. 8, no. 17, pp. 3516–3535, 2008.

250. B. Subramanian, V. K. Bansal, and N. N. V. Kav, "Proteome-level investigation of Brassica carinata-derived resistance to Leptosphaeria maculans," Journal of Agricultural and Food Chemistry, vol. 53, no. 2, pp. 313–324, 2005.

251. H. Konishi, K. Ishiguro, and S. Komatsu, "A proteomics approach towards understanding blast fungus infection of rice grown under different levels of nitrogen fertilization," Proteomics, vol. 1, no. 9, pp. 1162–1171, 2001.

252. S. U. N. T. A. E. Kim, K. Y. U. S. Cho, S. Yu, et al., "Proteomic analysis of differentially expressed proteins induced by rice blast fungus and elicitor in suspension-cultured rice cells," Proteomics, vol. 3, no. 12, pp. 2368–2378, 2003.

253. S. T. Kim, S. G. Kim, D. H. Hwang, et al., "Proteomic analysis of pathogen-responsive proteins from rice leaves induced by rice blast fungus, Magnaporthe grisea," Proteomics, vol. 4, no. 11, pp. 3569–3578, 2004. · ·

254. S. T. Kim, S. Yu, S. G. Kim, et al., "Proteome analysis of rice blast fungus (Magnaporthe grisea) proteome during appressorium formation," Proteomics, vol. 4, no. 11, pp. 3579–3587, 2004.

255. M. Liao, Y. Li, and Z. Wang, "Identification of elicitor-responsive proteins in rice leaves by a proteomic approach," Proteomics, vol. 9, no. 10, pp. 2809–2819, 2009. · ·

256. K. U. N. Yuan, B. O. Zhang, Y. Zhang, Q. Cheng, M. Wang, and M. Huang, "Identification of differentially expressed proteins in poplar leaves induced by Marssonina brunnea f. sp. Multigermtubi,"Journal of Genetics and Genomics, vol. 35, no. 1, pp. 49–60, 2008.

257. C. P. Pirovani, H. A. S. Carvalho, R. C. R. Machado, et al., "Protein extraction for proteome analysis from cacao leaves and meristems, organs infected by Moniliophthora perniciosa, the causal agent of the witches› broom disease," Electrophoresis, vol. 29, no. 11, pp. 2391–2401, 2008.

258. F. Wen, H. D. VanEtten, G. Tsaprailis, and M. C. Hawes, "Extracellular proteins in pea root tip and border cell exudates," Plant Physiology, vol. 143, no. 2, pp. 773–783, 2007.

259. Z. Chan, G. Qin, X. Xu, B. Li, and S. Tian, "Proteome approach to characterize proteins induced by antagonist yeast and salicylic acid in peach fruit," Journal of Proteome Research, vol. 6, no. 5, pp. 1677–1688, 2007.

260. M. A. Islam, R. N. Sturrock, and A. K. M. Ekramoddoullah, "A proteomics approach to identify proteins differentially expressed in Douglas-fir seedlings infected by Phellinus sulphurascens," Journal of Proteomics,

vol. 71, no. 4, pp. 425–438, 2008.

261. R. C. Amey, T. Schleicher, J. Slinn, et al., "Proteomic analysis of a compatible interaction between Pisum sativum (pea) and the downy mildew pathogen Peronospora viciae," European Journal of Plant Pathology, vol. 122, no. 1, pp. 41–55, 2008.

262. T. Cao, S. Srivastava, M. H. Rahman, et al., "Proteome-level changes in the roots of Brassica napus as a result of Plasmodiophora brassicae infection," Plant Science, vol. 174, no. 1, pp. 97–115, 2008.

263. J. Lee, T. M. Bricker, M. Lefevre, S. R. M. Pinson, and J. H. Oard, "Proteomic and genetic approaches to identifying defence-related proteins in rice challenged with the fungal pathogen Rhizoctonia solani,"Molecular Plant Pathology, vol. 7, no. 5, pp. 405–416, 2006. · ·

264. J. Lee, J. Feng, K. B. Campbell, et al., "Quantitative proteomic analysis of bean plants infected by a virulent and avirulent obligate rust fungus," Molecular and Cellular Proteomics, vol. 8, no. 1, pp. 19–31, 2009.

265. Y. U. Liang, S. Srivastava, M. H. Rahman, S. E. Strelkov, and N. N. V. Kav, "Proteome changes in leaves of Brassica napus L. as a result of Sclerotinia sclerotiorum challenge," Journal of Agricultural and Food Chemistry, vol. 56, no. 6, pp. 1963–1976, 2008.

266. V. Bhadauria, S. Banniza, L.-X. Wang, Y.-D. Wei, and Y.-L. Peng, "Proteomic studies of phytopathogenic fungi, oomycetes and their interactions with hosts," European Journal of Plant Pathology, vol. 126, no. 1, pp. 81–95, 2010.

267. U. Mathesius, "Comparative proteomic studies of root-microbe interactions," Journal of Proteomics, vol. 72, no. 3, pp. 353–366, 2009.

268. A. Mehta, A. C. M. Brasileiro, D. S. L. Souza, et al., "Plant-pathogen interactions: what is proteomics telling us?" FEBS Journal, vol. 275, no. 15, pp. 3731–3746, 2008.

269. Y. U. Liang, H. U. I. Chen, M. Tang, and S. Shen, "Proteome analysis of an ectomycorrhizal fungusBoletus edulis under salt shock," Mycological Research, vol. 111, no. 8, pp. 939–946, 2007.

270. A. M. Murad, E. F. Noronha, R. N. Miller, et al., "Proteomic analysis of Metarhizium anisopliae secretion in the presence of the insect pest Callosobruchus maculatus," Microbiology, vol. 154, no. 12, pp. 3766–3774, 2008.

271. W. U. N. Y. Lin, J. U. I. Y. Chang, C. H. Hish, and T. Z. U. M. Pan, "Profiling the Monascus pilosusproteome during nitrogen limitation," Journal of Agricultural and Food Chemistry, vol. 56, no. 2, pp. 433–441,

2008.

272. A. V. Wymelenberg, G. Sabat, D. Martinez, et al., "The Phanerochaete chrysosporium secretome: database predictions and initial mass spectrometry peptide identifications in cellulose-grown medium,"Journal of Biotechnology, vol. 118, no. 1, pp. 17–34, 2005. · ·

273. R. Marra, P. Ambrosino, V. Carbone, et al., "Study of the three-way interaction between Trichoderma atroviride, plant and fungal pathogens by using a proteomic approach," Current Genetics, vol. 50, no. 5, pp. 307–321, 2006.

274. S. C. Tseng, S. H. U. Y. Liu, H. H. Yang, C. T. Lo, and K. O. U. C. Peng, "Proteomic study of biocontrol mechanisms of Trichoderma harzianum ETS 323 in response to Rhizoctonia solani," Journal of Agricultural and Food Chemistry, vol. 56, no. 16, pp. 6914–6922, 2008.

275. V. Seidl, I. S. Druzhinina, and C. P. Kubicek, "A screening system for carbon sources enhancing β-N-acetylglucosaminidase formation in Hypocrea atroviridis (Trichoderma atroviride)," Microbiology, vol. 152, no. 7, pp. 2003–2012, 2006.

276. S. Nagendran, H. E. Hallen-Adams, J. M. Paper, N. Aslam, and J. D. Walton, "Reduced genomic potential for secreted plant cell-wall-degrading enzymes in the ectomycorrhizal fungus Amanita bisporigera, based on the secretome of Trichoderma reesei," Fungal Genetics and Biology, vol. 46, no. 5, pp. 427–435, 2009.

277. K. J. Welham, M. A. Domin, K. Johnson, L. Jones, and D. S. Ashton, "Characterization of fungal spores by laser desorption/ionization time-of-flight mass spectrometry," Rapid Communications in Mass Spectrometry, vol. 14, no. 5, pp. 307–310, 2000. ·

278. M. Both, M. Csukai, M. P. H. Stumpf, and P. D. Spanu, "Gene expression profiles of Blumeria graminisindicate dynamic changes to primary metabolism during development of an obligate biotrophic pathogen," Plant Cell, vol. 17, no. 7, pp. 2107–2122, 2005. · ·

279. M. Both, S. E. Eckert, M. Csukai, E. Muller, G. Dimopoulos, and P. D. Spanu, "Transcript profiles ofBlumeria graminis development during infection reveal a cluster of genes that are potential virulence determinants," Molecular Plant-Microbe Interactions, vol. 18, no. 2, pp. 125–133, 2005.

280. M. R. Remm, C. E. V. Storm, and E. L. L. Sonnhammer, "Automatic clustering of orthologs and in-paralogs from pairwise species comparisons," Journal of Molecular Biology, vol. 314, no. 5, pp. 1041–1052, 2001.

281. J. Grinyer, L. Kautto, M. Traini, et al., "Proteome mapping of the Trichoderma reesei 20S proteasome,"Current Genetics, vol. 51, no. 2, pp. 79–88, 2007.

282. P. S. Solomon, K. A. R. C. Tan, P. Sanchez, R. M. Cooper, and R. P. Oliver, "The disruption of a Gαsubunit sheds new light on the pathogenicity of Stagonospora nodorum on wheat," Molecular Plant-Microbe Interactions, vol. 17, no. 5, pp. 456–466, 2004. ·

283. J. C. Misas-Villamil and R. A. van der Hoorn, "Enzyme-inhibitor interactions at the plant-pathogen interface," Current Opinion in Plant Biology, vol. 11, no. 4, pp. 380–388, 2008.

284. P. Kankanala, K. Czymmek, and B. Valent, "Roles for rice membrane dynamics and plasmodesmata during biotrophic invasion by the blast fungus," Plant Cell, vol. 19, no. 2, pp. 706–724, 2007.

285. H. Tjalsma, A. Bolhuis, J. D. H. Jongbloed, S. Bron, and J. M. van Dijl, "Signal peptide-dependent protein transport in Bacillus subtilis: a genome-based survey of the secretome," Microbiology and Molecular Biology Reviews, vol. 64, no. 3, pp. 515–547, 2000. ·

286. R. P. De Vries, "Regulation of Aspergillus genes encoding plant cell wall polysaccharide-degrading enzymes; relevance for industrial production," Applied Microbiology and Biotechnology, vol. 61, no. 1, pp. 10–20, 2003. ·

287. F. M. Freimoser, S. Screen, G. Hu, and R. St. Leger, "EST analysis of genes expressed by the zygomycete pathogen Conidiobolus coronatus during growth on insect cuticle," Microbiology, vol. 149, no. 7, pp. 1893–1900, 2003. ·

288. J. D. Walton, "Deconstructing the cell wall," Plant Physiology, vol. 104, no. 4, pp. 1113–1118, 1994. ·

289. N. Brito, J. J. Espino, and C. González, "The endo-β-1,4-xylanase Xyn11A is required for virulence inBotrytis cinerea," Molecular Plant-Microbe Interactions, vol. 19, no. 1, pp. 25–32, 2006. · ·

290. H. Deising, R. L. Nicholson, M. Haug, R. J. Howard, and K. Mendgen, "Adhesion pad formation and the involvement of cutinase and esterases in the attachment of uredospores to the host cuticle," Plant Cell, vol. 4, no. 9, pp. 1101–1111, 1992. ·

291. A. Isshiki, K. Akimitsu, M. Yamamoto, and H. Yamamoto, "Endopolygalacturonase is essential for citrus black rot caused by Alternaria citri but not brown spot caused by Alternaria alternata," Molecular Plant-Microbe Interactions, vol. 14, no. 6, pp. 749–757, 2001.
·

292. B. Oeser, P. M. Heidrich, U. Müller, P. Tudzynski, and K. B. Tenberge, "Polygalacturonase is a pathogenicity factor in the Claviceps purpurea/ rye interaction," Fungal Genetics and Biology, vol. 36, no. 3, pp. 176–186, 2002.

293. A. ten Have, W. Mulder, J. Visser, and J. A. L. van Kan, "The endopolygalacturonase gene Bcpg1 is required to full virulence of Botrytis cinerea," Molecular Plant-Microbe Interactions, vol. 11, no. 10, pp. 1009–1016, 1998. ·

294. C. A. Voigt, W. Schäfer, and S. Salomon, "A secreted lipase of Fusarium graminearum is a virulence factor required for infection of cereals," Plant Journal, vol. 42, no. 3, pp. 364–375, 2005.

295. N. Yakoby, D. Beno-Moualem, N. T. Keen, A. Dinoor, O. Pines, and D. Prusky, "Colletotrichum gloeosporioides pelB is an important virulence factor in avocado fruit-fungus interaction," Molecular Plant-Microbe Interactions, vol. 14, no. 8, pp. 988–995, 2001. ·

296. J. W. Bennett, "The molds of Katrina. Update," New York Academy of Sciences Magazine, pp. 6–9, 2006.

297. G. L. F. Wallis, R. J. Swift, F. W. Hemming, A. P. J. Trinci, and J. F. Peberdy, "Glucoamylase overexpression and secretion in Aspergillus niger: analysis of glycosylation," Biochimica et Biophysica Acta, vol. 1472, no. 3, pp. 576–586, 1999.

298. D. Lim, P. Hains, B. Walsh, P. Bergquist, and H. Nevalainen, "Proteins associated with the cell envelope of Trichoderma reesei: a proteomic approach," Proteomics, vol. 1, no. 7, pp. 899–910, 2001. ·

299. M. B. Suárez, L. Sanz, M. I. Chamorro, et al., "Proteomic analysis of secreted proteins from Trichoderma harzianum: identification of a fungal cell wall-induced aspartic protease," Fungal Genetics and Biology, vol. 42, no. 11, pp. 924–934, 2005.

300. C. Tian, W. T. Beeson, A. T. Iavarone, et al., "Systems analysis of plant cell wall degradation by the model filamentous fungus Neurospora crassa," Proceedings of the National Academy of Sciences of the United States of America, vol. 106, no. 52, pp. 22157–22162, 2009.

301. J. M. Cork and M. D. Purugganan, "The evolution of molecular genetic pathways and networks,"BioEssays, vol. 26, no. 5, pp. 479–484, 2004.

302. T. Köcher and G. Superti-Furga, "Mass spectrometry-based functional proteomics: from molecular machines to protein networks," Nature Methods, vol. 4, no. 10, pp. 807–815, 2007.

303. W. F. Loomis and P. W. Sternberg, "Genetic networks," Science, vol.

269, no. 5224, p. 649, 1995. ·

304. M. E. Cusick, N. Klitgord, M. Vidal, and D. E. Hill, "Interactome: gateway into systems biology," Human Molecular Genetics, vol. 14, no. 2, pp. R171–R181, 2005. · ·

305. L. Kiemer and G. Cesareni, "Comparative interactomics: comparing apples and pears?" Trends in Biotechnology, vol. 25, no. 10, pp. 448–454, 2007.

306. L. Giot, J. S. Bader, C. Brouwer, et al., "A protein interaction map of Drosophila melanogaster," Science, vol. 302, no. 5651, pp. 1727–1736, 2003.

307. D. J. LaCount, M. Vignali, R. Chettier, et al., "A protein interaction network of the malaria parasitePlasmodium falciparum," Nature, vol. 438, no. 7064, pp. 103–107, 2005.

308. S. Li, C. M. Armstrong, N. Bertin, et al., "A map of the interactome network of the metazoan C. elegans,"Science, vol. 303, no. 5657, pp. 540–543, 2004.

309. J. R. Parrish, J. Yu, G. Liu, et al., "A proteome-wide protein interaction map for Campylobacter jejuni,"Genome Biology, vol. 8, no. 7, p. R130, 2007.

310. P. Uetz, L. Glot, G. Cagney, et al., "A comprehensive analysis of protein-protein interactions inSaccharomyces cerevisiae," Nature, vol. 403, no. 6770, pp. 623–627, 2000.

311. T. Berggård, S. Linse, and P. James, "Methods for the detection and analysis of protein-protein interactions," Proteomics, vol. 7, no. 16, pp. 2833–2842, 2007.

312. J. A. N. A. Miernyk and J. A. Y. J. Thelen, "Biochemical approaches for discovering protein-protein interactions," Plant Journal, vol. 53, no. 4, pp. 597–609, 2008.

313. B. Causier and B. Davies, "Analysing protein-protein interactions with the yeast two-hybrid system,"Plant Molecular Biology, vol. 50, no. 6, pp. 855–870, 2002. · ·

314. L. R. Matthews, P. Vaglio, J. Reboul, et al., "Identification of potential interaction networks using sequence-based searches for conserved protein-protein interactions or "interologs"," Genome Research, vol. 11, no. 12, pp. 2120–2126, 2001.

315. T. Ideker, O. Ozier, B. Schwikowski, and A. F. Siegel, "Discovering regulatory and signalling circuits in molecular interaction networks," Bioinformatics, vol. 18, supplement 1, pp. S233–S240, 2002. ·

316. X. Wu, L. E. I. Zhu, J. I. E. Guo, D. A. Y. Zhang, and K. U. I. Lin, "Prediction of yeast protein-protein interaction network: insights from the Gene Ontology and annotations," Nucleic Acids Research, vol. 34, no. 7, pp. 2137–2150, 2006.

317. S. E. E. K. Ng, Z. Zhang, and S. H. Tan, "Integrative approach for computationally inferring protein domain interactions," Bioinformatics, vol. 19, no. 8, pp. 923–929, 2003.

318. R. Jothi, M. G. Kann, and T. M. Przytycka, "Predicting protein-protein interaction by searching evolutionary tree automorphism space," Bioinformatics, vol. 21, supplement 1, pp. i241–i250, 2005. · ·

319. U. Ogmen, O. Keskin, A. S. Aytuna, R. Nussinov, and A. Gursoy, "PRISM: protein interactions by structural matching," Nucleic Acids Research, vol. 33, no. 2, pp. W331–W336, 2005.

320. Y. Qi, J. Klein-Seetharaman, and Z. I. V. Bar-Joseph, "A mixture of feature experts approach for protein-protein interaction prediction," BMC Bioinformatics, vol. 8, supplement 10, p. S6, 2007.

321. J. Shen, J. Zhang, X. Luo, et al., "Predicting protein-protein interactions based only on sequences information," Proceedings of the National Academy of Sciences of the United States of America, vol. 104, no. 11, pp. 4337–4341, 2007.

322. C. von Mering, L. J. Jensen, M. Kuhn, et al., "STRING 7—recent developments in the integration and prediction of protein interactions," Nucleic Acids Research, vol. 35, no. 1, pp. D358–D362, 2007.

323. D. R. Rhodes, S. A. Tomlins, S. Varambally, et al., "Probabilistic model of the human protein-protein interaction network," Nature Biotechnology, vol. 23, no. 8, pp. 951–959, 2005.

324. E. Hirsh and R. Sharan, "Identification of conserved protein complexes based on a model of protein network evolution," Bioinformatics, vol. 23, no. 2, pp. e170–e176, 2007.

325. S. P. Gygi, Y. Rochon, B. R. Franza, and R. Aebersold, "Correlation between protein and mRNA abundance in yeast," Molecular and Cellular Biology, vol. 19, no. 3, pp. 1720–1730, 1999.

Chapter 3

LIFESTYLE TRANSITIONS IN PLANT PATHOGENIC COLLETOTRICHUM FUNGI DECIPHERED BY GENOME AND TRANSCRIPTOME ANALYSES

Richard J O'Connell[1], Michael R Thon[2], Stéphane Hacquard[1], Stefan G Amyotte[3], Jochen Kleemann[1], Maria F Torres[3], Ulrike Damm[4], Ester A Buiate[3], Lynn Epstein[5], Noam Alkan[6], Janine Altmüller[7], Lucia Alvarado-Balderrama[8], Christopher A Bauser[9], Christian Becker[7], Bruce W Birren[8], Zehua Chen[8], Jaeyoung Choi[10], Jo Anne Crouch[11], Jonathan P Duvick[12,30], Mark A Farman[3], Pamela Gan[13], David Heiman[8], Bernard Henrissat[14], Richard J Howard[12], Mehdi Kabbage[15], Christian Koch[16], Barbara Kracher[1], Yasuyuki Kubo[17], Audrey D Law[3], Marc-Henri Lebrun[18], Yong-Hwan Lee[10], Itay Miyara[6], Neil Moore[19], Ulla Neumann[20], Karl Nordström[21], Daniel G Panaccione[22], Ralph Panstruga[1,23], Michael Place[24], Robert H Proctor[25], Dov Prusky[6], Gabriel Rech[2], Richard Reinhardt[26], Jeffrey A Rollins[27], Steve Rounsley[8], Christopher L Schardl[13], David C Schwartz[24], Narmada Shenoy[8], Ken Shirasu[13], Usha R Sikhakolli[28], Kurt Stüber[26], Serenella A Sukno[2], James A Sweigard[12], Yoshitaka Takano[29], Hiroyuki Takahara[1,30], Frances Trail[28], H Charlotte van der Does[1,30], Lars M Voll[16], Isa Will1, Sarah Young[8], Qiandong Zeng[8], Jingze Zhang[8], Shiguo Zhou[24], Martin B Dickman[15], Paul Schulze-Lefert[1], Emiel Ver Loren van Themaat[1], Li-Jun Ma[8,30] & Lisa J Vaillancourt[3]

[1]Department of Plant Microbe Interactions, Max Planck Institute for Plant Breeding Research, Cologne, Germany

[2]Centro Hispano-Luso de Investigaciones Agrarias, Departamento de Microbiología y Genética, Universidad de Salamanca, Villamayor, Spain

[3]Department of Plant Pathology, University of Kentucky, Lexington, Kentucky, USA

[4]Centraalbureau voor Schimmelcultures, Koninklijke Nederlandse Akademie van Wetenschappen, Fungal Biodiversity Centre, Utrecht, The Netherlands

[5]Department of Plant Pathology, University of California, Davis, California, USA

[6]Department of Postharvest Science of Fresh Produce, Agricultural Research Organization, The Volcani Center, Bet Dagan, Israel

[7]Cologne Center for Genomics, University of Cologne, Cologne, Germany

[8]Broad Institute, Cambridge, Massachusetts, USA

[9]GATC Biotech AG, Konstanz, Germany

[10]Department of Agricultural Biotechnology, Center for Fungal Genetic Resources, Seoul National University, Seoul, Korea

[11]Systematic Mycology and Microbiology Laboratory, US Department of Agriculture, Agricultural Research Service, Beltsville, Maryland, USA

[12]Pioneer Hi-Bred International, DuPont Agricultural Biotechnology, Wilmington, Delaware, USA

[13]Plant Immunity Research Group, RIKEN Plant Science Center, Yokohama, Japan

[14]Laboratoire Architecture et Fonction des Macromolécules Biologiques, Centre National de la Recherche Scientifique, Unité Mixte de Recherche 7257, Université Aix-Marseille, Marseille, France

[15]Department of Plant Pathology and Microbiology, Institute for Plant Genomics and Biotechnology, Borlaug Genomics and Bioinformatics Center, Texas A&M University, College Station, Texas, USA

[16]Department of Biology, Division of Biochemistry, Friedrich-Alexander-University ErlangenNuremberg, Erlangen, Germany

[17]Laboratory of Plant Pathology, Graduate School of Life and Environmental Sciences, Kyoto Prefectural University, Kyoto, Japan

[18]Institut National de la Recherche Agronomique, Biologie et Gestion des Risques en Agriculture—Champignons Pathogènes des Plantes, Thiverval-Grignon, France

[19]Department of Computer Science, University of Kentucky, Lexington, Kentucky, USA

[20]Central Microscopy, Max Planck Institute for Plant Breeding Research, Cologne, Germany

[21]Department of Plant Developmental Biology, Max Planck Institute for Plant Breeding Research, Cologne, Germany

[22]Division of Plant and Soil Sciences, West Virginia University, Morgantown, West Virginia, USA

[23]Institute for Biology I, Unit of Plant Molecular Cell Biology, Rheinisch-Westfälische Technische Hochschule Aachen University, Aachen, Germany

[24]Laboratory for Molecular and Computational Genomics, University of Wisconsin-Madison, Madison, Wisconsin, USA

[25]US Department of Agriculture, Agriculture Research Service, National Center for Agricultural Utilization Research, Peoria, Illinois, USA

[26]Max Planck Genome Centre Cologne, Cologne, Germany

[27]Department of Plant Pathology, University of Florida, Gainesville, Florida, USA

[28]Department of Plant Biology, Michigan State University, East Lansing, Michigan, USA

[29]Laboratory of Plant Pathology, Graduate School of Agriculture, Kyoto University, Kyoto, Japan

[30]Department of Genetics, Development and Cell Biology, Iowa State University, Ames, Iowa, USA (J.P.D.), Department of Bioproduction Science, Faculty of Bioresources and Environmental Sciences, Ishikawa Prefectural University, Ishikawa, Japan (H.T.), Department of Plant Pathology, Swammerdam Institute for Life Sciences, University of Amsterdam, Amsterdam, The Netherlands (H.C.v.d.D.) and The College of Natural Sciences, University of Massachusetts Amherst, Amherst, Massachusetts, USA (L.-J.M.)

Colletotrichum **species are fungal pathogens that devastate crop plants worldwide. Host infection involves the differentiation of specialized cell types that are associated with penetration, growth inside living host cells** (biotrophy) and tissue destruction (necrotrophy). We report here genome and transcriptome analyses of *Colletotrichum higginsianum* infecting *Arabidopsis thaliana* **and** *Colletotrichum graminicola* **infecting maize. Comparative genomics showed that both fungi have large sets of pathogenicity-related genes, but families of genes encoding secreted effectors, pectin-degrading enzymes, secondary metabolism enzymes, transporters and peptidases are expanded in** *C. higginsianum.* Genome-wide expression profiling revealed that these genes are transcribed in successive waves that are linked to pathogenic transitions: effectors and secondary metabolism enzymes are induced before penetration and during biotrophy, whereas most hydrolases and transporters are upregulated later, at the switch to necrotrophy. Our findings show that preinvasion perception of plant-derived signals substantially reprograms fungal gene expression and indicate previously unknown functions for particular fungal cell types.

The genus *Colletotrichum* (Sordariomycetes, Ascomycota; Fig. 1a) comprises ~600 species[1]attacking over 3,200 species of monocot and dicot plants. These pathogens use a multistage hemibiotrophic infection strategy[2]: dome-shaped appressoria first puncture host surfaces using a combination of mechanical force and enzymatic degradation, bulbous biotrophic hyphae enveloped by an intact host plasma membrane then develop inside living epidermal cells, and finally, the fungus switches to necrotrophy and differentiates thin, fast-growing hyphae that kill and destroy host tissues (Fig. 1b,c). We sequenced two *Colletotrichum* species with different host specificities and infection strategies: *C. higginsianum* attacks several members of *Brassicaceae*, including *Arabidopsis*, and has emerged as a tractable model for studying fungal pathogenicity and plant immune responses[3, 4, 5]. Biotrophy in this fungus is confined to the first invaded host cell and is followed by a complete switch to necrotrophy[5]. In contrast, *C. graminicola* primarily infects maize (*Zea mays*), causing annual losses of approximately 1 billion dollars in the United States alone[6]. In this species, biotrophy extends into many host cells and persists at the advancing colony margin while the center of the colony becomes necrotrophic[7].

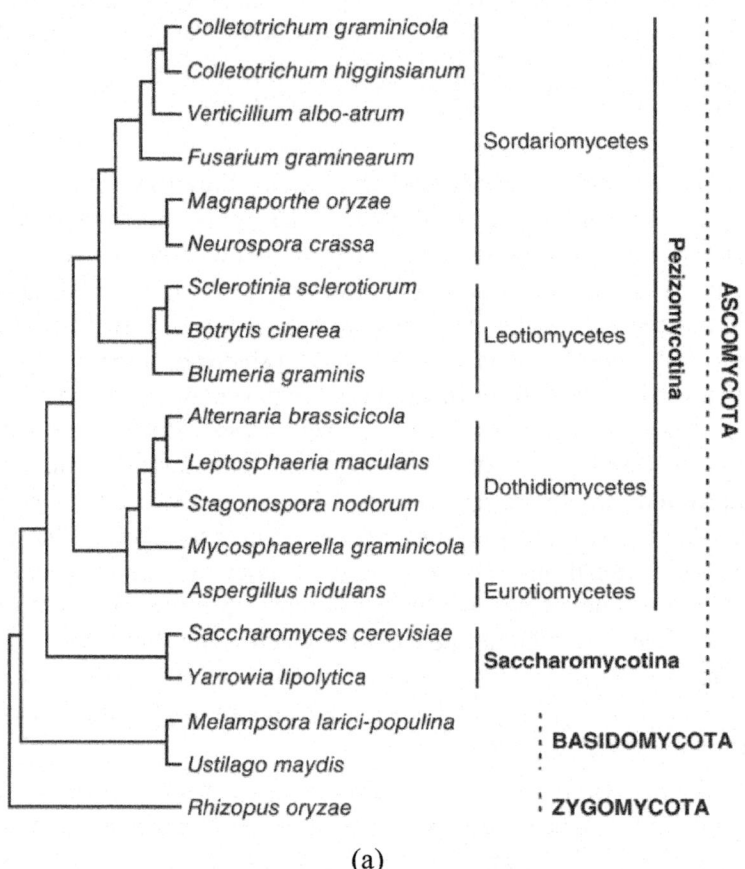

(a)

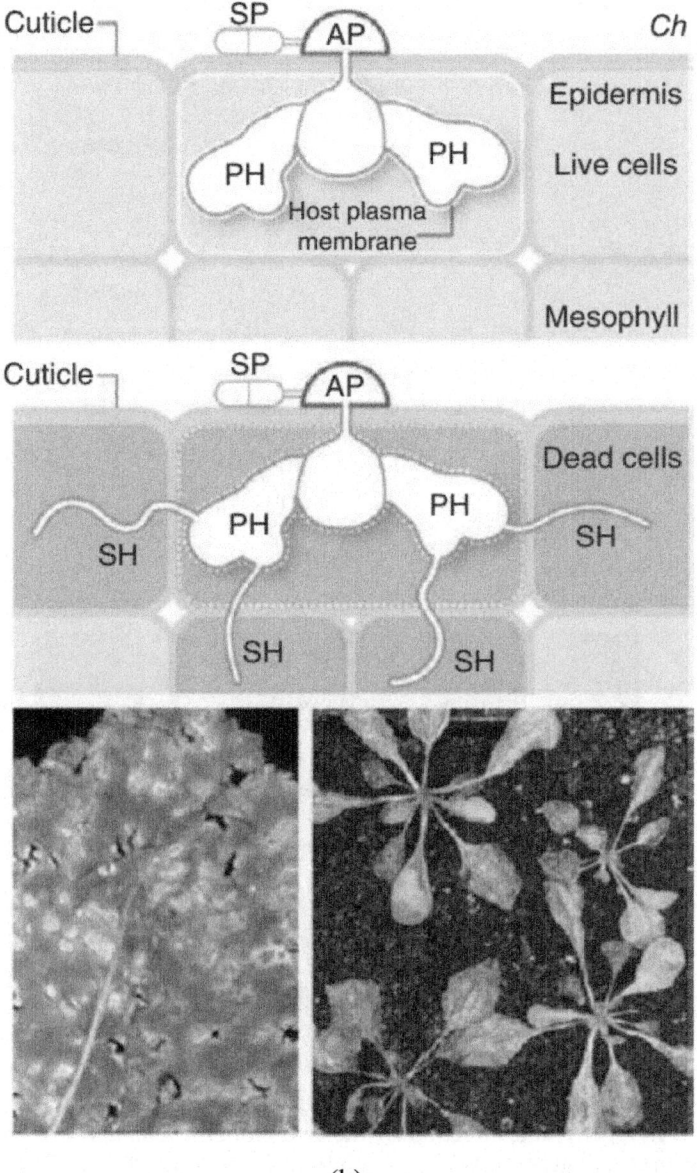

(b)

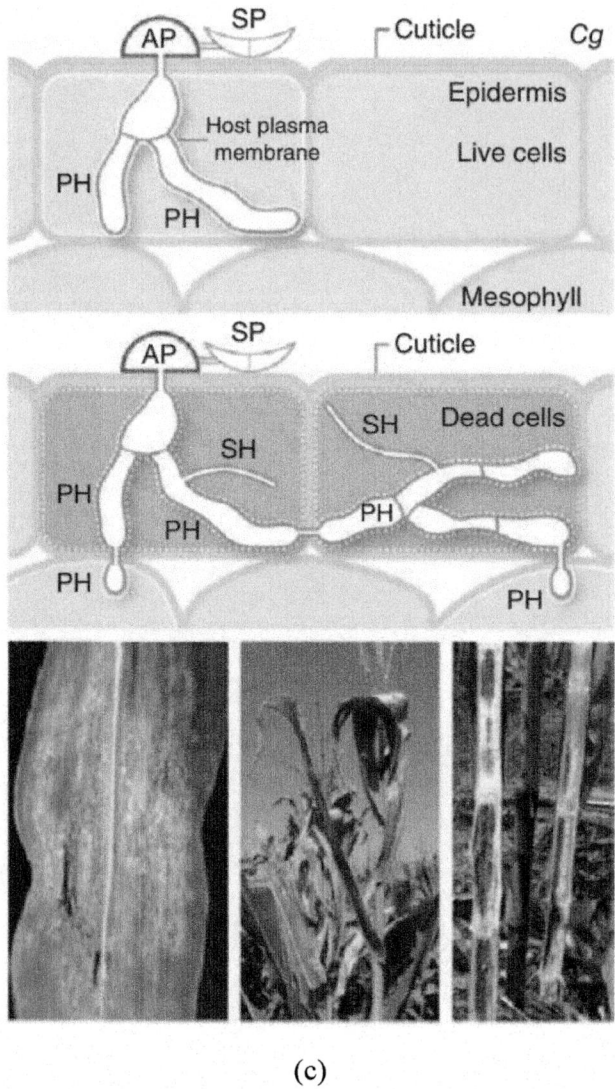

(c)

Figure 1: Phylogeny and infection of the two *Colletotrichum* **species analyzed in this study.** (a) Cladogram showing the phylogenetic relationship of *Colletotrichum* to other sequenced fungi, including 13 species used for comparative analyses (see Fig. 3). The unscaled tree was constructed using CVTree[34] with *Rhizopus oryzae* as the outgroup. (b) Infection process of *C. higginsianum* (*Ch*) and leaf anthracnose symptoms on *Brassica* and *Arabidopsis*. The *Brassica* image is reproduced with permission of University of Georgia Plant Pathology Archive (bugwood.org/). (c) Infection process of *C. graminicola*(*Cg*), and leaf-blight, top die-back and stalk-rot symptoms on maize. SP, spore; AP, appressorium; PH, biotrophic primary hyphae; SH, necrotrophic secondary hyphae.

Optical mapping showed that the genomes of the two species are similar in size and structure. *C. graminicola* has a 57.4-Mb genome that is distributed among 13 chromosomes, including three minichromosomes less than 1 Mb in size, whereas *C. higginsianum* has a 53.4-Mb genome comprising 12 chromosomes, including two minichromosomes. We sequenced the *C. graminicola* genome using Sanger and 454 platforms, which provided a high-quality reference assembly of 50.9 Mb. We sequenced the *C. higginsianum* genome using 454 and Illumina platforms, yielding an assembly of 49.3 Mb. Repetitive DNA comprises 12.2% of the *C. graminicola* genome assembly and 1.2% of the *C. higginsianum* assembly. The repeats clustered in genomic regions with low GC content in *C. graminicola*, similar to the AT-rich isochores found in *Leptosphaeria maculans*[8]. Including unassembled genomic regions (mostly repeats, such as ribosomal DNA, telomeres, centromeres and transposons), repetitive DNA was estimated to total 22.3% of the *C. graminicola* genome and 9.1% of the *C. higginsianum* genome. The two *Colletotrichum* species diverged relatively recently (~47 million years ago), after the separation of monocots and dicots 140–150 million years ago[9].

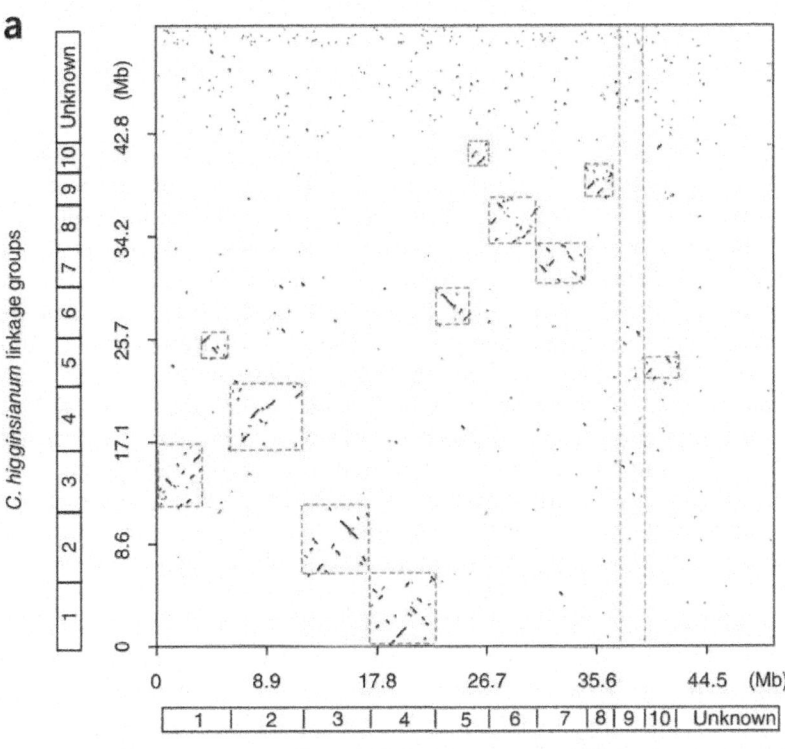

a

C. graminicola linkage groups (Mb)

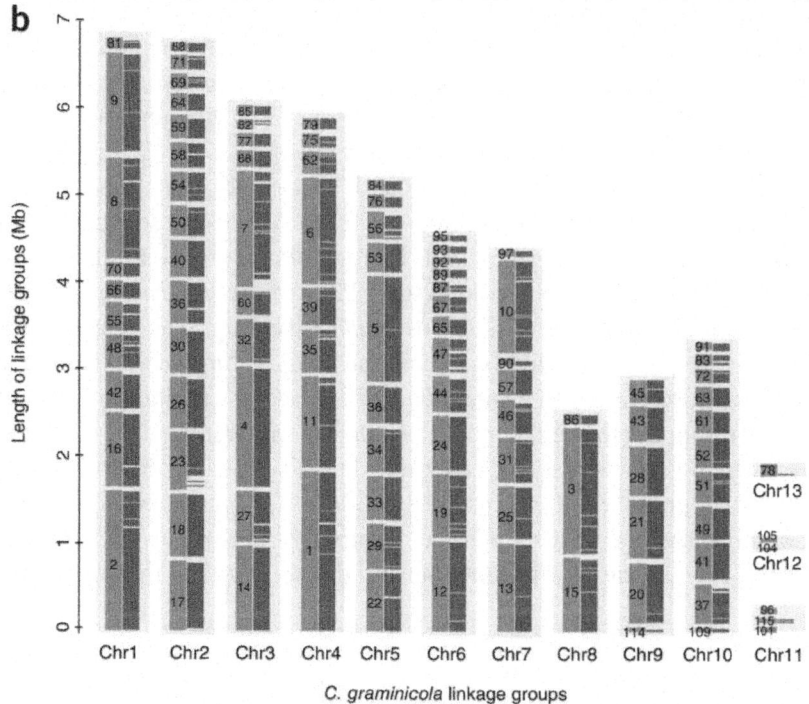

Figure 2: Conservation of synteny between the genomes of *C. graminicola* and *C. higginsianum*. (a) Dot plot showing the syntenic blocks between the 13 chromosomes (optical linkage groups) of *C. graminicola* (horizontal axis) and the 12 chromosomes of *C. higginsianum* (vertical axis). Homologies between chromosomes of each species are highlighted in red dashed boxes. Homologous sequences of *C. graminicola* chromosome 9, indicated between the blue dashed lines, are dispersed among many *C. higginsianum* chromosomes. (b) Global view of syntenic alignments between the genomes of *C. graminicola* and *C. higginsianum*. Linkage groups of *C. graminicola* are shown as the reference, with linkage group lengths defined by the *C. graminicola* optical map. For each chromosome, numbered genomic scaffolds (dark gray) positioned on the optical linkage groups are separated by scaffold breaks. The magenta blocks show syntenic mapping of the *C. higginsianum* sequences; notably, there is a near absence of homologous sequences among the minichromosomes.

Although *C. graminicola* and *C. higginsianum* belong to sister clades within the genus, only 35% of the two genomes are syntenic, which is less than the synteny between *Botrytis cinerea* and *Sclerotinia sclerotiorum*[10]. Nevertheless, an analysis of synteny between the two *Colletotrichum* genomes identified homologous chromosomes and revealed that major intrachromosomal rearrangements have occurred in one or both species (Fig.

2a). The minichromosomes do not contain homologous sequences (Fig. 2b), suggesting that they are lineage-specific innovations, and in *C. graminicola*, the minichromosomes are enriched with repetitive DNA (averaging 23%) compared to the core genome (averaging 5.5%).

We predicted the existence of 12,006 protein-coding genes in *C. graminicola* compared to 16,172 in *C. higginsianum*. Having been compiled from short-read data only, the *C. higginsianum* assembly is more fragmented than that of *C. graminicola*, resulting in some genes (5.2%) being split into two or more gene models, whereas others (4%) are truncated versions of the complete gene. After correcting for this fragmentation, the estimated gene content of *C. higginsianum* (15,331) is still markedly larger than that of *C. graminicola*. The two species share 9,795 orthologous genes. Using Markov clustering (MCL)[11] to analyze the proteomes, we found that 10,077 *C. higginsianum* genes belong to multicopy gene clusters, compared to 5,342 genes in *C. graminicola*, suggesting that the greater gene content of *C. higginsianum* results partly from gene duplication. The MCL analysis also revealed that gene clusters encoding serine proteases, methyl transferases, polyketide synthases, cytochrome P450 enzymes and small-molecule efflux pumps are expanded in *C. higginsianum* compared to *C. graminicola*, which we verified by manual inspection. Clusters that are expanded in *C. graminicola* relative to *C. higginsianum* include a family of genes encoding atypical cellulases (glycoside hydrolase GH61, described below) and another encoding secreted histidine acid phosphatases, which probably mobilize phytic acid, the main form of stored phosphorus in plants[12].

C. higginsianum and *C. graminicola* are particularly well equipped with genes encoding carbohydrate-active enzymes (CAZymes)[13] that potentially degrade the plant cell wall[14] (Fig. 3a) and modify the fungal cell wall. Both species encode more CAZymes than 13 other fungal genomes we examined. These expanded CAZyme arsenals are more similar to those of other hemibiotrophic and necrotrophic pathogens than to the highly reduced set found in biotrophs such as *Melampsora* and *Blumeria* (Fig. 3a). The exceptionally large and diverse inventory of CAZymes encoded by both *Colletotrichum* genomes provides a rich source of enzymes for potential commercial exploitation[15]. *C. higginsianum* encodes over twice as many pectin-degrading enzymes as does *C. graminicola* (Fig. 3b), the majority (62%) of which are activated during necrotrophy. Conversely, although both species encode similar numbers of cellulases and hemicellulases, *C. graminicola* activates many more of these genes during necrotrophy (48%) than does *C. higginsianum* (26%), including 22 GH61 copper-dependent oxygenases, which act in concert with classical cellulases to enhance lignocellulose hydrolysis[16, 17]. Thus, *C.*

graminicola and *C. higginsianum* use very different strategies to deconstruct plant cell walls, reflecting their host preferences: dicot cell walls are enriched with pectin (35% in dicots compared to 10% in maize), whereas the cell walls of grasses contain more hemicellulose (60% in grasses compared to 30% in dicots) and phenolics (up to 5%)[18].

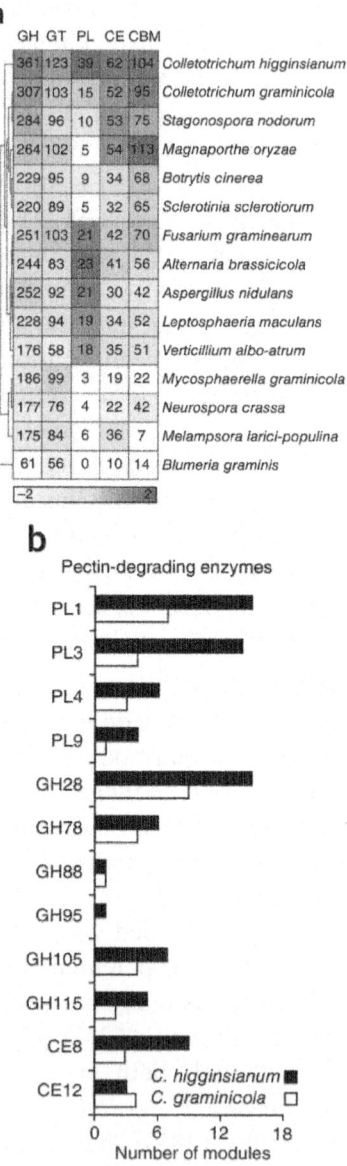

Figure 3: Comparison of fungal carbohydrate-active enzyme (CAZyme) repertoires. (a) Hierarchical clustering of CAZyme classes from *Colletotrichum* and 13 other

fungal genomes. GH, glycoside hydrolase; GT, glycosyltransferase; PL, polysaccharide lyase; CE, carbohydrate esterase; CBM, carbohydrate-binding module. The numbers of enzyme modules in each genome are shown. Overrepresented (orange to red) and underrepresented modules (pale yellow to white) are depicted as fold changes relative to the class mean. (b) Comparison of the pectin-degrading enzyme repertoires of *C. higginsianum* and *C. graminicola* shown as the number of modules in each CAZyme family[11]. In total, *C. higginsianum* encodes 86 such modules, whereas *C. graminicola* encodes only 42.

Many phytopathogens secrete proteins known as effectors that facilitate infection by reprogramming host cells and modulating plant immunity[19]. By defining candidate secreted effectors (CSEPs) as predicted extracellular proteins without any homology to proteins outside the genus *Colletotrichum*, we found 177 CSEP-encoding genes in *C. graminicola*, 85 (48%) of which were species specific. In contrast, *C. higginsianum* encodes twice as many CSEPs (365), including more species-specific proteins (264, or 72%). The CSEPs are mostly small proteins (averaging 110 residues and 175 residues in *C. higginsianum* and *C. graminicola*, respectively) and are more cysteine rich than the total proteome. CSEP-encoding genes are randomly distributed across the chromosomes of *C. graminicola*, with no evidence for clustering, enrichment on particular chromosomes or localization near transposable elements or telomeres, as has been reported for some other plant pathogens[8, 20,21, 22]. An MCL analysis revealed that relatively few *Colletotrichum* CSEPs (14% in both species) belong to small multigenic families with two to five members. The larger, more diversified CSEP repertoire of *C. higginsianum* might be an adaptation to invade a broader range of host plants than *C. graminicola*, which is restricted to infection of *Zea* under field conditions[23].

Both *Colletotrichum* species encode markedly more secondary metabolism enzymes (103 in *C. higginsianum* and 74 in *C. graminicola*) than other sequenced fungi (2–58 in ascomycetes[24, 25]). In fungi, secondary metabolism genes are typically located in clusters[26]; we found 42 of these clusters in *C. graminicola* and 39 in *C. higginsianum*, surpassing the numbers found in most other sequenced ascomycetes. Only 11 secondary metabolism gene clusters are shared between the two *Colletotrichum* species, and only 6 of these clusters show limited synteny (Fig. 4). This cluster diversity seems to result from gene duplication or loss and chromosomal rearrangements and may be related to the association of secondary metabolism gene clusters (71% in *C. graminicola*) with repetitive DNA. Because each secondary metabolism gene cluster is probably involved in the biosynthesis of a specific metabolite[24], each *Colletotrichum* species can be expected to produce unusually large and divergent spectra of secondary metabolites, some of which may be previously unknown bioactive molecules.

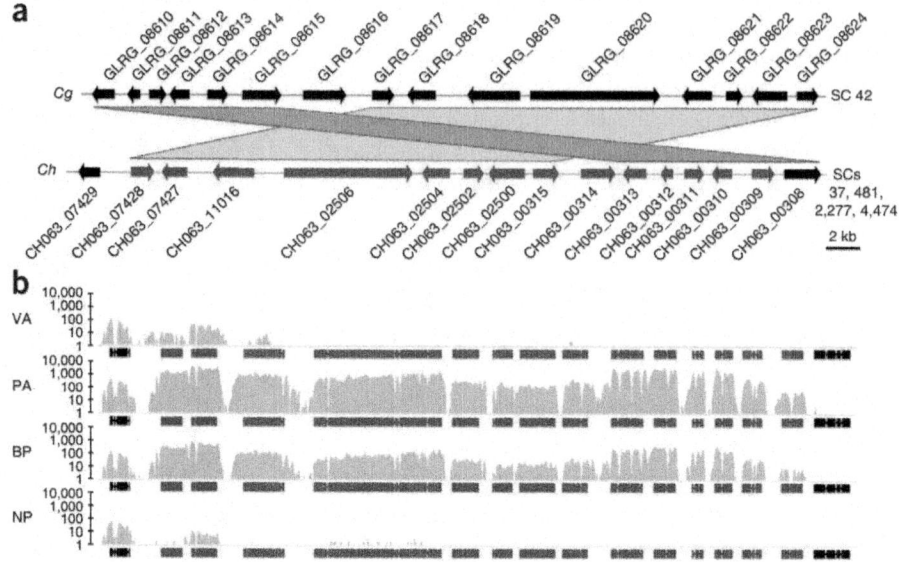

Figure 4: Structure and transcription of a secondary metabolism gene cluster. (a) Gene cluster 18 from *C. graminicola (Cg)* is orthologous to cluster 10 from *C. higginsianum (Ch)*. The latter is split between four small scaffolds in the Broad Institute genome annotation (supercontigs (SCs) 37, 481, 2,277 and 4,474) and was reconstructed based on an improved genome assembly. Microsynteny is indicated by gray bars. The 14 genes highlighted in red in the *C. higginsianum* cluster are co-regulated. (b) Visualization of RNA-Seq coverage across the *C. higginsianum* polyketide biosynthesis cluster. The gray curves indicate read coverage (log scale) for the four samples. Co-regulated gene models are highlighted in red. VA, *in vitro* appressoria; PA, *in planta* appressoria; BP, biotrophic phase; NP, necrotrophic phase.

To investigate how the fungal genetic program is deployed during host infection, we applied Illumina RNA sequencing to both pathosystems. We collected samples from infected *Arabidopsis* or maize leaves at intervals corresponding to pre-penetration appressoria, the early biotrophic phase and the transition to necrotrophy and from *C. higginsianum* appressoria formed *in vitro* (Fig. 5a). Almost all the gene models were transcribed *in planta* (14,972 *C. higginsianum* genes, or 92%, and 10,812 *C. graminicola* genes, or 90%). However, this transcription was highly dynamic, particularly in *C. higginsianum*, where 7,162 genes (44%) were differentially regulated (log2 fold change >2, *P* < 0.05) between one or more of the infection stages. Fewer genes (2,619, or 22%) were differentially regulated in *C. graminicola*, which may reflect the contrasting biology of this species, where biotrophic and necrotrophic growth occur simultaneously. The more clearly defined infection stages of *C. higginsianum* provided better temporal and spatial resolution of expression changes, and we therefore highlight our results for this species.

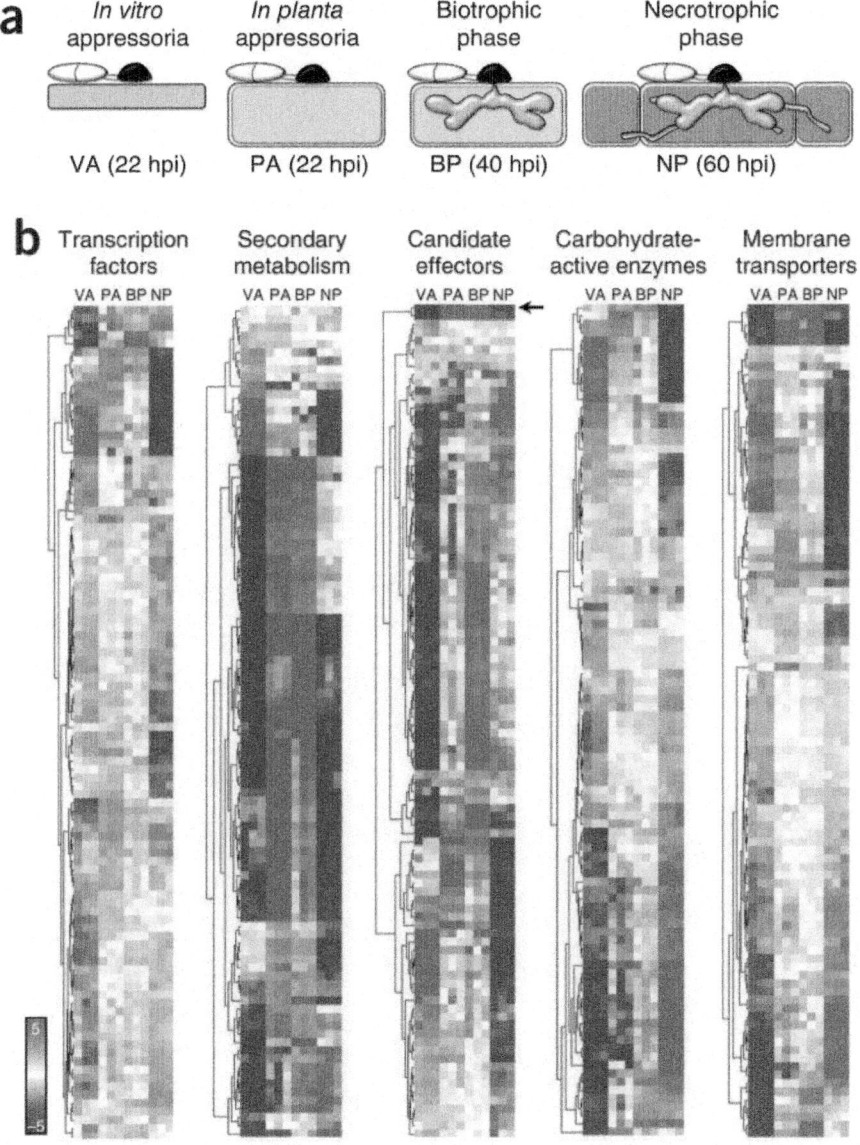

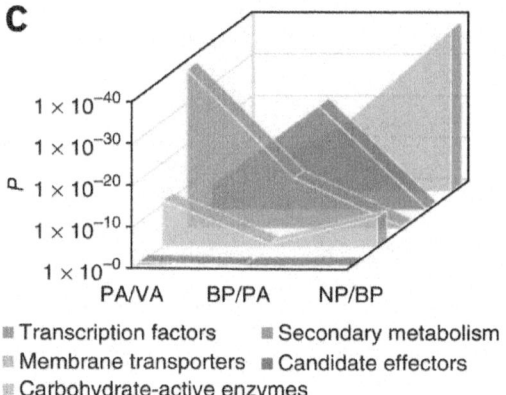

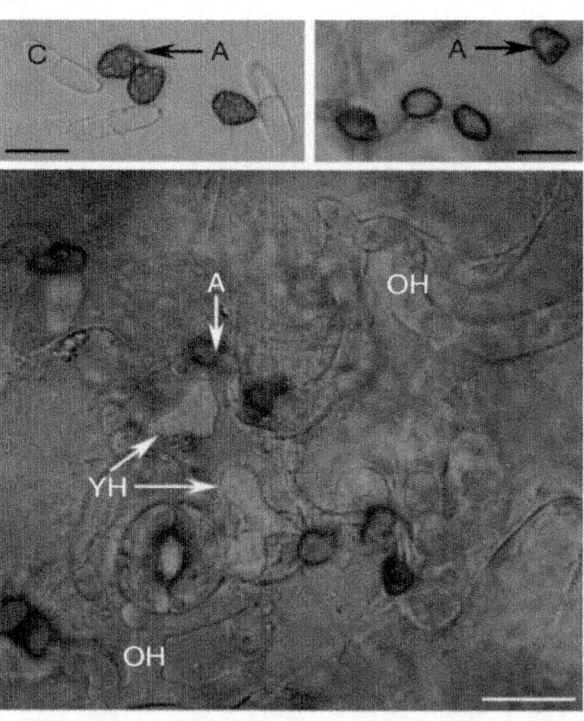

Figure 5: Expression profiling of pathogenicity-related genes in *C. higginsianum*. (**a**) Schematic representation of the four *C. higginsianum* developmental stages selected for RNA sequencing. Gray indicates polystyrene, green indicates living plant cell, and brown indicates dead plant cell. Hpi, hours post-inoculation. (**b**) Heatmaps of gene expression showing the 100 most highly expressed and significantly regulated genes (log2 fold change >2, $P < 0.05$) in five functional categories. Overrepresented (pale

red to dark red) and underrepresented transcripts (pale blue to dark blue) are shown as log2 fold changes relative to the mean expression measured across all four stages. The arrow indicates the CSEP-encoding gene *ChEC6* (CH063_01084). (**c**) The statistical significance of gene induction (*y* axis) in five functional categories during fungal developmental transitions (*x* axis). The *P*values were calculated using a one-sided Fisher's exact test and represent the probability of observing the number of significantly induced genes for a specific category during a transition given the total number of significantly induced genes during that transition (log2 fold change >2, *P* < 0.05) and the total number of genes in the category. (**d**) Transcriptional regulation of the effector gene *ChEC6* by plant-derived signals. Confocal micrographs showing *C. higginsianum* expressing the *mCherry* reporter gene under the native*ChEC6* promoter (overlays of bright-field and fluorescence channels). Appressoria (A) formed on polystyrene are unlabeled (top left), whereas those on the leaf surface (top right) have fluorescent cytoplasm. After host penetration, labeling is visible in young biotrophic hyphae (YH) but not older biotrophic hyphae (OH) (bottom). Scale bars, 10 μm. C, conidium.

Five gene categories relevant to pathogenicity (encoding transcription factors, secondary metabolism enzymes, CSEPs, CAZymes and transporters) had markedly different expression patterns during infection (Fig. 5b). We distinguished three waves of gene activation corresponding to pathogenic transitions (Fig. 5c). Among the genes upregulated at the appressorial phase were those encoding CAZymes that are predicted to degrade cutin, cellulose, hemicellulose and pectin, which may contribute to initial host penetration, together with a larger set of enzymes that potentially remodel the fungal cell wall (Fig. 5b). However, early during infection, the transcriptome of *C. higginsianum* was dominated by secondary metabolism genes, with 12 different secondary metabolism gene clusters being induced before penetration and during biotrophy (Fig. 5b,c). This indicates previously unsuspected roles for appressoria and biotrophic hyphae in synthesizing an array of small molecules for delivery to the first infected plant cells. Because these cells initially remain alive, such molecules are probably not toxins and may instead function in host manipulation, similar to protein effectors[27]. Remarkably, the *C. higginsianum* secondary metabolism gene cluster with the strongest activation at this stage was silent in *C. graminicola* at all the infection stages we examined (Fig. 4), suggesting that additional metabolite diversity is generated through transcriptional regulation.

Different sets of CSEP-encoding genes were expressed at each infection stage, but the majority of these genes were strongly induced during biotrophy (Fig. 5b,c). This suggests *Colletotrichum* requires a maximum capacity for host manipulation during intracellular colonization and that biotrophic hyphae provide a major interface for effector delivery to host cells. These specialized hyphae morphologically resemble the haustoria of obligate biotrophs, which function both as platforms for effector secretion and feeding structures for the uptake of sugars and amino acids[28, 29]. However, we found no evidence for specific transcriptional reprogramming of nutrient transporters in *C. higginsianum* during biotrophy (Fig. 5b,c), suggesting that the biotrophic hyphae of this pathogen function primarily to deliver protein effectors and secondary metabolites to the plant cell.

Transcripts encoding a vast array of lytic enzymes are induced at the transition to necrotrophy, when the pathogen uses dead and dying host cells as a nutrient source to support rapid colonization and sporulation (Fig. 5b,c). These enzymes include 44 putative secreted proteases and 146 CAZymes that potentially cleave all major polysaccharides in the host wall. Concomitantly, numerous genes encoding plasma membrane transporters that may be required for assimilating the products of this degradative activity, for example, oligopeptides, amino acids and sugars, are also induced (Fig. 5b,c). In fungi, genes encoding secreted proteases, CAZymes and permeases are often subject to pH regulation[30]. Consistent with this, we found evidence that necrotrophy in *C. higginsianum* is associated with local alkalinization of *Arabidopsis* tissue, probably resulting from fungal ammonia secretion[31], but tissue alkalinization was less pronounced in maize colonized by *C. graminicola*at this stage.

Notably, although appressoria *in vitro* are morphologically indistinguishable from those *in planta*, their transcriptomes are substantially different, with 1,515 genes significantly induced by host contact. One of these, the CSEP-encoding gene called *ChEC6*(CH063_01084)[32], is the most highly and significantly induced of all *C. higginsianum* genes (>50,000-fold) compared to appressoria *in vitro*. To experimentally verify this expression pattern at the cellular level, we generated transgenic *C. higginsianum* strains expressing a reporter gene under the control of the *ChEC6* promoter (Fig. 5d). Using this method, we confirmed that the transcription of *ChEC6* was plant specific, starting in the appressorium before penetration and continuing in young biotrophic hyphae, but it was switched off before the hyphae were fully expanded, indicating that its expression is transient and tightly regulated. The large-scale reprogramming of appressorial gene expression *in planta* shows that these specialized cells are highly responsive to host-derived cues that are perceived before penetration. Long regarded as organs of attachment and penetration[33], our findings assign a previously unsuspected sensory function to fungal appressoria, enabling the pathogen to prepare for the subsequent invasion of living host cells.

Major hemibiotrophic plant pathogens such as *Colletotrichum* and the rice blast fungus*Magnaporthe oryzae* undergo major transformations in cell morphology and infection mode when switching from growth on the plant surface to intracellular biotrophy and from biotrophy to necrotrophy. Genome sequencing combined with high-throughput transcriptome sequencing revealed the transcriptional dynamics underlying these transitions and led us to redefine the functions of appressoria and intracellular hyphae. Despite their similar morphologies, a genomic comparison of *C. higginsianum* and *C. graminicola* uncovered major differences in their gene content. We propose that the diversification of functions required for host interaction, notably, the secretion of small-molecule and protein effectors and the degradation of plant polymers, allows *C. higginsianum* to colonize a wider range of plant species. In contrast, *C. graminicola*, a pathogen that is adapted to a narrow range of hosts, has maintained a more targeted arsenal of virulence factors.

METHODS

Sequencing and Assembly

C. graminicola strain M1.001 (M2) was collected in Missouri from infected maize (Fungal Genetics Stock Center culture 10212). *C. higginsianum* strain IMI349063 was isolated from *Brassica campestris* in Trinidad and Tobago (CABI culture collection, Wallingford, UK). The genome assemblies of *C. graminicola* were generated at the Broad Institute by combining data from Sanger and 454 pyrosequencing using a Newbler hybrid approach. Paired-end reads from 468,734 plasmids and 67,151 fosmids improved the continuity of the assembly. In the assembled genome, more than 98.5% of the sequence bases had quality scores >40. The *C. higginsianum* genome assembly was generated by GATC Biotech AG (Konstanz, Germany) by combining 454 GS-FLX shotgun reads and Illumina GAII mate-pair reads. Additionally, 864 fosmids were end-sequenced with Sanger technology. After removing dinucleotide repeats, the 454 reads and the fosmid end sequences were coassembled using the SeqMan NGen assembler (DNAStar Inc., USA). Contigs were then sorted into scaffolds using the paired-end information derived from an Illumina 3-kb–insert mate-pair library (2×36 bp reads). Scaffolds were manually edited to correct falsely joined contigs and falsely arranged scaffolds. To correct homopolymer sequencing errors in the 454 data, the Illumina GA data (76-fold coverage) were mapped to the scaffolded contigs, and the depth of coverage was used to create a final corrected consensus sequence.

Gene Annotation

A total of 28,424 expressed sequence tags (ESTs) from two *C. graminicola* complementary DNA libraries and 828,592 ESTs from six *C. higginsianum* libraries were used to provide a training set for the gene-calling pipeline and for validating the gene models. Protein-coding genes were annotated in *C. graminicola* using multiple lines of evidence from BLAST, PFAM searches and EST alignments, as described previously[20]. Gene structures were predicted using the Broad Institute automated gene-calling pipeline[35] based on a combination of gene models predicted by the programs FGENESH (Softberry Inc., USA), GENEID[36], GeneMark[37], SNAP[38] and Augustus[39] together with EST-based and manually curated gene models. GENEID, FGENESH, SNAP and Augustus were trained using a set of high-confidence EST-based gene models generated by clustering Blat-aligned species-specific ESTs. By combining BLAST, EST and *ab initio* predictions, annotators manually built additional gene models that were otherwise missed by the automated annotation. *C. graminicola* was predicted to have 12,006 gene models, 39% of which were verified by the alignment of 13,600 Sanger EST reads. The *C. higginsianum* gene set was created similarly and was filtered using TBLASTN alignments from 10,661 of the *C. graminicola* gene models ($<1 \times 10^{-10}$). Another 1,564 gene models were based on evidence from *C. higginsianum* ESTs, and 600 were based on EVidenceModeler (EVM) models having BLAST hits to proteins in the UniRef90 database. *C. higginsianum* was predicted to have 16,172 protein-coding genes, 89% of which were validated by the alignment of 135,923 ESTs from 454 sequencing.

Optical Mapping

C. graminicola and *C. higginsianum* protoplasts[40] were lysed and prepared for optical mapping[41] using *Mlu*I (with an average fragment size for both genomes of 9.2 kb). Raw datasets comprising single DNA molecule maps (Rmaps; 300× coverage per genome) were assembled into genome-wide contig maps spanning each chromosome using divide-and-conquer[41] and iterative assembly strategies[42]. PROmer from the MUMmer package[43] was used to conduct pairwise comparisons between the *C. graminicola* and *C. higginsianum* genomes (Fig. 2a). The synteny map (Fig. 2b) was generated using the Argo browser[44].

Transposable Element Analysis

Repetitive DNA elements were identified by performing a self BLASTN of each genome and processing the output with a custom Perl script (available on request), which identified multicopy sequences and organized them into nonredundant families. Consensus sequences of these families were then used to generate a custom library for RepeatMasker to scan both genome assemblies. The distributions of the genes, the transposable elements and the GC content were examined within a 100-kb window, sliding 10 kb across each chromosome.

Orthology and Multigene Families

To identify differences in gene family size between *C. graminicola* and *C. higginsianum*, we clustered their proteomes using the Markov clustering program MCL[11]. An all-versus-all BLASTP search was performed using default parameters, followed by clustering with MCL using an inflation value of 2.0. We also included the proteomes of 13 additional fungal species (Fig. 3). Sequences were aligned using MAFFT[45], and phylogenetic trees were constructed using the neighbor-joining method, followed by a bootstrap test with 100 replications. Sequence editing and alignment and phylogenetic analyses were performed using Geneious Pro (version 5.5; see URLs).

Annotation of Specific Gene Categories

Secretomes of both *Colletotrichum* species were predicted using WoLF-PSORT[46]. CSEPs were defined as extracellular proteins with no significant BLAST homology (expect value $<1 \times 10^{-3}$) to sequences in the UniProt database (SwissProt and TrEMBL components). Homologs of proteins from outside the genus *Colletotrichum* were excluded. Genes encoding putative carbohydrate-active enzymes were identified using the CAZy annotation pipeline[13]. To identify secreted peptidase genes, sequences of predicted extracellular proteins were subjected to a MEROPS Batch BLAST analysis[47]. Membrane transporters were identified from BLAST searches against the Transporter Collection Database (see URLs) and *Saccharomyces* Genome Database (see URLs). Secondary metabolism genes were initially identified using MCL and gene family searches using the Broad Institute *Colletotrichum* database, BLAST searches against GenBank and InterProScan analysis (see URLs). The Secondary Metabolite Unknown Region Finder (SMURF)[26] was used to predict secondary metabolism gene clusters. SMURF was applied to the Velvet assembly of *C. higginsianum* . Candidate genes identified using automated searches were inspected manually, including protein sequence alignments to known enzymes and searches against the NCBI Conserved Domain Database. The Fungal Cytochrome P450 Database (see URLs) and Fungal Transcription Factor Database were used to annotate cytochrome P450 enzymes and transcription factors, respectively.

Whole-Genome Transcriptome Profiling

Arabidopsis leaves infected by *C. higginsianum* were obtained as described previously[48]. Sampling and RNA isolation of the pre-penetration stage (22 h hpi), the early biotrophic stage (40 hpi), the switch between biotrophy and necrotrophy (60 hpi) and *in vitro* appressoria (22 hpi) have been described previously[32, 49]. Each experimental repetition of the *in planta* stages was based on RNA extracted from ~300 leaves. Maize leaf sheaths infected by *C. graminicola* were obtained as described previously[50]. Sheaths from the maize inbred line Mo940 at the V3 stage were cut into 5-cm–long segments and inoculated with two 10-μl drops of spore suspension (5 × 10⁵ spores per ml). Sheaths containing mature pre-penetration appressoria (24 hpi), intracellular biotrophic hyphae (36 hpi) and necrotrophic hyphae with water-soaked lesions (60 hpi) were sampled. Each leaf sheath was trimmed to include only the inoculated area, and total RNA was extracted as described previously[51] (15 maize sheaths per experimental repetition). The RNA integrity of all samples was verified on an Agilent 2100 Bioanalyzer.

Twelve *C. higginsianum* libraries (four developmental stages and three biological replicates) and nine *C. graminicola* libraries (three developmental stages and three biological replicates) were prepared with the Illumina TruSeq RNA Sample Preparation Kit and sequenced using the Illumina Genome Analyzer IIx (single reads, 100 bp for *C. higginsianum* and 76 bp for *C. graminicola*). The RNA-Seq reads were mapped to the annotated genomes with TopHat ($a = 10$, $g = 5$)[52] and transformed into counts per annotated gene per sample with the 'coverageBed› function from the BEDtools suite[53] and custom R scripts. Differentially expressed genes between two developmental stages were detected using the 'exactTest› function from the R package EdgeR[54]. To calculate fold changes, the number of reads for each gene in each library was normalized by the total number of mapped reads for the library, and direct ratios (log2) were calculated between the different developmental stages. Transcripts with a significant *P* value (<0.05) and more than a twofold change (log2) in transcript level were considered to be differentially expressed. All *P* values were corrected for false discoveries resulting from multiple hypothesis testing using the Benjamini-Hochberg procedure. Heatmaps of gene expression profiles were generated with the Genesis expression analysis package[55]. All codes for the RNA-Seq processing are available upon request. The *C. higginsianum* RNA-Seq data were also mapped onto the unannotated Velvet genome assembly using bowtie[56] and visualized with SAMtools (see URLs) and the IGV browser. RNA-Seq expression profiles were validated by quantitative RT-PCR.

Molecular Phylogeny and Evolutionary Divergence Date Estimation

A whole-genome cladogram showing the phylogenetic relationships of *C. graminicola* and *C. higginsianum* to 17 other sequenced fungi was constructed with CVTree[34] (Fig. 1). A phylogeny was generated for the genus *Colletotrichum* based on sequencing five genes in 28 selected isolates. To estimate the evolutionary divergence date for *C. graminicola* and *C. higginsianum*, a phylogenetic analysis was performed using the 13 species. The proteomes were clustered using MCL, and proteins in each cluster were aligned using MUSCLE. Sixty-four clusters containing only one protein from each species and having at least 80% average pairwise nucleotide identity were used for further analyses. Sequence alignments were concatenated, and a phylogenetic tree was constructed with MrBayes[57] using the WAG amino acid substitution model. Date estimates were computed using the program r8s[58] with the nonparametric rate smoothing (NPRS) method using date estimates by Lücking *et al.*[59].

Fluorescent Reporter Gene Assay

The promoter of the CSEP-encoding gene *ChEC6* (CH063_01084) was fused to *mCherry*[60] and a transcriptional terminator by overlap fusion PCR[61] using the primer pairs shown. The genomic region between the *ChEC6* start codon and the stop codon of its upstream gene (1,198 bp) was amplified with primer pair 1. The *mCherry* gene was amplified with primer pair 2. The transcriptional terminator of *Aspergillus nidulans trpC* was amplified from the plasmid pBin-GFP-hph[5] with primer pair 3. After fusion, the insert was subcloned into the plasmid pENTR/D-TOPO (Invitrogen) and verified by sequencing. The insert was cut out with BamHI and EcoRI and ligated into the plasmid pBIGDR1, providing direct repeat recombination-mediated gene targeting[62]. A *ku70* mutant of *C. higginsianum* strain IMI349063 (ref. 62) was used for *Agrobacterium*-mediated transformation[3]. Confocal images of transformants were obtained using a Leica TCS SP2 confocal laser scanning microscope. Excitation for imaging mCherry fluorescence was at 563 nm, and emission was detected at 566–620 nm.

Host Tissue Alkalinization

The pH of the host cells during infection was measured using the cell-permeant pH-sensitive dye 2',7'-bis(carboxyethyl)-5(6)-carboxyfluorescein (BCECF) for analysis by epifluorescence microscopy[31]. Fluorescence intensity values were correlated with direct pH determinations obtained with a piercing-tip pH electrode (Eutech, Singapore). Ammonia concentrations in infected maize and *Arabidopsis* leaf tissues were measured using a photometric ammonium assay kit (Merck, Germany).

ACCESSION CODES

The *C. graminicola* and *C. higginsianum* genome assemblies have been deposited in NCBI's Whole-Genome Shotgun Project with accession numbers ACOD0100000000 and CACQ0200000000, respectively. The RNA-Seq data for *C. graminicola* and *C. higginsianum*have been deposited in the NCBI Gene Expression Omnibus under GEO Series accession numbers GSE34632 and GSE33683, respectively.

Primary Accessions

Gene Expression Omnibus

- GSE34632
- GSE33683

ACKNOWLEDGMENTS

This manuscript is dedicated to the memory of Robert Hanau. This work was primarily supported by US Department of Agriculture (USDA)–Cooperative State Research, Education and Extension Service (CSREES) grant 2007-35600-17829 (L.J.V., M.B.D., S.R. and M.R.T.), the Max Planck Society (R.J.O.) and Deutsche Forschungsgemeinschaft (SPP1212) grant OC 104/1-3 (R.J.O.). Other funding sources included USDA-CSREES grant 2009-34457-20125 (L.J.V.), the University of Kentucky College of Agriculture Research Office (L.J.V.), Ministerio de Ciencia e Innovación (MICINN) of Spain grants AGL2008-03177/AGR and AGL2011-29446/AGR (M.R.T.) and the Programme for Promotion of Basic and Applied Research for Innovations in Bio-oriented Industry (P.G. and K. Shirasu). This is manuscript number 12-12-011, published with the approval of the Director of the University of Kentucky Agricultural Experiment Station.

REFERENCES

1.   Crous, P.W., Gams, W., Stalpers, J.A., Robert, V. & Stegehuis, G. MycoBank: an online initiative to launch mycology into the 21st century. *Stud. Mycol.* **50**, 19–22 (2004).

2.   Perfect, S.E., Hughes, H.B., O'Connell, R.J. & Green, J.R. *Colletotrichum*—a model genus for studies on pathology and fungal-plant interactions. *Fungal Genet. Biol.* **27**, 186–198(1999).

3.   Huser, A., Takahara, H., Schmalenbach, W. & O'Connell, R. Discovery of pathogenicity genes in the crucifer anthracnose fungus *Colletotrichum higginsianum*, using random insertional mutagenesis. *Mol. Plant Microbe*

Interact. **22**, 143–156 (2009).

4.   Narusaka, M. *et al. RRS1* and *RPS4* provide a dual resistance-gene system against fungal and bacterial pathogens. *Plant J.* **60**, 218–226 (2009).

5.   O'Connell, R. *et al.* A novel *Arabidopsis-Colletotrichum* pathosystem for the molecular dissection of plant-fungal interactions. *Mol. Plant Microbe Interact.* **17**, 272–282 (2004).

6.   Frey, T.J., Weldekidan, T., Colbert, T., Wolters, P.J.C.C. & Hawk, J.A. Fitness evaluation of*Rcg1*, a locus that confers resistance to *Colletotrichum graminicola* (Ces.) G.W. Wils. Using Near-Isogenic Maize Hybrids. *Crop Sci.* **51**, 1551–1563 (2011).

7.   Mims, C.W. & Vaillancourt, L.J. Ultrastructural characterization of infection and colonization of maize leaves by *Colletotrichum graminicola*, and by a *C. graminicola* pathogenicity mutant. *Phytopathology* **92**, 803–812 (2002).

8.   Rouxel, T. *et al.* Effector diversification within compartments of the *Leptosphaeria maculans*genome affected by repeat-induced point mutations. *Nat. Commun.* **2**, 202 (2011).

9.   Chaw, S.M., Chang, C.C., Chen, H.L. & Li, W.H. Dating the monocot-dicot divergence and the origin of core Eudicots using whole chloroplast genomes. *J. Mol. Evol.* **58**, 424–441(2004).

10. Amselem, J. *et al.* Genomic analysis of the necrotrophic fungal pathogens *Sclerotinia sclerotiorum* and *Botrytis cinerea. PLoS Genet.* **7**, e1002230 (2011).

11. Enright, A.J., Van Dongen, S. & Ouzounis, C.A. An efficient algorithm for large-scale detection of protein families. *Nucleic Acids Res.* **30**, 1575–1584 (2002).

12. Oh, B.C., Choi, W.C., Park, S., Kim, Y.O. & Oh, T.K. Biochemical properties and substrate specificities of alkaline and histidine acid phytases. *Appl. Microbiol. Biotechnol.* **63**,362–372 (2004).

13. Cantarel, B.L. *et al.* The Carbohydrate-Active EnZymes database (CAZy): an expert resource for glycogenomics. *Nucleic Acids Res.* **37**, D233–D238 (2009).

14. van den Brink, J. & de Vries, R.P. Fungal enzyme sets for plant polysaccharide degradation. *Appl. Microbiol. Biotechnol.* **91**, 1477–1492 (2011).

15. King, B.C. *et al.* Arsenal of plant cell wall degrading enzymes reflects host preference among plant pathogenic fungi. *Biotechnol. Biofuels* **4**, 4 (2011).

16. Quinlan, R.J. *et al*. Insights into the oxidative degradation of cellulose by a copper metalloenzyme that exploits biomass components. *Proc. Natl. Acad. Sci. USA* **108**,15079–15084 (2011).

17. Beeson, W.T., Phillips, C.M., Cate, J.H. & Marletta, M.A. Oxidative cleavage of cellulose by fungal copper-dependent polysaccharide monooxygenases. *J. Am. Chem. Soc.* **134**,890–892 (2012).

18. Vogel, J. Unique aspects of the grass cell wall. *Curr. Opin. Plant Biol.* **11**, 301–307 (2008).

19. Stergiopoulos, I. & de Wit, P.J.G.M. Fungal effector proteins. *Annu. Rev. Phytopathol.* **47**,233–263 (2009).

20. Ma, L.-J. *et al*. Comparative genomics reveals mobile pathogenicity chromosomes in*Fusarium*. *Nature* **464**, 367–373 (2010).

21. Farman, M.L. Telomeres in the rice blast fungus *Magnaporthe oryzae*: the world of the end as we know it. *FEMS Microbiol. Lett.* **273**, 125–132 (2007).

22. Kämper, J. *et al*. Insights from the genome of the biotrophic fungal plant pathogen *Ustilago maydis*. *Nature* **444**, 97–101 (2006).

23. Crouch, J.A. & Beirn, L.A. Anthracnose of cereals and grasses. *Fungal Divers.* **39**, 19–44(2009).

24. Collemare, J., Billard, A., Bohnert, H.U. & Lebrun, M.H. Biosynthesis of secondary metabolites in the rice blast fungus *Magnaporthe grisea*: the role of hybrid PKS-NRPS in pathogenicity. *Mycol. Res.* **112**, 207–215 (2008).

25. Spanu, P.D. *et al*. Genome expansion and gene loss in powdery mildew fungi reveal functional trade-offs in extreme parasitism. *Science* **330**, 1543–1546 (2010).

26. Khaldi, N. *et al*. SMURF: genomic mapping of fungal secondary metabolite clusters. *Fungal Genet. Biol.* **47**, 736–741 (2010).

27. Collemare, J. & Lebrun, M.-H. Fungal secondary metabolites: ancient toxins and novel effectors in plant-microbe interactions. in *Effectors in Plant-Microbe Interactions* (eds. Martin, F. & Kamoun, S.) 379–402 (Wiley-Blackwell, Oxford, 2011).

28. Voegele, R.T. & Mendgen, K.W. Nutrient uptake in rust fungi: how sweet is parasitic life?*Euphytica* **179**, 41–55 (2011).

29. Catanzariti, A.-M., Dodds, P.N., Lawrence, G.J., Ayliffe, M.A. & Ellis, J.G. Haustorially expressed secreted proteins from flax rust are highly enriched for avirulence elicitors. *Plant Cell* **18**, 243–256 (2006).

30. Peñalva, M.A. & Arst, H.N. Regulation of gene expression by ambient

pH in filamentous fungi and yeasts. *Microbiol. Mol. Biol. Rev.* **66**, 426–446 (2002).

31. Alkan, N., Fluhr, R., Sherman, A. & Prusky, D. Role of ammonia secretion and pH modulation on pathogenicity of *Colletotrichum coccodes* on tomato fruit. *Mol. Plant Microbe Interact.* **21**, 1058–1066 (2008).

32. Kleemann, J. *et al.* Sequential delivery of host-induced virulence effectors by appressoria and intracellular hyphae of the phytopathogen. *Colletotrichum higginsianum. PLoS Pathog.* **8**, e1002643 (2012).

33. Frank, B. Ueber einige neue und weniger bekannte Pflanzenkrankheiten. *Ber. Deutsch. Bot. Gesell.* **1**, 29–34 (1883).

34. Wang, H., Xu, Z., Gao, L. & Hao, B. A fungal phylogeny based on 82 complete genomes using the composition vector method. *BMC Evol. Biol.* **9**, 195 (2009).

35. Stanke, M., Steinkamp, R., Waack, S. & Morgenstern, B. AUGUSTUS: a web server for gene finding in eukaryotes. *Nucleic Acids Res.* **32**, W309–W312 (2004).

36. Haas, B.J., Zeng, Q., Pearson, M.D., Cuomo, C.A. & Wortman, J.R. Approaches to fungal genome annotation. *Mycology* **2**, 118–141 (2011).

37. Parra, G., Blanco, E. & Guigó, R. GeneID in *Drosophila. Genome Res.* **10**, 511–515 (2000).

38. Borodovsky, M. & McIninch, J. GeneMark: parallel gene recognition for both DNA strands. *Comput. Chem.* **17**, 123–133 (1993).

39. Korf, I. Gene finding in novel genomes. *BMC Bioinformatics* **5**, 59 (2004).

40. Thon, M.R., Nuckles, E.M. & Vaillancourt, L.J. Restriction enzyme-mediated integration used to produce pathogenicity mutants of *Colletotrichum graminicola. Mol. Plant Microbe Interact.* **13**, 1356–1365 (2000).

41. Zhou, S. *et al.* Validation of rice genome sequence by optical mapping. *BMC Genomics* **8**, 278 (2007).

42. Zhou, S. *et al.* A single molecule scaffold for the maize genome. *PLoS Genet.* **5**, e1000711 (2009).

43. Kurtz, S. *et al.* Versatile and open software for comparing large genomes. *Genome Biol.* **5**, R12 (2004).

44. Engels, R. *et al.* Combo: a whole genome comparative browser. *Bioinformatics* **22**, 1782–1783 (2006).

45. Katoh, K., Misawa, K., Kuma, K. & Miyata, T. MAFFT: a novel method for rapid multiple sequence alignment based on fast Fourier

transform. *Nucleic Acids Res.* **30**, 3059–3066(2002).

46. Horton, P. *et al.* WoLF PSORT: protein localization predictor. *Nucleic Acids Res.* **35**,W585–W587 (2007).

47. Rawlings, N.D., Barrett, A.J. & Bateman, A. MEROPS: the peptidase database. *Nucleic Acids Res.* **38**, D227–D233 (2010).

48. Takahara, H., Dolf, A., Endl, E. & O'Connell, R. Flow cytometric purification of*Colletotrichum higginsianum* biotrophic hyphae from *Arabidopsis* leaves for stage-specific transcriptome analysis. *Plant J.* **59**, 672–683 (2009).

49. Kleemann, J., Takahara, H., Stüber, K. & O'Connell, R. Identification of soluble secreted proteins from appressoria of *Colletotrichum higginsianum* by analysis of expressed sequence tags. *Microbiology* **154**, 1204–1217 (2008).

50. Kankanala, P., Czymmek, K. & Valent, B. Roles for rice membrane dynamics and plasmodesmata during biotrophic invasion by the blast fungus. *Plant Cell* **19**, 706–724(2007).

51. Metz, R.P., Kwak, H.-I., Gustafson, T., Laffin, B. & Porter, W.W. Differential transcriptional regulation by mouse single-minded 2s. *J. Biol. Chem.* **281**, 10839–10848 (2006).

52. Trapnell, C., Pachter, L. & Salzberg, S.L. TopHat: discovering splice junctions with RNA-Seq. *Bioinformatics* **25**, 1105–1111 (2009).

53. Quinlan, A.R. & Hall, I.M. BEDTools: a flexible suite of utilities for comparing genomic features. *Bioinformatics* **26**, 841–842 (2010).

54. Robinson, M.D., McCarthy, D.J. & Smyth, G.K. EdgeR: a Bioconductor package for differential expression analysis of digital gene expression data. *Bioinformatics* **26**, 139–140(2010).

55. Sturn, A., Quackenbush, J. & Trajanoski, Z. Genesis: cluster analysis of microarray data.*Bioinformatics* **18**, 207–208 (2002).

56. Langmead, B., Trapnell, C., Pop, M. & Salzberg, S.L. Ultrafast and memory-efficient alignment of short DNA sequences to the human genome. *Genome Biol.* **10**, R25 (2009).

57. Ronquist, F. & Huelsenbeck, J.P. MrBayes 3: Bayesian phylogenetic inference under mixed models. *Bioinformatics* **19**, 1572–1574 (2003).

58. Sanderson, M.J. r8s: inferring absolute rates of molecular evolution and divergence times in the absence of a molecular clock. *Bioinformatics* **19**, 301–302 (2003).

59. Lücking, R., Huhndorf, S., Pfister, D.H., Plata, E.R. & Lumbsch,

H.T. Fungi evolved right on track. *Mycologia* **101**, 810–822 (2009).

60. Shaner, N.C. *et al.* Improved monomeric red, orange and yellow fluorescent proteins derived from *Discosoma* sp. red fluorescent protein. *Nat. Biotechnol.* **22**, 1567–1572 (2004).

61. Szewczyk, E. *et al.* Fusion PCR and gene targeting in *Aspergillus nidulans*. *Nat. Protoc.* **1**, 3111–3120 (2006).

62. Ushimaru, T. *et al.* Development of an efficient gene targeting system in *Colletotrichum higginsianum* using a non-homologous end-joining mutant and *Agrobacterium tumefaciens*–mediated gene transfer. *Mol. Genet. Genomics* **284**, 357–371 (2010).

Chapter 4

CONTROL OF FOLIAR PATHOGENS OF SPRING BARLEY USING A COMBINATION OF RESISTANCE ELICITORS

Dale R. Walters[1], Neil D. Havis[1], Linda Paterson[1], Jeanette Taylor[1], David J. Walsh[2] and Cecile Sablou[1]

[1]Crop and Soil Systems Research Group, Scotland's Rural College, Edinburgh, UK

[2]Engineering, Science and Technology Department, Scotland's Rural College, Edinburgh, UK

The ability of the resistance elicitors acibenzolar-S-methyl (ASM), β-aminobutyric acid (BABA), *cis*-jasmone (CJ), and a combination of the three products, to control infection of spring barley by *Rhynchosporium commune* was examined under glasshouse conditions. Significant control of *R. commune* was provided by ASM and CJ, but the largest reduction in infection was obtained with the combination of the three elicitors. This elicitor combination was found to up-regulate the expression of *PR-1b*, which is used as a molecular marker for systemic acquired resistance (SAR). However, the elicitor combination also down-regulated the expression of *LOX2*, a gene involved in the biosynthesis of jasmonic acid (JA). In field experiments over 3 consecutive years, the effects of the elicitor combination were influenced greatly by crop variety and by year. For example, the elicitor combination applied on its own provided significant control of powdery mildew (*Blumeria graminis* f.sp. *hordei*) and *R. commune* in 2009, whereas no control on either variety was observed in 2007. In contrast, treatments involving both the elicitor combination and fungicides provided disease control and yield increases which were equal to, and in some cases better than that provided by the best fungicide-only treatment. The prospects for the use of elicitor plus fungicide treatments to control foliar pathogens of spring barley in practice are discussed.

INTRODUCTION

Application of various agents to plants can lead to the induction of resistance to subsequent pathogen attack, both locally, and systemically (Walters et al., 2013). Such induced resistance can be split into systemic acquired resistance (SAR) and induced systemic resistance (ISR). SAR is characterized by a restriction of pathogen growth and a suppression of disease symptom development compared to non-induced plants infected with the same pathogen. Its onset is associated with an accumulation of salicylic acid (SA) at sites of infection and systemically, and with the coordinated activation of a specific set of genes encoding pathogenesis-related (PR) proteins. Treatment of plants with SA or one of its functional analogs e.g., acibenzolar-S-methyl (ASM; marketed in Europe as Bion® and in North America as Actigard®), induces SAR and activates the same set of *PR* genes. ISR develops as a result of colonization of plant roots by plant growth-promoting rhizobacteria (PGPR) and has been shown to function independently of SA and activation of *PR* genes, requiring instead, jasmonic acid (JA), and ethylene (ET) (Pieterse et al., 2012; Spoel and Dong, 2012).

Research over the past decade suggests that SA induces defenses against biotrophic pathogens, while JA mediates defenses against necrotrophic pathogens (Glazebrook, 2005). It is thought that cross-talk between the two signaling pathways might help to fine-tune defense responses against a particular pathogen according to its mode of infection (Beckers and Spoel, 2006). Interestingly, Spoel et al. (2007) found that infection with the biotrophic pathogen *Pseudomonas syringae*, which induces SA-mediated defense, rendered *Arabidopsis thaliana* more susceptible to the necrotrophic pathogen *Alternaria brassicicola*, although this trade-off was restricted to plant tissue adjacent to the initial infection site. In terms of plant-insect interactions, JA is known to play a major role in mediating defenses against insect herbivory (Bostock, 2005), although the situation is rather more complex than was once thought, since both SA- and JA-responsive gene expression can be elicited by aphids and whiteflies. In terms of insect herbivory, although there are examples of negative cross-talk i.e., SA-mediated suppression of JA-inducible gene expression, there are also examples of no trade-offs, and even of positive effects (see Walters et al., 2013).

Because induced resistance offers the prospect of broad spectrum disease control using the plant's own resistance mechanisms, there has been great interest in the development of agents which can mimic natural inducers of resistance (Lyon, 2007). These include elicitor molecules released during the early stages of the plant-pathogen interaction and the signaling pathways used to trigger defenses locally and systemically. Examples include ASM, which has been shown to elicit SAR in a wide range of plant-pathogen interactions (Leadbeater and Staub, 2007), the non-protein amino acid β-aminobutyric acid (BABA), and the oxylipin, *cis*-jasmone (CJ) (Walters et al., 2007).

BABA is known to induce resistance against pathogens in various systems, including tomato, potato, grapevine, and pea (Cohen et al., 1999; Jakab et al., 2001). In field experiments, Cohen (2002) found that BABA provided significant control of late blight of potato, while Liljeroth et al. (2010) showed that BABA used together with a reduced fungicide dose gave the same level of late blight control as a full dose of the standard fungicide treatment.

CJ is structurally related to JA and methyl jasmonate (MeJA), both of which are well known to activate plant defenses (Farmer and Ryan, 1990; Thaler et al., 1996), although CJ activates a unique and more limited set of genes than does treatment with MeJA (e.g., Pickett et al., 2007). CJ is released naturally from plants damaged by insects and when applied artificially, can activate defense against insects (Birkett et al., 2000; Bruce et al., 2003).

The efficacy of induced resistance under field conditions is variable, representing a major obstacle to its use in practical crop protection (Reglinski and Walters, 2009). Induced resistance is a complex plant response to pathogen attack and as such, will be modified by many factors including genotype. However, insufficient attention has been paid to investigating the mechanisms underlying variable efficacy and approaches that might be adopted to incorporate elicitors into crop protection practice, such as use of elicitors and fungicides together in the same disease control program, and use of combinations of elicitors. The latter aspect has received little attention to date, probably because of the trade-offs that might be associated with using elicitors which activate different signaling pathways, as mentioned above. In this paper, we report the results of field experiments over 3 consecutive years, undertaken to determine the potential for use of an elicitor combination to control foliar pathogens of spring barley. Some preliminary data from this study have appeared previously in a conference paper (Walters et al., 2010).

MATERIALS AND METHODS

Plant Growth and Pathogen Inoculation under Glasshouse Conditions

The spring barley (*Hordeum vulgare* L.) variety Cellar was used for glasshouse studies. Cellar was chosen since it exhibits moderate susceptibility to *Rhynchosporium commune* (HGCA, 2009). Seeds were sown in pots in Fisons Levington compost and grown in a walk-in growth room at 18°C with a 16 h photoperiod (190 μmol m^{-2}s^{-1} provided by 400 W mercury vapor lamps). The experiment was laid out in a randomized block design, with each of the five treatment groups consisting of 15 plants, with 10 plants used for disease assessment and 5 plants for gene expression analysis. Plants were used for efficacy experiments when the sixth leaf was fully formed and the seventh leaf emerging. Leaves 1–4 were sprayed with elicitors using a hand-held sprayer. Two days later, plants were inoculated with the leaf scald pathogen, *Rhynchosporium commune*, by spraying with a suspension of spores (1 × 10^5 spores/ml) in distilled water containing 0.01% Tween 20. Inoculated plants were then covered with plastic bags for 48 h (the first 24 h in the dark) and kept at 16°C to provide the conditions necessary for spore germination and early fungal development. Thereafter, the temperature of the growth room was increased to 18°C for the remainder of the experiment. Infection intensity on leaves 5–7 was assessed 21 days after inoculation by determining the % leaf area exhibiting symptoms on each of 10 plants. For gene expression experiments, leaves three and four were treated with elicitor and 2 days later were inoculated with *R. commune*. Leaves were harvested 2 days later and frozen at −80°C for gene expression analysis. Data presented are the means of three replicates.

The elicitors used in these experiments were ASM, BABA, and CJ. ASM (Bion®) was a gift from Syngenta, Basel, Switzerland; BABA was purchased from Sigma, Poole, Dorset, UK; CJ was purchased from Sigma Aldrich, Dorset, UK. ASM (1 mM), BABA (1 mM) and CJ (0.625 g/l) were made up in distilled water containing 0.01% Tween 20.

Field EXPERIMENTS

Field experiments were conducted in 2007 at Tibbermore, Perthshire, Scotland, and in 2008 and 2009 at Drumalbin, Lanark, Scotland. Total rainfall and average air temperatures at these sites during the period 1 June—1 September were:

Perthshire, 2007: rainfall = 219 mm; average air temperature = 17°C

Lanark, 2008: rainfall = 317 mm; average air temperature = 16°C

Lanark, 2009: rainfall = 338 mm; average air temperature = 17°C.

Two spring barley varieties (Cellar and Optic) used in all field experiments reported here. Cellar has a resistance rating (RR) of 9 for powdery mildew and 4 for *Rhynchosporium commune*, and Optic has a RR of 5 for powdery mildew and 4 for *R. commune* (RR scale: 10 = high resistance, 1 = low resistance; HGCA, 2009). Each variety was sown in a randomized block design at a seed rate of 360 seeds m^{-2} and an individual plot size of 10 × 2 m, using three plots per treatment. For each barley variety, the factor tested was the applied treatment (i.e., elicitor, fungicide, elicitor + fungicide). Plots received standard fertilizer and herbicide regimes and 16 treatment programs were compared (Table 1). Spray dates for treatments were based on plant growth stage as described by Zadoks et al. (1974) and were applied with a knapsack sprayer using an equivalent spray volume of 200 1 ha-1. Disease symptoms (% leaf area infected) and % GLA were assessed using 10 plants per plot at spray dates and at 14 day intervals after the final spray. Area under the disease progress curves (AUDPC) were calculated using the formula:

$$\Sigma(y_i + y(i+1))/2 \times (t(i+1) - t_i)\Sigma(y_i + y(i+1))/2 \times (t(i+1) - t_i)$$

where y_i is the disease rating at time t_i.

Table 1. Elicitor and fungicide treatments applied in field experiments in 2007–2009

Treatment	GS24	GS31	GS39
1	Nil	Nil	Nil
2	Elicitors	Nil	Nil
3	Nil	Elicitors	Nil
4	Nil	Nil	Elicitors
5	Elicitors	Elicitors	Nil
6	Elicitors	Nil	Elicitors
7	Fungicide[1]	Nil	Nil
8	Nil	Fungicide[1]	Nil
9	Nil	Nil	Fungicide[2]
10	Nil	Fungicide[1]	Fungicide[2]
11	Fungicide[1] + Elicitors	Nil	Nil
12	Nil	Fungicide[1] + Elicitors	Nil
13	Nil	Nil	Fungicide[2] + Elicitors
14	Elicitors	Fungicide[1]	Fungicide[2]
15	Elicitors	Fungicide[3]	Fungicide[4]
16	Elicitors	Elicitors + Fungicide[3]	Fungicide[4]

Treatments were applied at different growth stages (GS) as described in the Materials and Methods.

GS = growth stage (see Zadoks et al., 1974).

Elicitors = ASM + BABA + C.

Fungicide[1] = prothioconazole at full-rate x 0.4 + cyprodinil and picoxystyrobin at full-rate x 0.5.

Fungicide[2] = prothioconazole at full-rate x 0.4 + chlorothalonil at full-rate.

Fungicide[3] = prothioconazole at full-rate x 0.2 + cyprodinil and picoxystrobin at full-rate x 0.3.

Fungicide[4] = prothioconazole at full-rate x 0.2 + chlorothalonil at full-rate x 0.5 Plots were harvested at the end of the trial and yields expressed as tonnes/hectare at 85% dry matter content.

Gene Expression

Total RNA was extracted from barley leaves using a RNeasy™ kit (Qiagen, West Sussex, UK) and RNA yield determined using a Nanodrop 1000 spectrophotometer (Nanodrop Technologies, Wilmington, DE, USA). In order to remove any remaining trace of DNA likely to interfere with measurements, samples were treated with desoxyribonuclease enzymes using the DNA-*free*™ kit from Applied Biosystems (California, USA). The final quantity and quality of the RNA was tested using a RNA 6000 Nano Chip kit (Agilent Technologies, Santa Clara, CA, USA).

Primer sequences for *PR1-b, LOX2*, and the cyclophilin gene (internal control) are listed in Table 2. All sequences were purchased from Eurofins MWG Operon (Ebersburg, Germany) and all primers were designed using Beacon Designer software (Premier Biosoft International, Palo Alto, California, USA).

Table 2. Primer sequences for genes used in this study

Gene	Primer sequence (F)	Primer sequence (R)
PR1-b	CTACGACTACGGCTCCAACAC	GCATCACGGTTAGTATGGTTTCTG
LOX2	CGGCAGACTCCCTCATCACTAAAG	GGCAGCAACAGGTCGTGGTAG
Cyclophilin	CCTGTCGTGTCGTCGGTCTAAA	ACGCAGATCCAGCAGCCTAAAG

Amplicon lengths: PR1-b = 190 base pairs. LOX2 = 121 base pairs. Cyclophilin = 122 base pairs.

Following RNA extraction, cDNA was generated using a SuperScript™ first-strand cDNA synthesis kit (Invitrogen, USA). Quantitative real-time PCR (qRT-PCR) was then performed with a MX3000P system (Stratagene, CA, USA) using a Brilliant 11SYBR Green QPCR master mix with ROX (Agilent Technologies, Santa Clara, CA, USA). In order to construct standard curves for the genes, six data points were used with a 5-fold dilution series (1:10–1:31,250). A 25 μl reaction for PCR amplification contained 12.5 μl of SYBR Green master mix (see above), 0.75 μl forward primer, 0.75 μl reverse primer, 6 μl sterile distilled water, and 5 μl cDNA. All PCR reactions were performed in duplicate. The cycling conditions were as follows: pre-incubation for 10 min at 95°C, followed by 40 cycles, each consisting of 30 s denaturing at 95°C, 60 s annealing at 57°C, 30 s at 72°C for new strand synthesis. The standard curves were used to calculate the absolute quantity of the product in each sample, Relative expression values were then calculated by normalizing against the cyclophilin gene as an internal control.

Statistical Analysis

All data were subjected to One-Way ANOVA using the GenStat Release 11.1 statistical program. The effect of blocks was considered random and the effect of applied treatments was defined as fixed. % leaf area infected values from glasshouse experiments and % GLA data from field experiments were log-transformed prior to analysis. Comparison of treatment means was performed using Fisher's protected least significant difference (LSD) Test.

RESULTS

Effects of Elicitors under Glasshouse Conditions

Initial experiments were conducted under glasshouse conditions to examine the effects of Bion®, BABA, and CJ, singly and in combination, on infection of the barley variety Cellar by *R. commune*. Although BABA reduced infection compared to the untreated control, this difference was not significant. In contrast, treatment with Bion® and CJ led to significant reductions in *R. commune* infection, with Bion® reducing infection by 70% and CJ by 64% (Figure 1). The largest reduction in infection (96%) was obtained using a combination of Bion®, BABA, and CJ.

Application of the elicitor combination to leaves 1 and 2 of the barley variety Cellar led to changes in the expression of two defense-related genes in leaves 3 and 4. Thus, elicitor treatment resulted in significant increases in expression of *PR1b* in both leaves 3 and 4 (4.3-fold and 3.8-fold, respectively; Figures 2A,B). Expression of *PR1b* was also increased significantly in leaf 3 following inoculation of untreated leaves (2.6-fold increase), but was not affected in leaf 4 (Figures 2A,B). However, the largest increases in *PR1b* expression were obtained when leaves 3 and 4 were first treated with elicitor and subsequently inoculated with *R. commune*. Here, *PR1b* expression was increased 7-fold in leaves 3 and 4 compared to the untreated control (Figures2A,B). This suggests that the elicitor combination primes the plant for enhanced expression of *PR1b*.

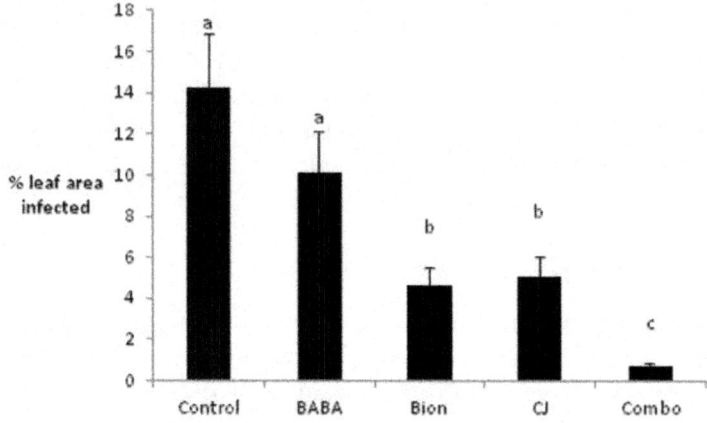

Figure 1. Effects of elicitors, singly and in combination, on infection of barley with *R. commune*. Leaves 1–4 were sprayed with elicitor and inoculated with *R. commune* 2 days later. Infection intensity was assessed 21 days later on leaves 5–7. Bars with a different letter are significantly different at $P < 0.05$ (Fisher's LSD).

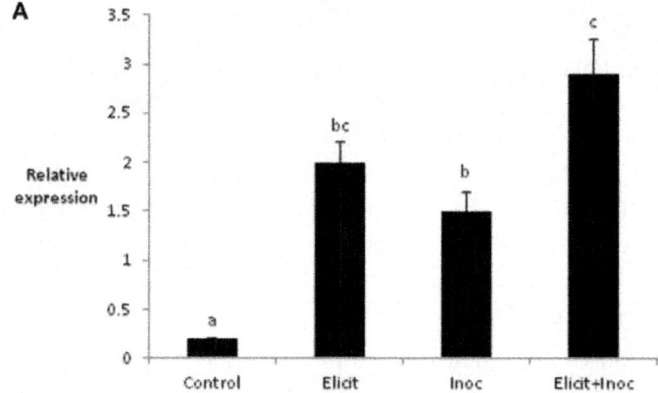

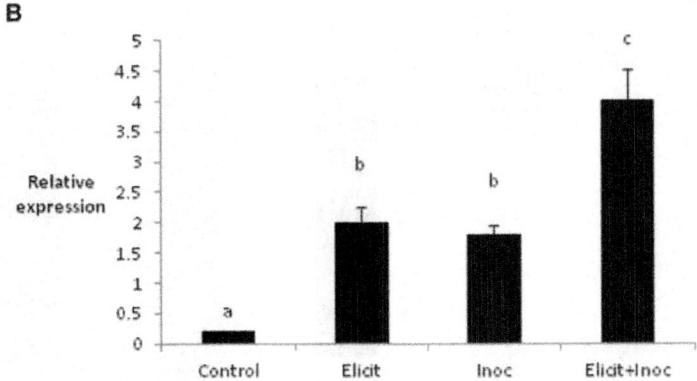

Figure 2. Relative quantity of *PR1-b* **in leaf 3 (A) and leaf 4 (B) of barley plants treated with the elicitor combination.** Leaves three and four were treated with elicitor and inoculated with *R. commune* 2 days later. Leaves were harvested 2 days after inoculation for analysis of gene expression. Bars with a different letter are significantly different at $P < 0.05$ (Fisher's LSD).

In contrast to *PR1b*, expression of *LOX2* was reduced by treatment with the elicitor combination, compared to the untreated control. Indeed, all three treatments led to substantial and significant decreases in expression of *LOX2* in leaves 3 and 4 (Figures 3A,B).

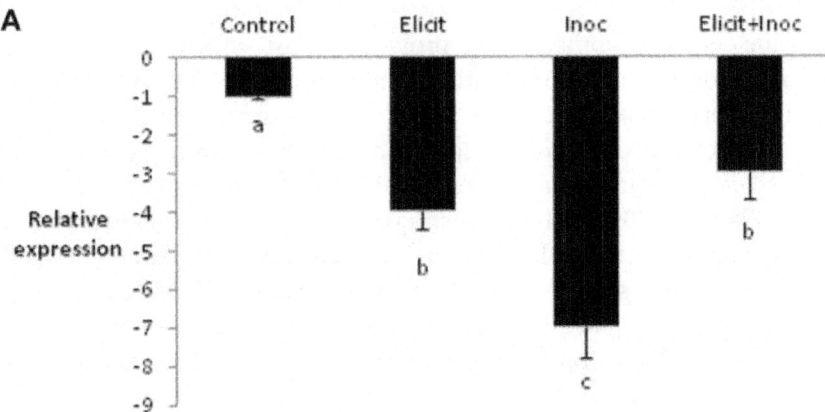

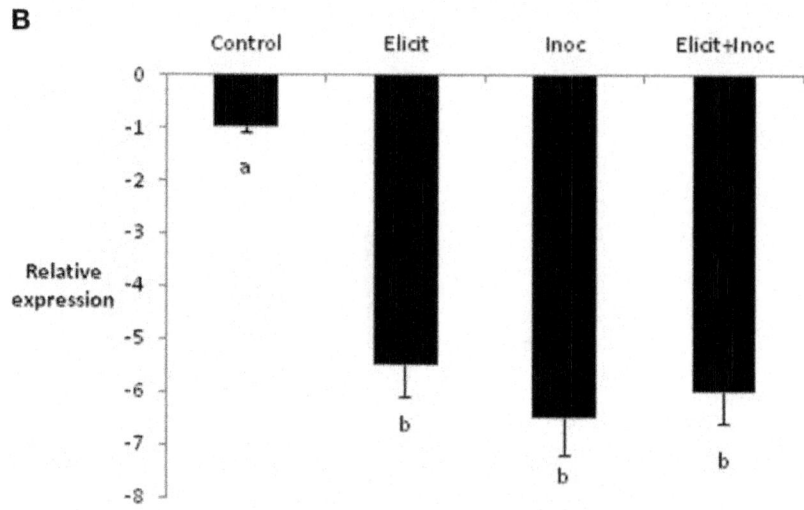

Figure 3. Relative quantity of *LOX2* **in leaf 3 (A) and leaf 4 (B) of barley plants treated with the elicitor combination**. Leaves three and four were treated with elicitor and inoculated with *R. commune* 2 days later. Leaves were harvested 2 days after inoculation for analysis of gene expression. Bars with a different letter are significantly different at $P < 0.05$ (Fisher's LSD).

Field Experiments

Since the elicitor combination provided most effective control of *R. commune* under glasshouse conditions, this treatment was chosen for inclusion in field experiments. Over the 3 years of field experiments, the two major foliar diseases detected on the spring barley crops were powdery mildew and *R. commune*. However, disease levels varied between years and in some years (2008 and 2009), only *R. commune* was observed on the variety Cellar. This probably reflected differences in weather at the different sites, since, for example, the 2007 season at the Perth site was considerably drier than the Lanark sites in 2008 and 2009. The 3 years of field experimentation, involving 17 different treatments on two varieties generated too much data to be shown here. Instead, only data for selected treatments are shown. These treatments are untreated (1), one fungicide-only treatment (10), one elicitor-only treatment (5), and one elicitor + fungicide (reduced rate) treatment (13). The data for all treatments are provided in Supplementary Material.

The efficacy of the elicitor combination was dependent on crop variety and year. Thus, in 2009, the elicitor combination provided significant control of both powdery mildew and *R. commune* on the variety Optic and of *R. commune* on variety Cellar (Figure 4). However, on both varieties, most effective disease control was achieved using a combination of the elicitor combination and fungicide. Indeed, the level of disease control achieved with the elicitor and fungicide treatment was at least as good as that obtained using the fungicide only treatment (Figure 4). Interestingly, although the elicitor combination on its own provided significant disease control on the two varieties, grain yield remained unchanged compared to untreated plants (Figure 4). In contrast, grain yield was increased significantly ($P < 0.05$) in both barley varieties treated with the elicitor plus fungicide treatment (Figure 4).

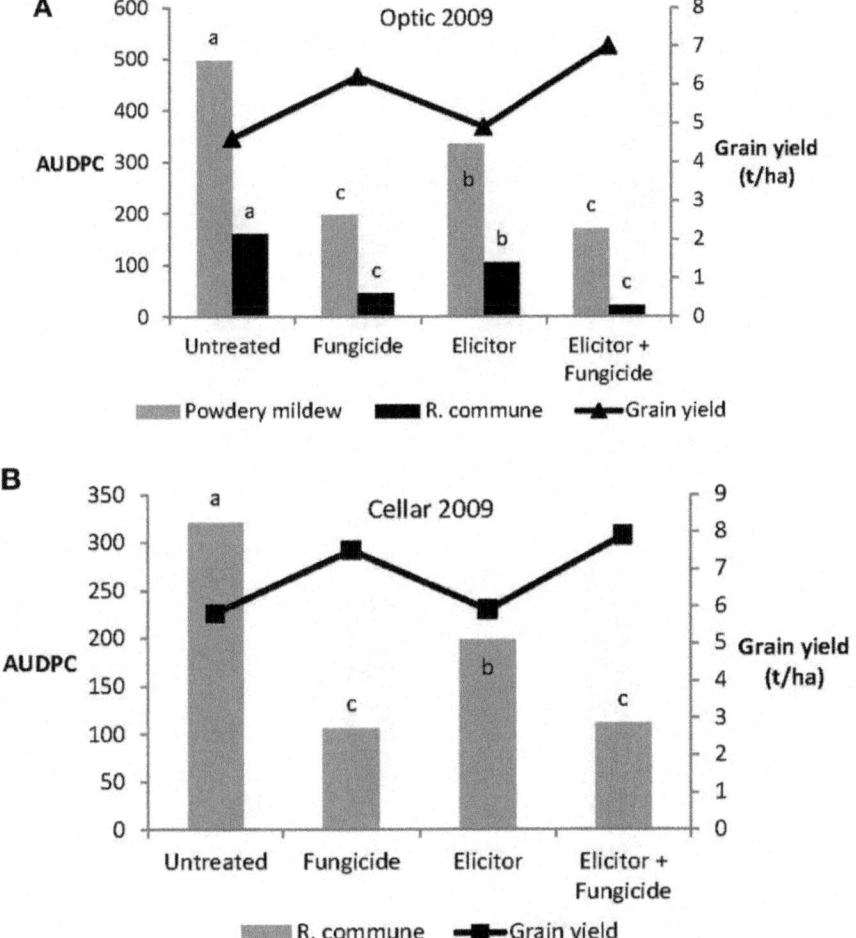

Figure 4. Effects of the elicitor combination and fungicides on (A) AUDPC for powdery mildew and *R. commune*, and grain yield in the spring barley variety Optic, and (B) on AUDPC for *R. commune* **and grain yield in the variety Cellar, in a field**

experiment in 2009. Bars with different letters are significantly different at $P < 0.05$ (Fisher's LSD).

In 2008, the elicitor combination applied on its own reduced levels of powdery mildew significantly on variety Optic, but provided no control of *R. commune* on either Optic or Cellar (Figure 5). Treatment with elicitor plus fungicide provided significant control of both diseases on Optic and of *R. commune* on Cellar. In both varieties, highest grain yields were obtained from plants receiving the elicitor plus fungicide treatment (Figure 5).

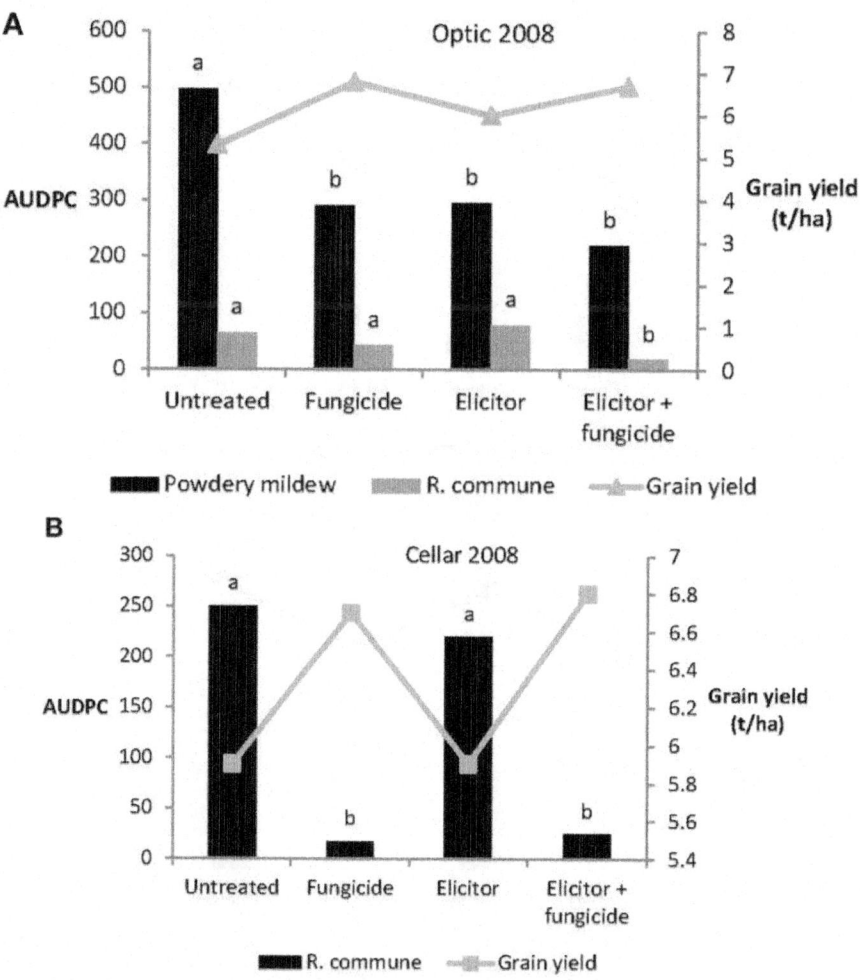

Figure 5. Effects of the elicitor combination and fungicides on (A) AUDPC for powdery mildew and *R. commune*, and grain yield in the spring barley variety Optic, and (B) on AUDPC for *R. commune* **and grain yield in the variety Cellar, in a field**

experiment in 2008. Bars with different letters are significantly different at $P < 0.05$ (Fisher's LSD).

Levels of both powdery mildew and *R. commune* were not significantly affected by treatment of either variety with the elicitor combination on its own in 2007 (Figure 6). In contrast, on both varieties, treatment with elicitor plus fungicide provided significant control of both diseases and in most cases the level of disease control achieved was as good as that obtained using the fungicide only treatment (Figure 6). Grain yields of both varieties were significantly increased in plants receiving the elicitor plus fungicide treatment compared to the other treatments, while in Optic, the elicitor combination on its own actually reduced grain yield (Figure 6).

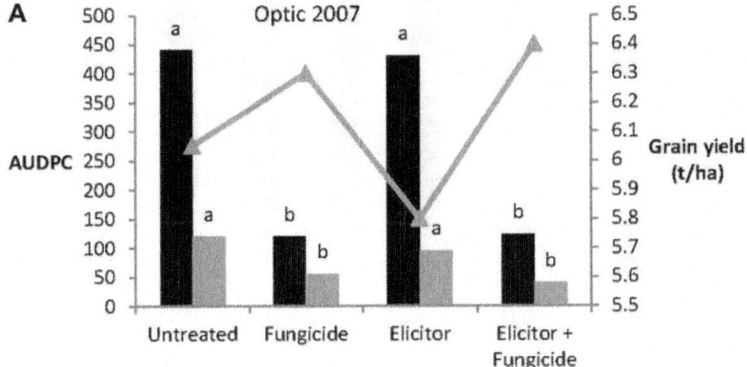

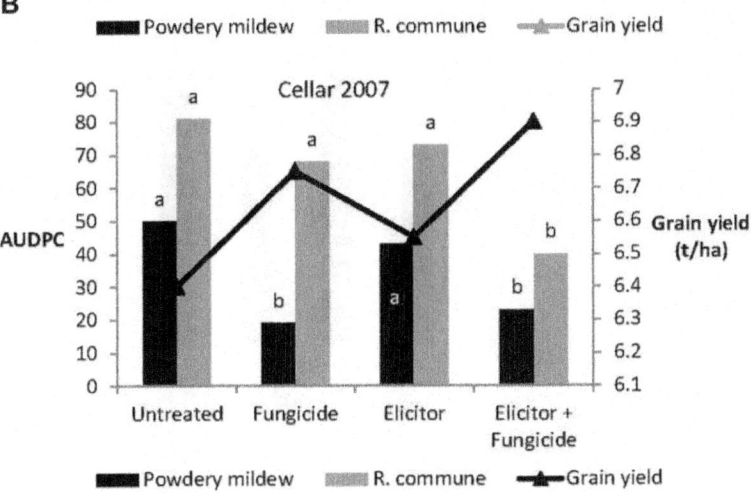

Figure 6. Effects of the elicitor combination and fungicides on (A) AUDPC for powdery mildew and *R. commune*, and grain yield in the spring barley variety Optic, and

(B) in the variety Cellar, in a field experiment in 2007. Bars with different letters are significantly different at $P< 0.05$ (Fisher's LSD).

The absence of any significant effect of the elicitor-only treatment on grain yields in the 2 varieties across the years probably reflects, in part, the lack of a significant effect of this treatment on green leaf area (GLA) (Figure7). This is supported by the fact that the significant reduction in grain yield in the variety Optic in 2007 (Figure6) was associated with a significant reduction in GLA (Figure 7E).

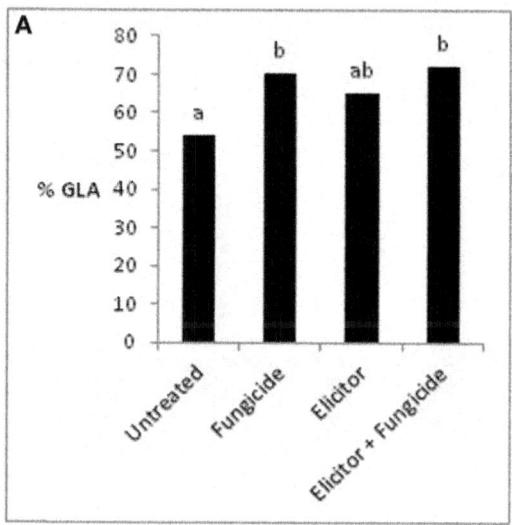

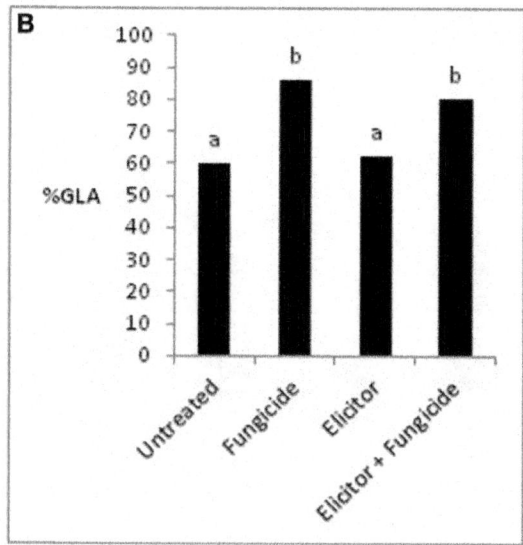

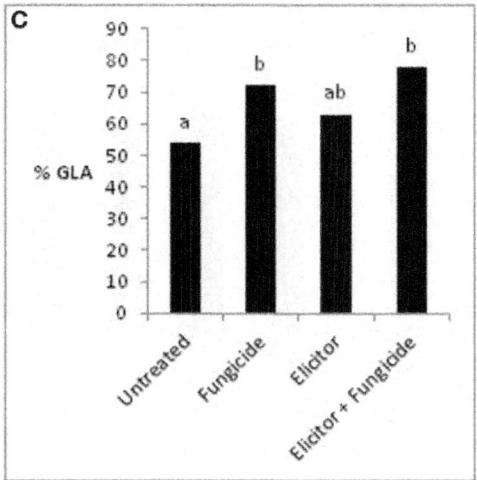

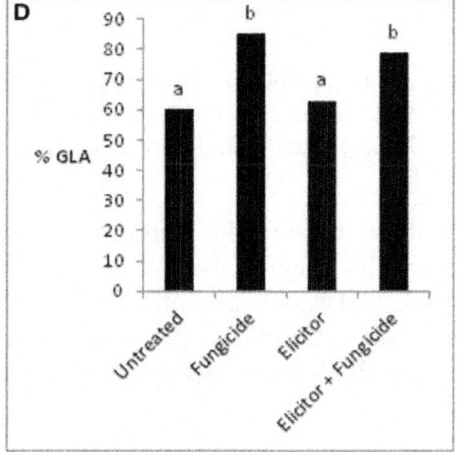

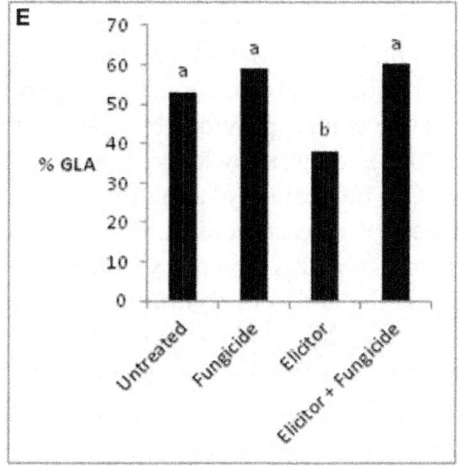

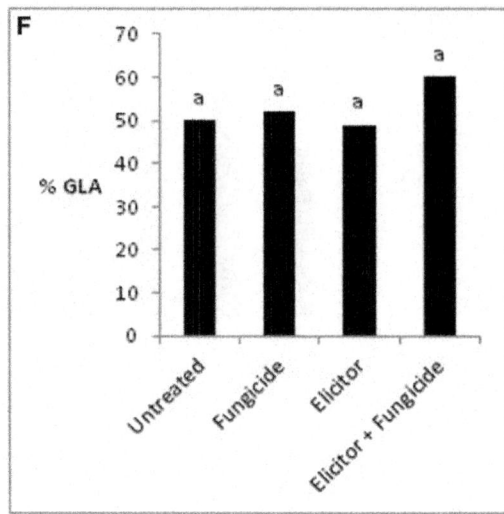

Figure 7. Effects of the elicitor combination and fungicides on percentage green leaf area (% GLA) in Optic and Cellar in 2009 (A,B), Optic and Cellar in 2008 (C,D), and Optic and Cellar in 2007 (E,F). Bars with different letters are significantly different at $P < 0.05$ (Fisher›s LSD).

This contrasts with the elicitor plus fungicide treatment, where, in most cases, increased grain yields (Figures 4–6) were associated with significantly increased GLAs (Figure 7). The only exception was the variety Cellar in 2007, where increased grain yield in the elicitor plus fungicide treatment (Figure 6B) was accompanied by a small, but statistically non-significant increase in GLA (Figure 7F).

DISCUSSION

Under glasshouse conditions, ASM and CJ reduced *R. commune* infection of barley by 64–70%, while BABA had much less effect. Although there are many examples of disease control provided by ASM and BABA (Cohen et al., 2010; Walters et al., 2013), to our knowledge, this is the first report of disease control provided by CJ. Interestingly, applying the three elicitors together provided the best levels of disease control, reducing infection by 96%, and confirms previous reports from this laboratory (Walters et al., 2011a,b). Reports of the use of elicitor combinations to control plant disease are rare, although combinations of ASM and Milsana® (extract of *Reynoutria sachalinensis*) were found to control powdery mildew on cucumber (Bokshi et al., 2008).

Molecular studies on tobacco and *Arabidopsis* have shown that ASM activates the SAR pathway by mimicking the activity of SA (Gaffney et al.,

1993; Friedrich et al., 1996; Lawton et al., 1996). The situation with BABA is more complex and seems to involve SA-dependent, SA-independent, and abscisic acid (ABA)-dependent mechanisms, with the relative importance of the different signaling pathways depending on the particular host-pathogen interaction (Ton et al., 2005). As indicated earlier, CJ is structurally related to JA and MeJA, although it up-regulates a unique set of genes compared to MeJA (Birkett et al., 2000; Pickett et al., 2007). Treatment of leaves one and two of barley with the combination of ASM, BABA, and CJ led to an approximately 6-fold up-regulation of *PR-1b* in leaves three and four, confirming previous reports on the effects of the elicitor combination on defense-related gene expression (Walters et al., 2011a,b). Expression of a *PR-1* gene is usually considered to be a molecular marker for SAR (van Loon et al., 2006) and therefore, these data suggest that the elicitor combination activated SAR in barley. *PR-1b* was also up-regulated by inoculation with *R. commune*(~4-fold increase in leaf 3, for example). However, the largest up-regulation of *PR-1b* was obtained when elicitor-treated plants were subsequently inoculated with *R. commune*. This suggests that the elicitor combination primed *PR-1b* gene expression in these plants. In contrast to *PR-1b*, the expression of *LOX2*exhibited a completely different trend. Treatment of barley plants with the elicitor combination resulted in a substantial down-regulation of *LOX2* (4-fold in leaf 3 and 5.5-fold in leaf 4). Inoculation with *R. commune* led to a greater down-regulation of *LOX2* in the treated leaves, while application of the elicitor combination followed by inoculation had no further effect on gene expression than the elicitor combination only. The *LOX2*gene, which is involved in the octadecanoid pathway, is auto-regulated by JA, thereby controlling a feed-forward loop in JA biosynthesis (Bell et al., 1995). Suppression of *LOX2* in transgenic *Arabidopsis* was shown to block JA biosynthesis during pathogen infection (Spoel et al., 2003), and the present data on *LOX2* in barley suggests that the elicitor combination might suppress activity of the JA pathway in barley. Concomitant activation of SAR and suppression of the JA pathway might be expected to enhance defense against biotrophic pathogens and increase susceptibility to necrotrophic pathogens (Glazebrook, 2005). *R. commune* is a hemibiotrophic pathogen, possessing an initial biotrophic phase followed by a prolonged necrotrophic phase (Walters et al., 2008). Perhaps it is no surprise therefore, that the elicitor combination reduces infection by *R. commune*. It is possible that the elicitor combination might compromise the ability of the plant to defend itself against necrotrophic pathogens and indeed, the elicitor combination was shown to provide control of powdery mildew and *R. commune*, but was associated with increased symptoms of leaf spot caused by the necrotrophic pathogen *Ramularia collo-cygni* on spring barley in the field (Walters et al., 2011b). Plant defense against chewing insects is mediated by

JA signaling (Pieterse et al., 2012), as is establishment of functional arbuscular mycorrhizal (AM) symbioses (Pozo and Azcón-Aguilar, 2007). In view of the down-regulation of *LOX2* in barley treated with the elicitor combination, it would be prudent to examine the effects of treated plants on defense against herbivorous insects and on the establishment of AM symbiosis.

In all 3 years of field experiments, the elicitor combination applied on its own was either partially effective or ineffective at controlling powdery mildew and *R. commune* on the two barley cultivars. In 2009, the elicitor combination provided significant control of both diseases on both spring barley varieties, and also controlled powdery mildew on Optic in 2008. However, the elicitor treatment did not control *R. commune* on either variety in 2008 and provided no disease control in 2007. Perhaps unsurprisingly therefore, given the poor levels of disease control provided by the elicitor-only treatment, grain yield was not significantly affected, apart from 2007, when grain yield of Optic was significantly reduced compared to the untreated control. Here, the reduced grain yield probably reflected the significantly reduced GLA in the elicitor-only treatment. Infection by many pathogens, including biotrophic pathogens such as powdery mildews, results in chlorosis and reduced photosynthetic rates (e.g., Walters and McRoberts, 2007) and the failure of the elicitor-only treatment to control infection would probably have affected rates of photosynthesis. Whether the elicitor combination affects photosynthesis in barley is not known, but should be examined.

The data on disease control presented in this paper highlight two typical problems associated with the practical use of elicitors on certain crops under field conditions, namely inconsistency and poor levels of disease control (Walters and Fountaine, 2009). In contrast to the elicitor-only treatments, the performance of the elicitor plus fungicide treatment was better both in terms of disease control and consistency. The acid test for such combined treatments is whether the performance of the combination is superior to that of the fungicide treatment on its own. Unfortunately, in most cases, although the combined elicitor and fungicide treatment performed as well as the fungicide-only treatment in terms of disease control, only in one case (control of *R. commune* on Optic in 2008) did the elicitor plus fungicide treatment out-perform the fungicide-only treatment. A similar situation was found for grain yield. Here, although grain yield tended to be increased by application of the elicitor plus fungicide treatment, most of these increases were not significantly different from the fungicide-only treatment. The lack of consistency in terms of disease control shown by the elicitor combination in barley, contrasts with the situation in oilseed rape (*Brassica napus*). Here, application of the elicitor combination to winter oilseed rape provided better control of light leaf spot caused by *Pyrenopeziza brassicae*, than standard fungicide treatments (Oxley and Walters, 2012).

Over the 3 years of field experiments, the elicitor plus fungicide combinations providing most consistent disease control were treatments 13, 14, and 16. As indicated earlier, only data for treatment 13 are shown, since this treatment performed most consistently throughout the study. Treatment 13 involved application of a combination of fungicide at reduced rate plus elicitor at GS39, with no control treatments applied at earlier growth stages. It has been suggested that application of elicitors earlier in the season might reduce inoculum levels, thereby requiring less fungicide to be applied later (Walters et al., 2013). On the basis of the results obtained in the present paper, this suggestion does not appear to work for spring barley. This suggests that, at least for spring barley, protecting later stages of crop growth is important in maintaining grain yield.

In some crops, use of elicitors and fungicides (or bactericides) can be effective. For example, the use of ASM (as Actigard®) in combination with fungicides and bactericides was recommended in tomato spray programs in North Carolina, USA (Ivors and Louws, 2007). The rationale here was that the elicitor would increase plant resistance, while the fungicides and bactericides would reduce pathogen inoculum levels. On mandarins (variety Murcott), tank-mixing ASM with azoxystrobin improved the efficacy of the fungicide by more than 50% (Miles et al., 2005), although this effect was clearly variety-specific, since no extra benefit of tank-mixing the elicitor and fungicide was obtained with the mandarin variety Imperial (Miles et al., 2004).

It has been suggested that one of the reasons for the relatively poor performance of elicitors is due to the likelihood that under field conditions, plants are already induced (Walters, 2009). Indeed, Herman et al. (2007) found that in tomato under field conditions, some defense genes were already expressed prior to ASM application. Nevertheless, the expression of these genes was increased further following ASM application. In preliminary work, examination of CAD activity in leaves from the field experiment in 2007 indicated that activity of the enzyme was already high prior to elicitor treatment (Paterson and Walters, unpublished results). Although it is tempting to suggest that CAD activity was increased further following elicitor application, any increases observed were not significant. It is possible that in barley, unlike tomato, prior induction of resistance compromises the ability of the plant to respond effectively to elicitors. Indeed, this was reported for barley treated with the elicitor combination, where prior inoculation with *R. commune* compromised the ability of the plant to respond to subsequent elicitor treatment (Walters et al., 2011a). It was suggested that this might help to explain the relatively poor performance of induced resistance in the field, particularly in cereals, compared to plants grown under controlled conditions (Walters et al., 2011a).

The results presented in this paper indicate quite clearly that use of a combination of elicitors alone does not provide effective disease control in spring barley. In contrast, using the elicitor combination and fungicides, even at half-rate, can provide levels of disease control and yield increases that are equal to the best fungicide-only treatment. From a practical perspective, an elicitor plus fungicide program is only likely to be attractive to a grower if it is cost-effective i.e., it provides levels of disease control and yield increases above that achieved using the fungicide on its own. This suggests that barley growers are unlikely to find the elicitor plus fungicide treatments examined in this work an attractive proposition. This might change however, if fungicide availability is further reduced through legislation. Indeed, elicitor/fungicide combinations could be valuable in reducing fungicide use, and prolonging the useful life of certain fungicides.

ACKNOWLEDGMENTS

We are grateful to Duncan McKenzie of Syngenta, Switzerland, for the gift of Bion®. This research was supported by the Scottish Government (Work Packages 1.4 and 6.4).

SUPPLEMENTARY MATERIAL

The Supplementary Material for this article can be found online at:http://www.frontiersin.org/journal/10.3389/fpls.2014.00241/abstract

REFERENCES

1. Beckers, G. J. M., and Spoel, S. H. (2006). Fine-tuning plant defence signalling: salicylate versus jasmonate. *Plant Biol.* 8, 1–10. doi: 10.1055/s-2005-872705

2. Bell, E., Creelman, R. A., and Mullett, J. E. (1995). A chloroplast lipoxygenase is required for wound-induced accumulation of jasmonic acid in*Arabidopsis*. *Proc. Natl. Acad. Sci. U.S.A.* 92, 8675–8679. doi: 10.1073/pnas.92.19.8675

3. Birkett, M. A., Campbell, C. A., Chamberlain, K., Guerrieri, E., Hick, A. J., Martin, J. L., et al. (2000). New roles for *cis*-jasmone as an insect semiochemical and in plant defense. *Proc. Natl. Acad. Sci. U.S.A.* 97, 9329–9334. doi: 10.1073/pnas.160241697

4.   Bokshi, A. I., Jobling, J., and McConchie, R. (2008). A single application of Milsana® followed by Bion® assists in the control of powdery mildew in cucumber and helps overcome yield losses. *J. Hort. Sci. Biotech.* 83, 701–706.

5.   Bostock, R. M. (2005). Signal crosstalk and induced resistance: straddling the line between cost and benefit. *Annu. Rev. Phytopathol.* 43, 545–580. doi: 10.1146/annurev.phyto.41.052002.095505

6.   Bruce, T. J. A., Pickett, J. A., and Smart, L. E. (2003). *cis*-jasmone switches on plant defence against insects. *Pest. Out.* 14, 96–98. doi: 10.1039/b305499n

7.   Cohen, Y. (2002). β-aminobutyric acid-induced resistance against plant pathogens. *Plant Dis.* 86, 448–457. doi: 10.1094/PDIS.2002.86.5.448

8.   Cohen, Y., Reuveni, M., and Baider, A. (1999). Local and systemic activity of BABA (DL- 3-aminobutyric acid) against *Plasmopara viticola* in grapevines. *Eur. J. Plant Pathol.* 105, 351–361. doi: 10.1023/A:1008734019040

9.   Cohen, Y., Rubin, A. E., and Kilfin, G. (2010). Mechanisms of induced resistance in lettuce against *Bremia lactucae* by DL-β-aminobutyric acid (BABA). *Eur. J. Plant Pathol.* 126, 553–573. doi: 10.1007/s10658-009-9564-6

10. Farmer, E. E., and Ryan, C. A. (1990). Interplant communication: airborne methyl jasmonate induces synthesis of proteinase inhibitors in plant leaves. *Proc. Natl. Acad. Sci. U.S.A.* 87, 7713–7716. doi: 10.1073/pnas.87.19.7713

11. Friedrich, L., Lawton, K., Ruess, W., Masner, P., Specker, N., Gut Rella, M., et al. (1996). A benzothiadiazole derivative induces systemic acquired resistance in tobacco. *Plant J.* 10, 61–70. doi: 10.1046/j.1365-313X.1996.10010061.x

12. Gaffney, T., Friedrich, L., Vernooij, B., Negrotto, D., Nye, G., Uknes, S., et al. (1993). Requirement of salicylic acid for the induction of systemic acquired resistance. *Science* 261, 754–756. doi: 10.1126/science.261.5122.754

13. Glazebrook, J. (2005). Contrasting mechanisms of defense against biotrophic and necrotrophic pathogens. *Annu. Rev. Phytopathol.* 43, 205–227. doi: 10.1146/annurev.phyto.43.040204.135923

14. Herman, M. A. B., Restrepo, S., and Smart, C. D. (2007). Defense gene expression patterns of three SAR-induced tomato cultivars in the field.*Physiol. Mol. Plant Pathol.* 71, 192–200. doi: 10.1016/j.

pmpp.2008.02.002

15. Home-Grown Cereals Authority (HGCA). (2009). *Recommended List for Spring Barley, 2009.* © *Agriculture and Horticulture Development Board.* Stoneleigh, UK: HGCA publications.

16. Ivors, K., and Louws, F. (2007). *Foliar Fungicide Spray Guide for Tomatoes in NC.* Available online at:www.ces.ncsu.edu/fletcher/programs/plantpath/tomato-spray-guide

17. Jakab, G., Cottier, V., Toquin, V., Rigoli, G., Zimmerli, L., Metraux, J.-P., et al. (2001). β- aminobutyric acid-induced resistance in plants. *Eur. J. Plant Pathol.* 107, 29–37. doi: 10.1023/A:1008730721037

18. Lawton, K. A., Friedrich, L., Hunt, M., Weymann, K., Delaney, T. P., Kessmann, H., et al. (1996). Benzothiadiazole induces disease resistance in*Arabidopsis* by activation of Systemic acquired resistance signal transduction pathway. *Plant J.* 10, 71–82. doi: 10.1046/j.1365-313X.1996.10010071.x

19. Leadbeater, A., and Staub, T. (2007). "Exploitation of induced resistance: a commercial Perspective," in *Induced Resistance For Plant Defence: A Sustainable Approach To Crop Protection,* eds D. Walters, A. Newton, and G. Lyon (Oxford: Blackwell Publishing), 229–242.

20. Liljeroth, E., Bengtsson, T., Wiik, L., and Andreasson, E. (2010). Induced resistance in potato to *Phytophthora infestans* – effects of BABA in greenhouse and field tests with different potato varieties. *Eur. J. Plant Pathol.* 127, 171–183. doi: 10.1007/s10658-010-9582-4

21. Lyon, G. (2007). "Agents that can elicit induced resistance," in *Induced Resistance For Plant Defence: A Sustainable Approach To Crop Protection,* eds D. Walters, A. Newton, and G. Lyon (Oxford: Blackwell Publishing), 9–29.

22. Miles, A. K., Willingham, S. L., and Cooke, A. W. (2004). Field evaluation of strobilurins and a plant activator for the control of citrus black spot. *Aust. Plant Pathol.* 33, 371–378. doi: 10.1071/AP04025

23. Miles, A. K., Willingham, S. L., and Cooke, A. W. (2005). Field evaluation of a plant activator, captan, chlorothalonil, copper hydroxide, iprodione, mancozeb and strobilurins for the control of citrus brown spot of mandarin. *Aust. Plant Pathol.* 34, 63–71. doi: 10.1071/AP04085

24. Oxley, S. J. P., and Walters, D. R. (2012). Control of light leaf spot (*Pyrenopeziza brassicae*) on winter oilseed rape (*Brassica napus*) with resistance elicitors. *Crop Prot.* 40, 59–62. doi: 10.1016/j.cropro.2012.04.028

25. Pickett, J. A., Birkett, M. A., Moraes, M. C. B., Bruce, T. J. A., Chamberlain, K., Gordon-Weeks, R., et al. (2007). *cis*-jasmone as an allelopathic agent in inducing plant defence. *Allelopathy J.* 19, 109–117.

26. Pieterse, C. M. J., van der Does, D., Zamioudis, C., Leon-Reyes, A., and van Wees, S. C. M. (2012). Hormonal modulation of plant immunity. *Annu. Rev. Plant Cell Dev. Biol.* 28, 489–521. doi: 10.1146/annurev-cellbio-092910-154055

27. Pozo, M. J., and Azcón-Aguilar, C. (2007). Unraveling mycorrhiza-induced resistance. *Curr. Opin. Plant Biol.* 10, 393–398. doi: 10.1016/j.pbi.2007.05.004

28. Reglinski, T., and Walters, D. (2009). "Induced resistance for plant disease control," in *Disease Control in Crops: Biological And Environmentally Friendly Approaches*, ed D. Walters (Oxford: Wiley-Blackwell), 62–92.

29. Spoel, S. H., and Dong, X. (2012). How do plants achieve immunity? Defence without specialized immune cells. *Nat. Rev. Immunol.* 12, 89–100. doi: 10.1038/nri3141

30. Spoel, S. H., Johnson, J. S., and Dong, X. (2007). Regulation of tradeoffs between plant defences against pathogens with different lifestyles. *Proc. Natl. Acad. Sci. U.S.A.* 104, 18842–18847. doi: 10.1073/pnas.0708139104

31. Spoel, S. H., Kornneef, A., Claessens, S. M. C., Korzrlius, J. P., van Pelt, J. A., Mueller, J. A., et al. (2003). NPR1 modulates cross-talk between salicylate and jasmonate dependent defense pathways through a novel function in the cytosol. *Plant Cell* 15, 760–770. doi: 10.1105/tpc.009159

32. Thaler, J. S., Stout, M. J., Karban, R., and Duffey, S. S. (1996). Exogenous jasmonates simulate insect wounding in tomato plants (*Lycopersicon esculentum*) in the laboratory and field. *J. Chem. Ecol.* 22, 1767–1781. doi: 10.1007/BF02028503

33. Ton, J., Jakab, G., Toquin, V., Flors, V., Iavicoli, A., Maeder, M. N., et al. (2005). Dissecting the β-aminobutyric acid induced priming phenomenon in *Arabidopsis*. *Plant Cell* 17, 987–999. doi: 10.1105/tpc.104.029728

34. van Loon, L. C., Rep, M., and Pieterse, C. M. J. (2006). Significance of inducible defense-related proteins in infected plants. *Annu. Rev. Phytopathol.* 44, 135–162. doi: 10.1146/annurev.phyto.44.070505.143425

35. Walters, D., Newton, A., and Lyon, G. (2007). *Induced Resistance for Plant Defence: A Sustainable Approach to Crop Protection*. Oxford: Blackwell Publishing. doi: 10.1002/9780470995983

36. Walters, D. R. (2009). Are plants in the field already induced? Implications for practical disease control. *Crop Prot.* 28, 459–465. doi: 10.1016/j.

cropro.2009.01.009

37. Walters, D. R., and Fountaine, J. M. (2009). Practical application of induced resistance to plant diseases: an appraisal of effectiveness under field conditions. *J. Agric. Sci.* 147, 523–535. doi: 10.1017/ S0021859609008806

38. Walters, D. R., Havis, N. D., Sablou, C., and Walsh, D. J. (2011b). Possible trade-off associated with use of a combination of resistance elicitors.*Physiol. Mol. Plant Pathol.* 75, 188–192. doi: 10.1016/j. pmpp.2011.02.001

39. Walters, D. R., and McRoberts, N. (2007). Plants and biotrophs: a pivotal role for cytokinins? *Trends Plant Sci.* 11, 581–586. doi: 10.1016/j. tplants.2006.10.003

40. Walters, D. R., McRoberts, N., and Fitt, B. D. L. (2008). Are green islands red herrings? Significance of green islands in plant interactions with pathogens and pests. *Biol. Rev.* 83, 79–102. doi: 10.1111/j.1469-185X.2007.00033.x

41. Walters, D. R., Paterson, L., and Havis, N. D. (2010). "Control of foliar diseases of spring barley using resistance elicitors," in *Proceedings Crop Protection in Northern Britain 2010* (Dundee), 91–96.

42. Walters, D. R., Paterson, L., Sablou, C., and Walsh, D. J. (2011a). Existing infection with *Rhynchosporium secalis* compromises the ability of barley to express induced resistance. *Eur. J. Plant Pathol.* 130, 73–82. doi: 10.1007/s10658-010-9733-7

43. Walters, D. R., Ratsep, J., and Havis, N. D. (2013). Controlling crop diseases using induced resistance: challenges for the future. *J. Exp. Bot.* 64, 1263–1280. doi: 10.1093/jxb/ert026

44. Zadoks, J. C., Chang, T. T., and Konzak, C. F. (1974). A decimal code for the growth stages of cereals. *Weed Res.* 14, 415–421. doi: 10.1111/ j.1365-3180.1974.tb01084.x

Chapter 5

A REVIEW ON BIOLOGICAL CONTROL OF FUNGAL PLANT PATHOGENS USING MICROBIAL ANTAGONISTS

Asghar Heydari[1] and Mohammad Pessarakli[2]

[1]Department of Plant Pathology, Iranian Research Institute of Plant Protection, Tehran, Iran

[2]Department of Plant Sciences, University of Arizona, Tucson, AZ, 85721, USA

ABSTRACT

The objective of this study was to review the published research works on biological control of fungal plant diseases during past 50 years. Fungal plant pathogens are among the most important factors that cause serious losses to agricultural products every year. Biological control of plant diseases including fungal pathogens has been considered a viable alternative method to chemical control. In plant pathology, the term biocontrol applies to the use of microbial antagonists to suppress diseases. Throughout their lifecycle, plants and pathogens interact with a wide variety of organisms. These interactions can significantly affect plant health in various ways. Different mode of actions of biocontrol-active microorganisms in controlling fungal plant diseases include hyperparasitism, predation, antibiosis, cross protection, competition for site and nutrient and induced resistance. Successful application of biological control strategies requires more knowledge-intensive management. Various methods for application of biocontrol agents include: application directly to the infection court at a high population level to swamp the pathogen, application at one place in which biocontrol microorganisms are applied at one place (each crop year) but at lower populations which then multiply and spread to other plant parts and give protection against pathogens and one time or occasional application that maintain pathogen populations below threshold levels. Commercial use and application of biological disease control have been slow mainly due to their variable performances under different environmental conditions in the field. To overcome this problem and in order to take the biocontrol technology to the field and improve the commercialization of biocontrol, it is important to develop new formulations of biocontrol microorganisms with higher degree of stability and survival. Majority of biocontrol products are applied against seed borne and soil borne fungal pathogens, including the causal agents of seed rot, damping-off and root rot diseases. These products are mostly used as seed treatment and have been effective in protecting several major crops such as wheat, rice, corn, sugar beet and cotton against fungal pathogens. However, in some cases, biocontrol microorganisms have also been tested as spray application on foliar diseases, including powdery mildew, downy mildew, blights and leaf spots. A few post-harvest fungal diseases have also been controlled by the use of antagonistic fungi and bacteria. Biocontrol microorganisms are also being used as the form of composts in some plants. Research data and observations in nurseries have shown that addition of composted organic matter to potting mixes results in suppression of soil borne diseases. A significant improvement have been made in different aspects of biological control of fungal plant diseases, but this area still need much more development and investigations to solve the existing problems. In order to have more effective biological control strategies in the future, it is critical to carry out more research studies on some less developed aspects of biocontrol, including development of novel formulations, understanding the impact of environmental factors on biocontrol agents, mass production of biocontrol microorganisms and the use of biotechnology and nano-technology in improvement of biocontrol mechanisms and strategies. Future outlooks of biocontrol of plant diseases is bright and promising and with the growing demand for biocontrol products among the growers, it is possible to use the biological control as an effective strategy to manage plant diseases, increase yield, protect the environment and biological resources and approach a sustainable agricultural system.

INTRODUCTION

Plant pests (harmful insects, parasitic weeds and pathogens) are among the most important biotic agents causing serious losses and damages to agricultural products. Plant pests need to be controlled to ensure food, feed and fiber production quantitatively and qualitatively. A number of different strategies are currently being employed to manage and control plant pests (Agrios, 1988; Baker, 1987; Cook, 1993; Bargabus et al., 2002, 2004; Benhamou, 2004; Chisholm et al., 2006; Heydari, 2007; Heydari et al., 2007; Islam et al., 2005; Kloepper et al., 2004). Beyond good agronomic and cultural practices, growers often rely heavily on chemical pesticide application (Agrios, 1988; Baker, 1987). However, the environmental pollution caused by excessive use of agrochemicals has led to considerable changes in people's attitudes towards the use of pesticides in agriculture. Today, there are strict regulations on chemical pesticide use and there is political pressure to remove the most hazardous chemicals from the market. In addition to the above-mentioned issues, the spread of plant diseases in natural ecosystems may preclude successful application of chemicals, because of the scale to which such applications might have to be applied. Consequently, some pest management researchers have focused their efforts on developing alternative inputs to synthetic chemicals for controlling pests and diseases (Baker, 1987; Cook, 1993).

Plant diseases are mostly controlled by the use of chemical pesticides and in some cases by cultural practices (Agrios, 1988; Cook, 1993). However, the widespread use of chemicals in agriculture has been a subject of public concern and scrutiny due to the potential harmful effects on the environment, their undesirable effects on non-target organisms and possible carcinogenicity of some chemicals (Agrios, 1988;Cook, 1993; Heydari, 2007; Heydari et al., 2007). Other problems include development of resistant races of pathogens, a gradual elimination and phasing out of some available pesticides and the reluctance of some chemical companies to develop and test new chemicals due to the problems with registration process and cost (Cook, 1993). The need for the development of non-chemical alternative methods to control plant diseases is therefore clear.

Biological control of plant diseases has been considered a viable alternative method to manage plant diseases (Cook, 1993). Biological control is the inhibition of growth, infection or reproduction of one organism using another organism (Cook, 1993; Baker, 1987). Biocontrol is environmentally safe and in some cases is the only option available to protect plants against pathogens (Cook, 1993). Biological controlemploys natural enemies of pests or pathogens to eradicate or control their population. This can involve the introduction of exotic species, or it can be a matter of harnessing whatever form of biological control exists naturally in the ecosystem. The induction of plant resistance using non-pathogenic or incompatible microorganisms is also a form of biological control (Cook, 1993; Schouten et al., 2004). Fungal plant diseases are considered the most important microbial agents causing serious losses in the agriculture annually (Agrios, 1988). Some fungal diseases that have successfully been controlled using biological agents are pathogens of pruning wounds and other cut surfaces, diseases of leaves and flowers, such as powdery mildew, diseases of fruits and vegetables, such as Botrytis and fungal pathogens in the soil (Agrios, 1988;Baker, 1987; Cook, 1993; Heydari, 2007; Heydari et al., 2004, 2007; Heydari and Misaghi, 1998, 1999, 2003). A variety of biological controls are available for use, but further development and effective adoption will require a greater understanding of the complex interactions among plants, people and the environment. To that end, the objectives of this review chapter is to present an advanced survey of the nature and practice of biological control as it is applied to the suppression of plant diseases. In this review, different aspects of biological control of fungal plant diseases including definitions, modes of action, application strategies, current status and future development and outlooks will be discussed.

TERMINOLOGY

The term biological control and its abbreviated synonym biocontrol have been used in different fields of biology, most notably entomology and plant pathology. In plant pathology, the term applies to the use of microbial antagonists to suppress diseases as well as the use of host-specific pathogens to control weed populations (Cook, 1993). In both fields, the organism that suppresses the pest or pathogen is referred to as the Biological Control Agent (BCA). More broadly, the term biological control also has been applied to the use of the natural products extracted or fermented from various sources (Cook, 1993). These formulations may be very simple mixtures of natural ingredients with specific activities or complex mixtures with multiple effects on the host as well as the target pest or pathogen. While such inputs may mimic the activities of living organisms, non-living inputs should more properly be referred to as biopesticides or biofertilizers, depending on the primary benefit provided to the host plant (Cook, 1993).

The various definitions offered in the scientific literature have sometimes caused confusion and controversy. For example, members of the United States National Research Council took into account modern biotechnological developments and referred to biological control as the use of natural or modified organisms, genes, or gene products, to reduce the effects of undesirable organisms and to favor desirable organisms such as crops, beneficial insects and microorganisms, but this definition spurred much subsequent debates and it was frequently considered too broad by many scientists who worked in the field.

Published definitions of biocontrol differ depending on the target of suppression; number, type and source of biological agents and the degree and timing of human intervention (Cook, 1993). Most broadly, biological control is the suppression of damaging activities of one organism by one or more other organisms, often referred to as natural enemies. With regards to plant diseases, suppression can be accomplished in many ways. If grower's activities are considered relevant, cultural practices such as the use of rotations and planting of disease resistant cultivars (whether naturally selected or genetically engineered) would be included in the definition (Cook, 1993).

Because the plant host responds to numerous biological factors, pathogenic and non-pathogenic, induced host resistance might be considered a form of biological control (Cook, 1993). More narrowly, biological control refers to the purposeful utilization of introduced or resident living organisms, other than disease resistant host plants, to suppress the activities and populations of one or more plant pathogens. This may involve the use of microbial inoculants to suppress a single type or class of plant diseases. This may also involve managing soils to promote the combined activities of native soil and plant-associated organisms that contribute to general suppression (Cook, 1993). Most narrowly, biological control refers to the suppression of a single pathogen by a single antagonist, in a single cropping system. Most specialists in the field would concur with one of the narrower definitions presented above.

INTERACTIONS BETWEEN PLANTS AND BENEFICIAL MICROBES

Throughout their lifecycle, plants and pathogens interact with a wide variety of organisms. These interactions can significantly affect plant health in various ways (Agrios, 1988; Bull et al., 2002; Katska, 1994; Chisholm et al., 2006; Fitter and Garbaye, 1994; McSpadden-Gardener and Weller, 2001). In order to understand the mechanisms of biological control, it is helpful to appreciate the different ways that organisms interact. Note too, in order to interact, organisms must have some form of direct or indirect contact. The types of interactions between plants and microorganisms have been referred to as mutualism, protocooperation, commensalisms, neutralism, competition, amensalism, parasitism and predation (Bankhead et al., 2004; Bull et al., 2002; Katska, 1994; Chisholm et al., 2006; Fitter and Garbaye, 1994; Hoitink and Boehm, 1999). While the terminology has been developed for macroecology, examples of all of these types of interactions can be found in the natural world at both the macroscopic and microscopic level. And, because the development of plant diseases involves both plants and microbes, the interactions that lead to biological control take place at multiple levels of scale (Bull et al., 2002; Katska, 1994; Chisholm et al., 2006; Fitter and Garbaye, 1994).

From the plant's point of view, biological control may be considered a positive result arising from different specific and non-specific interactions (Cook, 1993; Weller et al., 2002). We can begin to classify and functionally delineate the diverse components of ecosystems that contribute to biological control. Mutualism is an association among several species where all of them are benefited from this association (Biermann and Linderman, 1983; Bull et al., 2002; Katska, 1994; Chisholm et al., 2006; Duchesne, 1994; Fitter and

Garbaye, 1994; Garcia-Garrido and Ocampo, 1989; Kerry, 2000). Sometimes, it can be an obligatory relation involving close physical and biochemical contact between two organisms, such as those between plants and mycorrhizal fungi (Bull et al., 2002; Katska, 1994; Chisholm et al., 2006; Fitter and Garbaye, 1994). However, they are generally facultative and opportunistic.

For example, Rhizobium bacteria reproduce either in the soil or, to a much greater degree, through their mutualistic association with legume plants. These types of mutualism can contribute to biological control, by providing plant with improved nutrition and/or by stimulating host defense mechanism and ability (Chisholm et al., 2006; Fitter and Garbaye, 1994). Many of the microorganisms isolated and classified as biocontrol agents (BCA) can be considered facultative mutualists, because host and disease suppression by them will vary depending on the prevailing environmental conditions (Cook, 1993).

Commensalism is also a symbiotic interaction between two living organisms, where one organism benefits and the other is neither harmed nor benefited (Fitter and Garbaye, 1994). Most plant-associated microorganisms are assumed to be commensals with regards to the host plant, because their presence, individually or in total, rarely results in positive or negative consequences to the plant (Katska, 1994; Chisholm et al., 2006). While the presence of these microorganisms may present a variety of challenges to an infecting pathogen, their absence decreases pathogen infection or disease severity and is indicative of commensal interactions (Cook, 1993).

Biological interactions in which the population density of one species has absolutely no effect on the other are called neutralism (Berg et al., 2005; Chisholm et al., 2006). Related to biological control, an inability to associate the population dynamics of pathogen with that of another organism would indicate neutralism (Chisholm et al., 2006). In contrast, antagonism between organisms results in a negative outcome for one or both. Competition within and between species caused a decreased growth, activity, and/or fecundity of the interacting organisms (Cook, 1993). Biocontrol can occur when non-pathogens compete with pathogens for nutrients and sites in host plant. Direct interactions that benefit one population at the expense of another also affect our understanding of biological control (Cook, 1993).

Parasitism is also a symbiotic relation in which two organisms coexist over a prolonged period of time (Cook, 1993; Chisholm et al., 2006; Lo et al., 1997). In this type of interaction, one organism, usually the physically smaller (parasite) benefits and the other (host) is harmed. The activities of various hyperparasites, for example those agents that parasitize plant pathogens, can result in biocontrol (Lo et al., 1997). Another interesting contribution to biocontrol is when host infection and parasitism by relatively avirulent pathogens may lead to biocontrol of more virulent pathogens through the stimulation of host defense systems (Cook, 1993). Finally predation refers to the hunting and killing of one organism by another for consumption and sustenance. While the term predator typically refers to animals that feed at higher trophic levels in the macroscopic world, it has also been applied to the actions of microorganisms such as protists and mesofauna, e.g. fungal feeding nematodes and microarthropods, that consume pathogen biomass for sustenance (Cook, 1993).

Biological control can result in various forms of these types of interactions, depending on the environmental conditions within which they occur. Significant biological control, as was described above, generally arises from manipulating mutualisms between microorganisms and their plant hosts or from manipulating antagonisms between microbes and pathogens (Bull et al., 2002; Katska, 1994; Chisholm et al., 2006;Fitter and Garbaye, 1994).

MECHANISMS OF BIOLOGICAL CONTROL

Since biological control is a result of many different types of interactions among microorganisms, scientists have concentrated on characterization of mechanisms occurring in different experimental situations (Audenaert et al., 2002; De Meyer and Hofte, 1997; Elad and Baker, 1985; Heydari et al., 1997; Homma et al., 1989; Howell et al., 1988; Islam et al., 2005; Meziane et al., 2005; Ryu et al., 2004; Van Dijk and Nelson, 2000). In all cases, pathogens are antagonized by the presence and activities of other microorganisms that they encounter.

Direct antagonism results from physical contact and/or a high-degree of selectivity for the pathogen by the mechanism(s) expressed by the biocontrol active microorganisms. In this type of interaction, Hyperparasitism by obligate parasites of a plant pathogen would be considered the most direct type of mechanism because the activities of no other organism would be required to exert a suppressive effect (Harman et al., 2004; Linderman, 1994). In contrast, indirect antagonism is resulted from the activities that do not involve targeting a pathogen by a biocontrol active microorganism. Improvement and stimulation of plant host defense mechanism by non-pathogenic microorganisms is the most indirect form of antagonism (Kloepper et al., 1980; Lafontaine and Benhamo, 1996; Leeman et al., 1995; Maurhofer et al., 1994; Silva et al., 2004). While many studies have concentrated on the establishment of the importance of specific mechanisms of biocontrol to particular pathosystems, all of the mechanisms described below are likely to be operating to some extent in all natural and managed ecosystems.

The most effective biocontrol active microorganisms studied appear to antagonize plant pathogens employing several modes of actions (Cook, 1993). For example, pseudomonads known to produce the antibiotic 2, 4-diacetylphloroglucinol (DAPG) may also induce host defenses (Kloepper et al., 1980; Lafontaine and Benhamou, 1996; Leeman et al., 1995; Maurhofer et al., 1994; Silva et al., 2004). Additionally, DAPG-producers bacterial antagonists can aggressively colonize roots, a trait that might further contribute to their ability to suppress pathogen activity in the rhizosphere of plant through competition for organic nutrients. However, the most important modes of actions of biocontrol active microorganisms are as follows:

Mycoparasitism

In Hyperparasitism, the pathogen is directly attacked by a specific biocontrol agent (BCA) that kills it or its propagules. Four major groups of hyperparasites have generally been identified which include hypoviruses, facultative parasites, obligate bacterial pathogens and predators. An example of hypoparasites is the virus that infects Cryphonectria parasitica, the fungal causal agent of chestnut blight, which causes hypovirulence, a reduction in pathogenicity of the pathogen. This phenomenon has resulted in the control of chestnut blight in many places (Milgroom and Cortesi, 2004). However, the interaction of virus, fungus, tree and environment determines the success or failure of hypovirulence.

In addition to hypoviruses several fungal hypoparasites have also been identified including those that attack sclerotia (e.g., Coniothyrium minitans) or others that attack fungal hyphae (e.g. Pythium oligandrum). In some cases, a single fungal pathogen can be attacked by multiple hyperparasites. For example, Acremonium alternatum, Acrodontium crateriforme, Ampelomyces quisqualis, Cladosporium oxysporum and Gliocladium virens are just a few of the fungi that have the capacity to parasitize powdery mildew pathogens (Milgroom and Cortesi, 2004).

In contrast to hyperparasitism, microbial predation is more general, non-specific and generally provides less predictable levels of disease control. Some biocontrol agents exhibit predatory behavior under nutrient-limited conditions. Such as Trichoderma, a fungal antagonist that produces a range of enzymes that are directed against cell walls of pathogenic fungi. However, when fresh bark is used in composts,Trichoderma sp. does not directly attack the plant pathogen, Rhizoctonia solani. But, in decomposing bark, the concentration of readily available cellulose decreases and this activates the chitinase genes of Trichoderma sp. Which, in turn, produce chitinase to parasitize R. solani (Benhamou and Chet, 1997).

Antibiosis

Many microbes produce and secrete one or more compounds with antibiotic activity (Homma et al., 1989; Howell and Stipanovic, 1980; Islam et al., 2005; Leclére et al., 2005; Shahraki et al., 2009; Shanahan et al., 1992; Thomashow et al., 1990; Thomashow and Weller, 1988). In a general definition antibiotics are microbial toxins that can, at low concentrations, poison or kill other microorganisms. It has been shown that some antibiotics produced by microorganisms are particularly effective against plant pathogens and the diseases they cause (Homma et al., 1989; Howell and Stipanovic, 1980; Islam et al., 2005; Shanahan et al., 1992; Thomashow et al., 1990, 2002; Thomashow and Weller, 1988). In all cases, the antibiotics have been shown to be particularly effective at suppressing growth of the target pathogen in vitro and/or in situconditions. An effective antibiotic must be produced in sufficient quantities (dose) near the pathogen. In situ production of antibiotics by several different biocontrol agents has been studied (Thomashow et al., 1990). While several procedures have been developed to ascertain when and where biocontrol agents may produce antibiotics detecting expression in the infection court is difficult because of the heterogenous distribution of plant-associated microbes and the potential sites of infection (Thomashow et al., 1990).

However, in some cases, the relative importance of antibiotic production by biocontrol bacteria has been demonstrated. For example, mutant strains incapable of producing phenazines (Thomashow and Weller, 1988) or phloroglucinols (Keel et al., 1989) have been shown to be equally capable of colonizing the rhizosphere, but much less capable of suppressing soil borne root diseases than the corresponding wild-type and complemented mutant strains. Many biocontrol strains have been shown to produce multiple antibiotics which can suppress one or more pathogens (Homma et al., 1989; Howell and Stipanovic, 1980; Islam et al., 2005; Shanahan et al., 1992; Thomashow et al., 1990; Thomashow and Weller, 1988). The ability of production of several antibiotics probably results in suppression of diverse microbial competitors and plant pathogens.

Metabolite Production

Many biocontrol active microorganisms produce other metabolites that can interfere with pathogen growth and activities. Lytic enzymes are among these metabolites that can break down polymeric compounds, including chitin, proteins, cellulose, hemicellulose and DNA (Anderson et al., 2004; Howell et al., 1988; Loper and Buyer, 1991; Ordentlich et al., 1988; Press et al., 2001; Wilhite et al., 2001). Studies have shown that some of these metabolites can sometimes directly result in the suppression of plant pathogens. For example, control of Sclerotium rolfsii by Serratia marcescens appeared to be mediated by chitinase expression (Ordentlich et al., 1988). It seems more likely that antagonistic activities of these metabolites are indicative of the need to degrade complex polymers in order to obtain carbon nutrition. Microorganisms that show a preference in colonizing and suppression of plant pathogens might be classified as biocontrol agents. For example, Lysobacter and Myxobacteria that produce lytic enzymes have been shown to be effective against some plant pathogenic fungi (Bull et al., 2002).

Studies have shown that some products of lytic enzyme activity may have indirect efficacy against plant pathogens. For example, oligosaccharides derived from fungal cell walls have been shown to induce plant host defenses (Howell et al., 1988). It is believed that the effectiveness of the above compounds against plant pathogens is dependent on the composition and carbon and nitrogen sources of the soil and rhizosphere. For example, in post-harvest disease control, addition of chitosan which is a non-toxic and biodegradable polymer of beta-1, 4-glucosamine produced from chitin by alkaline deacylation stimulated microbial degradation of pathogens (Benhamou, 2004). Amendment of plant growth substratum with chitosan suppressed the root rot caused by Fusarium oxysporum f. sp. radicis-lycopersici in tomato (Lafontaine and Benhamou, 1996).

In addition to the above-mentioned metabolites, other microbial byproducts may also play important roles in plant disease biocontrol (Phillips et al., 2004). For example, Hydrogen cyanide (HCN) effectively blocks the cytochrome oxidase pathway and is highly toxic to all aerobic microorganisms at picomolar concentrations (Ramette et al., 2003). The production of HCN by certain fluorescent pseudomonads is believed to be effective against plant pathogens. Results of some research studies in this regard have shown that P. fluorescens CHA0, an antagonistic bacterium, produces antibiotics including siderophores and HCN, but suppression of black rot of tobacco caused by Thielaviopsis basicola appeared to be due primarily to HCN production. In another study Howell et al. (1988) reported that volatile compounds such as ammonia produced by Enterobacter cloacae were involved in the suppression of cotton seedling damping-off caused by Pythium ultimum.

Competition

The nutrient sources in the soil and rhizosphere are frequently not sufficient for microorganisms. For a successful colonization of phytosphere and rhizosphere a microbe must effectively compete for the available nutrients (Elad and Baker, 1985; Keel et al., 1989; Loper and Buyer, 1991). On plant surfaces, host-supplied nutrients include exudates, leachates, or senesced tissue. In addition to these, nutrients can also be obtained from waste products of other organisms such as insects and the soil. This is a general believe that competition between pathogens and non-pathogens for nutrient resources is an important issue in biocontrol (Elad and Baker, 1985; Keel et al., 1989; Loper and Buyer, 1991). Ii is also believed that competition for nutrients is more critical for soil borne pathogens, including Fusarium and Pythiumspecies that infect through mycelial contact than foliar pathogens that germinate directly on plant surfaces and infect through appressoria and infection pegs (Elad and Baker, 1985; Keel et al., 1989; Loper and Buyer, 1991). Results of a study by Anderson et al. (1988) revealed that production of a particular plant glycoprotein called agglutinin was correlated with potential of Pseudomonas putida to colonize the root system. P. putida mutants deficient in this ability exhibited reduced capacity to colonize the rhizosphere and a corresponding reduction in Fusarium wilt suppression in cucumber (Tari and Anderson, 1988).

It has been shown that non-pathogenic plant-associated microrganisms generally protect the plant by rapid colonization and thereby exhausting the limited available substrates so that none are available for pathogens to grow. For example, effective catabolism of nutrients in the spermosphere has been identified as a mechanism contributing to the suppression of Pythium ultimum by Enterobacter cloacae (Van Dijk and Nelson, 2000; Kageyama and Nelson, 2003). At the same time, these microbes produce metabolites that are effective in suppression of pathogens. These microbes colonize the sites where water and carbon-containing nutrients are most readily available, such as exit points of secondary roots, damaged epidermal cells and nectaries and utilize the root mucilage.

Competition for rare but essential micronutrients, such as iron, has also been shown to be important in biological disease control. Iron is extremely limited in the rhizosphere, depending on soil pH. In highly oxidized and aerated soil, iron is present in ferric form (Kageyama and Nelson, 2003; Shahraki et al., 2009), which is insoluble in water and the concentration may be extremely low. This very low concentration cannot support the growth of microorganisms. To survive in such environment, organisms were found to secrete iron-binding ligands called Siderophores having high ability to obtain iron from the micro-organisms (Shahraki et al., 2009). Almost all microorganisms produce siderophores, of either the catechol type or hydroxamate type (Kageyama and Nelson, 2003).

A direct correlation was established in vitro between siderophore synthesis in fluorescent pseudomonads and their capacity to inhibit germination of chlamydospores of F. oxysporum (Elad and Baker, 1985). It was shown that mutants incapable of producing some siderophores, such as pyoverdine, were reduced in their capacity to suppress different plant pathogens (Keel et al., 1989; Loper and Buyer, 1991). The increased efficiency in iron uptake of the commensal microorganisms is thought to be a critical factor in their root colonization ability which is a major factor in biocontrol performance of bacterial antagonists.

Induction of resistance: Plants actively respond to a variety of environmental stimulating factors, including gravity, light, temperature, physical stress, water and nutrient availability and chemicals produced by soil and plant associated microorganisms (Audenaert et al., 2002; De Meyer and Hofte, 1997; Kloepper et al., 1980; Leeman et al., 1995; Moyne et al., 2000; Vallad and Goodman, 2004; Van Loon et al., 1998; Van Peer and Schippers, 1992; Van Wees et al., 1997). Such stimuli can either induce or condition plant host defenses through biochemical changes that enhance resistance against subsequent infection by a variety of pathogens. Induction of host defenses can be local and/or systemic in nature, depending on the type, source and amount of stimulation agents (Audenaert et al., 2002; De Meyer and Hofte, 1997; Kloepper et al., 1980; Leeman et al., 1995; Moyne et al., 2000; Vallad and Goodman, 2004; Van Loon et al., 1998; Van Peer and Schippers, 1992; Van Wees et al., 1997).

Recently, plant pathologists have begun to characterize the determinants and pathways of induced resistance stimulated by biological control agents and other non-pathogenic microorganisms (Audenaert et al., 2002; Moyne et al., 2000; Vallad and Goodman, 2004). The first pathway called Systemic Acquired Resistance (SAR), is mediated by Salicylic Acid (SA), a chemical compound which is usually produced after pathogen infection and typically leads to the expression of Pathogenesis-related (PR) proteins (Vallad and Goodman, 2004). These PR proteins include a variety of enzymes some of which may act directly to lyse invading cells, reinforce cell wall boundaries to resist infections, or induce localized cell death (Vallad and Goodman, 2004).

Second pathway, called Induced Systemic Resistance (ISR), is mediated by Jasmonic Acid (JA) and/or ethylene, which are produced following applications of some nonpathogenic rhizobacteria (Audenaert et al., 2002; De Meyer and Hofte, 1997; Kloepper et al., 1980; Leeman et al., 1995;Moyne et al., 2000; Van Loon et al., 1998; Van Peer and Schippers, 1992; Van Wees et al., 1997). Interestingly, the SA- and JA- dependent defense pathways can be mutually antagonistic and some bacterial pathogens take advantage of this to overcome the SAR. For example, pathogenic strains of Pseudomonas syringae produce coronatine, which is similar to JA, to overcome the SA-mediated pathway (Vallad and Goodman, 2004). Since the various host-resistance pathways can be activated to variable degrees by different microorganisms and insect feeding, it is therefore possible that multiple stimuli are constantly being received and processed by the plant. Thus, the magnitude and duration of host defense induction will likely vary over time. Only if induction can be controlled, i.e., by overwhelming or synergistically interacting with endogenous signals, will host resistance be increased (Audenaert et al., 2002; De Meyer and Hofte, 1997; Kloepper et al., 1980; Leeman et al., 1995; Moyne et al., 2000).

Some strains of root-colonizing microorganisms have been identified as potential elicitors of plant host defenses. For example, some biocontrol active strains of Pseudomonas sp. and Trichoderma sp. are known to strongly induce plant host defenses (Haas and Defago, 2005;Harman et al., 2004). In other instances, inoculation with Plant Growth Promoting Rhizobacteria (PGPR) have been shown to be effective in controlling multiple diseases caused by different fungal pathogens, including anthracnose (Colletotrichum lagenarium). A number of chemical elicitors of SAR and ISR such as salicylic acid, siderophore, lipopolysaccharides and 2, 3-butanediol may be produced by the PGPR strains upon inoculation (Ryu et al., 2004; Van Loon et al., 1998).

A substantial number of microbial products have been reported to elicit host defenses, indicating that host defenses are likely stimulated continually during the plant's lifecycle (Ryu et al., 2004; Van Loon et al., 1998). These inducers include lipopolysaccharides and flagellin from Gram-negative bacteria; cold shock proteins of diverse bacteria; transglutaminase, elicitins and a-glucans in Oomycetes; invertase in yeast; chitin and ergosterol in all fungi; and xylanase in Trichoderma (Ryu et al., 2004). These findings indicate that plants would detect the composition of their plant-associated microbial communities and respond to changes in the quantity, quality and localization of many different signals. The importance of such interactions is indicated by the fact that further induction of host resistance pathways, by chemical and microbiological inducers, is not always effective in improving plant health or productivity in the field (Vallad and Goodman, 2004).

METHODS OF APPLICATION OF ANTAGONISTS

Overall Application

Successful application of biological control strategies requires more knowledge-intensive management (Baker, 1987; Cook, 1993; Heydari et al., 2004; Shah-Smith and Burns, 1997). Understanding when and where biological control of plant pathogens can be profitable, requires an appreciation of its place within integrated pest management systems (Cook, 1993; Heydari et al., 2004; Shah-Smith and Burns, 1997).

In general, the foundation of a sound pest and disease management program in an annual cropping system begins with cultural practices that alter the farm landscape to promote crop health (Cook, 1993; Heydari et al., 2004; Shah-Smith and Burns, 1997). These include crop rotations that limit the availability of host material used by plant pathogens (Cook, 1993). Proper use of tillage can disrupt pathogen life cycles and prepare seed beds of optimal moisture and bulk density. Careful management of soil fertility and moisture can also limit plant diseases by minimizing plant stress (Cook, 1993). In nurseries and greenhouses environmental control can be more tightly regulated in terms of temperature, light, moisture and soil composition, but the design of such systems cannot wholly eliminate disease problems (Paulitz and Belanger, 2001).

The second layer of defense against pests consists of the quality of crop germplasm. Breeding for pathogen resistance including fungal pathogens contributes substantially to crop success in most regions (Cook, 1993). Newer technologies that directly incorporate genes into crop genomes, commonly referred to as genetic modification or genetic engineering, are bringing new traits into crop. Other technologies, such as seed washing, testing for pathogens and treatments are also used to keep germplasm pathogen-free. In perennial cropping systems, such as orchards and forests, germplasm quality may be more important than cultural practices, because rotation and tillage cannot be used as regularly (Agrios, 1988; Cook, 1993). Upon these two layers, growers can further reduce pathogen pressure by considering both biological and chemical inputs.

Biologically based inputs such as microbial fungicides can be used to interfere with pathogen activities. Registered biofungicides are generally labeled with short reentry intervals and pre-harvest intervals, giving greater flexibility to growers who need to balance their operational requirements and disease management goals. When living microorganisms are introduced, they may also augment natural beneficial populations to further reduce the damage caused by targeted pathogens (Cook, 1993; Heydari et al., 2004; Shah-Smith and Burns, 1997).

Applying to the Infection Site

Application directly to the infection court at a high population level to swamp the pathogen (inundate application), seed coating and treatment with antagonistic fungi and bacteria, e.g., *Trichoderma harzianum* and *Psudomonas fluorescens*(Cook, 1993; Heydari and Misaghi, 2003; Heydari et al., 2004), antagonists applied to fruit for protection in storage, e.g., *Pseudomonas fluorescens* (De Capdeville et al., 2002; El-Ghaouth et al., 2000; Janisiewicz and Korsten, 2002; Janisiewicz and Peterson, 2004) and application to soil at the site of seed placement (Heydari and Misaghi, 2003). These types of applications are the most commonly used procedures which have resulted in the successful control of several fungal plant pathogens.

One place application: in this procedure, biocontrol microorganisms are applied at one place (each crop year), but at lower populations which then multiply and spread to other plant parts and give protection (augmentative application) against fungal pathogens. An Example of this method is Plant Growth Promoting Rhizobacteria (PGPR) and atoxigenic *Aspergillus flavus* on wheat seed scattered on the soil to spread to cotton flowers where they displace aflatoxin producing strains of A. flavus and fungal antagonists added to soil (Islam et al., 2005; Kloepper et al., 2004).

Occasional application: One time or occasional application maintains pathogen populations below threshold levels. In theory, parasites of the pathogen, or hypovirulent (disease carrying) strains of the pathogen, might be used and not require yearly repetition (e.g., hypovirulent strains of the chestnut blight pathogen) in which host plant is inoculated with attenuated strains of pathogenic that protects the host plant against the virulent strains of pathogen (Milgroom and Cortesi, 2004).

BIOCONTROL OF DIFFERENT FUNGAL PATHOGENS

Microorganisms naturally present in the plants ecosystem will help reduce disease potential or disease damage, but only if they are allowed to grow vigorously (Cook, 1993). They accomplish these tasks by competing with the pathogens for food sources, producing metabolites that inhibit the growth of the pathogens and physically eliminating the pathogens from the plant by occupying the space and sites first. Microorganisms not naturally present in plant environment can be introduced in an attempt to control diseases (Cook, 1993). This can be done by application of organic materials that contain natural microbial populations such as composts or natural microbial populations added to them including natural organic fertilizers with microbial supplements. In both cases, the products must be applied prior to disease development as they are preventive and not curative (Baker, 1987; Cook, 1993). Natural organic fertilizers should be used for their nutritional value (nitrogen and potassium) and not for any possible secondary effects.

Fungal plant pathogens are very diverse and cause diseases on different parts of plants such as root, stem, leaf, fruit, etc. In this section, application of biological control strategies for controlling fungal diseases on different parts of plants will be discussed.

The majority of research on biocontrol of fungal diseases have focused on soil borne diseases rather than foliar or post-harvest. According to the results of numerous research projects, several fungal and bacterial biocontrol agents have been used as seed and soil application to reduce the incidence of plant diseases caused by soil borne fungal pathogens (Cook, 1993; Heydari, 2007; Heydari et al., 2004; Heydari and Misaghi, 2003; Lo et al., 1995, 1996, 1997; McSpadden-Gardener, 2001; Naraghi et al., 2004; Ramette et al., 2003; Scheuerell et al., 2005). Since many plant pathogens can spread readily in the foliar parts, control of these diseases requires both suppression of initial plant infection and reduction of the infection rate (Lo et al., 1997). Granular applications of strain 1295-22 of *Trihoderma harzianum* has been shown to significantly inhibit disease severity of some plant diseases during the initial stage of disease development, most likely by reducing levels of the pathogen inoculum in the soil (Lo et al., 1995, 1996, 1997). It is apparent, therefore, that soil applications alone cannot effectively control the foliar phases of this disease.

Additives have been commonly used with fungicides to improve efficacy and they also may enhance the ability of biocontrol agents to reduce plant diseases. For example, it was reported that seed treatment using 10% Pelgel with solid matrix priming markedly enhanced the efficacy of Trichoderma strains to control Pythium sp. on various crops (Lo et al., 1997). Research has indicated that for control of multiple fungal plant diseases, greater control was obtained when Triton X-100 was included than when no additives, Pelgel, or Tween 20 were used (Lo et al., 1997). The use of specific surfactants with Trichoderma strains seems essential to obtain levels of control equivalent to those achieved with chemical fungicides. Detergents such as Triton X-100 may have several functions in biocontrol systems. They may slow the growth of pathogens more than that of the biocontrol agents or they may enhance wetting and adhesion of spores to infection courts (Lo et al., 1997). In preliminary experiments, both Tween 20 and Triton X-100 slowed the growth of both *T. harzianum* and the pathogens, but the ratio of the growth rates of *T. harzianum* and pathogens was greater with Triton X-100 than with Tween 20 (Lo et al., 1997).

Living organisms, in addition to yielding a large quantity of biomass of the bioprotectant fungus, must perform effectively in each application. To examine this, different spore formulations of *Trihoderma harzianum* were compared in a study for controlling plant diseases (Lo et al., 1996). It was found that all formulations provided equivalent levels of control, indicating that the method of spore production may not be a key factor in the efficacy of this fungal biocontrol agent in controlling these diseases (Lo et al., 1996). To predictably and successfully use biological control agents for fungal disease control, it is critical that their biology and ecology be more completely understood. Therefore, effective antagonists must become established in plant ecosystems and remain active against target pathogens during periods favorable for plant infection.

Broadcast application of granules of Trichoderma to control plant diseases has resulted in establishment of stable and effective populations of plants in soils (Lo et al., 1995, 1996, 1997). Similarly, it was shown that the populations of *T. harzianum* in soils treated with spray applications were as high as those in soils treated with granular formulations (Lo et al., 1996). Population levels of strain 1295-22 in about 5x10 5 cfu g-1 of soil significantly reduced Pythium blight, root rot and brown patch diseases (Lo et al., 1997). However, spray applications, even though resulted in numerically similar levels of root colonization, did not provide the same benefit. This may reflect the differences in inoculum potential of granules versus spray applications. Granules are applied as a several-millimeter-diameter particle that is completely colonized by the fungus. Conidial inoculum, on the other hand, is much smaller and would therefore be expected to possess lower inoculum potential than the granular formulation (Lo et al., 1997).

Conversely, in greenhouse and field experiments, it was found that *Trihoderma harzianum* significantly reduced some foliar phases of plant diseases when spray applications of conidial suspensions containing Triton X-100 were used (Lo et al., 1995, 1996, 1997). Weekly spray applications were as effective as the standard (monthly) fungicide applications. These results indicate that the efficacy of *T. harzianum* against plant diseases, especially those involving secondary infections, is very strongly affected by the method of application (Lo et al., 1997).

The ability to survive on the plant phylloplane is also a desirable trait for strains of fungal and bacterial antagonists used as biocontrol agents against foliar diseases (Lo et al., 1997). Spray applications of strain 1295-22 of *T. harzianum* has resulted in disease suppressive population levels on leaf (Lo et al., 1997). These populations were sufficient to suppress Pythium root rot, brown patch and dollar spot over the entire season. Thus, *T. harzianum* 1295-22 may possess a measure of phylloplane competence on the plants. The ideal biocontrol strategy attempts to introduce or promote the activity of biocontrol agents only when and where they are needed or are most effective and minimizes wasteful application of inoculum to non-target habitats (Lo et al., 1997). Thus, for effective delivery, it is necessary to consider plant–pathogen–antagonist interactions in terms of time and space.

Pythium, Rhizoctonia and *Sclerotinia* are important soil borne pathogens of many plant species and their survival structures in soil serve as primary inoculum. Consequently, suppression of the initial inoculum will be the first step in managing these pathogens (Lo et al., 1997). The granular application of biocontrol agents should be followed by monthly spray applications to suppress foliar phases of these diseases. Inhibition of the secondary infection and dissemination of these pathogens is also important for disease management (Lo et al., 1997). Monthly spray applications of *T. harzianum* could provide a second step in protection of plant foliage from attack by preventing these pathogens from initially infecting leaves and by reducing the spread of disease or other methods of inoculum dissemination. Finally, results of Lo et al. (1997) study have indicated that it will be necessary to apply weekly sprays for highly effective control of these pathogens under severe disease conditions.

In addition to Trihoderma and other fungal antagonists, several antagonistic bacterial species including Pseudomonas fluorescens, P. putida,P. aerofaciens, Burkholderia cepacia, Bacillus subtillis, B. Polymyxa and B. cerrues have also been used successfully in biological control of different soil borne fungal diseases (Heydari et al., 1997, 2004, 2007; Heydari and Misaghi, 2003; Kloepper et al., 2004; Leeman et al., 1995;Shahraki et al., 2009; Shishido et al., 2005; Weller and Cook, 1983; Zaki et al., 1998). By application of these bacterial antagonists, various fungal pathogens including Rhizoctonia solani, Fusarium moxysporium, F. solani, Verticillium dahliae, Gaummannomuces graminis and soil borne diseases caused by them such as seed rot, damping-off, root rot, vascular wilt and take-all have been biologically controlled on major agricultural crops including cotton, sugar beet, wheat, rice and different vegetables.

Although the majority of biological control research have been concentrated on soil borne fungal diseases, a number of studies have focused on fungal pathogens causing diseases and disorders in above-ground parts of plants (Kessel et al., 2005; Khodakaramian et al., 2008; Kovach et al., 2000; Milgroom and Cortesi, 2004; Smith et al., 1993). For example, Anderson et al. (2004) studied the possibility of biological control of fungal pathogens in the phylosphere and proposed that it may be possible to reduce the incidence and development of these diseases using fungal and bacterial antagonists.

In another study, biological control of powdery mildew disease on different crops using antagonistic fungi was investigated and it was found that biocontrol-active microorganisms can potentially be applied against this very important foliar diseases. Botrytis cinera which is the causal agent of gray mold on many plants (Agrios, 1988) was successfully controlled by the use of biocontrol-active microorganism on strawberry (Kovach et al., 2000). In another study conducted by Smith et al. (1993) biological control of cotton leak of cucumber caused by a fungal foliar pathogen was studied. It was found that

Bacillus cerrues, a bacterial antagonist was capable of reducing the incidence of the disease significantly (Smith et al., 1993).

Another example of using biocontrol-active microorganisms against foliar fungal pathogen is the study in which chestnut blight was successfully controlled by the virus that infects Cryphonectria parasitica, the fungal causal agent of the disease through the mechanism of hypovirulence, a reduction in pathogenicity of the pathogen. This phenomenon has resulted in control of the chestnut blight in many places (Milgroom and Cortesi, 2004). However, the interactions of virus, fungus, tree and environment play very important role in the success of disease control.

In addition to soil borne and foliar diseases some studies have also tested the efficacy of biocontrol-active microorganisms on post harvest fungal pathogens which cause losses to fruits and vegetables during post harvest and storage periods (Janisiewicz and Korsten, 2002). Spray applications of fungal and bacterial antagonists have resulted in significant reduction in the infection caused by some fungal pathogens in the storage.

THE USE OF COMPOST AS BIOFERTILIZER

Research data and observations in nurseries have shown that addition of composted organic matter to potting mixes results in suppression of soil borne fungal diseases (McKellar and Nelson, 2003; Paulitz and Belanger, 2001). The concentration of suppressive microorganisms in compost amended substrates is very high, but greatly reduced in soils or potting mixes after the amendment (McKellar and Nelson, 2003; Paulitz and Belanger, 2001). As a result, predictive disease suppression models have been developed based on the composition and concentration of microbial biomass.

The effectiveness of composts in suppression of soil borne diseases is dependent on heat kill, organic matters decomposition, recolonization of compost by suppressive microorganisms following heat kill and physical and chemical factors (McKellar and Nelson, 2003). Although previous works have focused on plant soil borne diseases, current research indicates that potting mixes containing composted organic materials which also have been inoculated with Trichoderma hamatum can be effective as a biocontrol alternative to foliar fungicides; however, the mechanism of this systemic type of induced resistance is not yet understood (McKellar and Nelson, 2003). Although the growers have traditionally relied on aged pine bark and composted biosolids to provide the potential for disease suppression, research indicates that composted animal manure have the potential to replace some of these components, but a consistent quantity and quality of these materials will need to be incorporated (McKellar and Nelson, 2003; Paulitz and Belanger, 2001). The maturity (stability) of the composted manure and its salinity largely determine its ability to induce suppression.

COMMERCIALIZATION OF BIOCONTROL

Commercial use and application of biological disease control have been slow mainly due to their variable performances under different environmental conditions in the field (Fravel, 2005; Mercier and Lindow, 2001; Paulitz and Belengar, 2001; Wang et al., 2003). Many biocontrol agents perform well in the laboratory and green house conditions but fail to do so in the field. This problem can only be solved by better understanding of the environmental parameters that affect biocontrol agents (Fravel, 2005; Mercier and Lindow, 2001; Paulitz and Belengar, 2001; Wang et al., 2003). In addition to this problem, there has also been relatively little investment in the development and production of commercial formulation of biocontrol-active microorganisms probably due to the cost of developing, testing, registering and marketing of these products (Heydari et al., 2007; Ardakani et al., 2009).

Biological control agents are generally formulated as wetable powders, dusts, granules and aqueous or oil-based liquid products using different mineral and organic carriers (Ardakani et al., 2009).

Currently in the market, a number of biologically based products are being sold for the control of fungal plant diseases (Ardakani et al., 2009). A growing number of companies are also developing new products that are in the process of registration. Many of these companies are small, privately owned firms with a limited product-line. Others are publicly traded and have substantial capitalization values. In addition, larger companies with more diverse product lines that include a variety of agrochemicals and biotechnological products have played a significant role in the development and marketing of products for the control of plant pathogens (Ardakani et al., 2009).

Biocontrol products are either marketed as stand-alone products or formulated as mixtures with other microbials. Some products with biocontrol properties may not be registered, but are sold instead as plant strengtheners or growth promoters without any specific claims regarding disease control (Ardakani et al., 2009). To help improve the global market perception of biopesticides as effective products, the biopesticide Industry Alliance is establishing a certification process to ensure industry standards for efficacy, quality and consistency. To improve commercial use and application of biological disease control it is extremely important to emphasize and concentrate on several factors including training of growers, formulation of biocontrol microorganisms and studying the role of environmental factors.

FUTURE OUTLOOK

Biological control really developed as an academic discipline during the 1970s and is now a mature science supported in both the public and private sector. Research related to biological control is published in many different scientific journals, particularly those related to plant pathology and entomology. Additionally, there are some academic journals specifically devoted to this disipline. In the United States, research funds for the biological control are provided primarily by several USDA programs. These include the Section 406 programs, regional IPM grants, Integrated Organic Program, IR-4 and several programs funded as part of the National Research Initiative (Bloom et al., 2003). Monies also exist to stimulate the development of commercial ventures through the small business innovation research programs. Such ventures are intended to be conduits for academic research that can be used to develop new companies (Spadaro and Gullino, 2005).

Much has been learned from the biological control research conducted over the past forty years. But, in addition to learning the lessons of the past, biocontrol researchers need to look forward to define new and different questions, the answers to which will help facilitate new biocontrol technologies and applications. Currently, fundamental advances in computing, molecular biology, analytical chemistry and statistics have led to new research aimed at characterizing the structure and functions of biocontrol agents, pathogens and host plants at the molecular, cellular and ecological levels (Spadaro and Gullino, 2005). Some of the research criteria that will advance our understanding of biological control and the conditions under which it can be most fruitfully applied are as follows:

Ecology of Antagonistic Microbes

Ecological factors play very important roles in the performance and activity of biocontrol-active microorganisms. In this regard, the following criteria need to be clarified and studied:

- The distribution of fungal pathogens and their antagonists in the environment
- The optimum conditions in which biocontrol microorganisms exert their suppressive capacities
- The response of native and introduced populations to different management practices
- The determinants factor of successful colonization and expression of biocontrol traits

- The components and dynamics of plant host defense induction

Application Methods

In regard with application strategies still there are some areas which should be investigated and developed for the enhancement of the effectiveness of biocontrol microorganisms. These areas are as follows:

- The search for more effective strains or strain variants for current applications
- The use of genetic engineering of microbes and plants for enhancing biocontrol application methods
- The development of proper formulations to enhance activities of known biocontrol agents

Introducing New Strains and Mechanisms

Since fungal plant pathogens are very diverse and their pathogenicity is different on host plants, it is therefore very important to look for new and novel biocontrol microorganisms with different mechanisms. In this regard, the following criteria need further investigation:

- The use of previously uncharacterized microbes as biological control agents
- Study on the roles of other genes and gene products which are involved in pathogen suppression
- The efficacy of using novel strain combinations in comparison with individual agents
- Study on the signal molecules of plant and microbial origin which regulate the expression of biocontrol traits by different agents

Integrated Pest Management

Since the ultimate goal of biological control of plant diseases is to assist the growers to combat and control plant pathogens in the field which is the real agricultural environment, it is therefore important to practically integrate biocontrol strategies into agricultural system. In this regard, the following criteria should be considered and followed carefully:

- Selection of production systems that can most benefit from biocontrol for disease management
- Application and use of biocontrol strategies which best fit with other IPM system components

- Development of effective biocontrol-cultivar combinations by plant breeders

Research and development: Nowadays, growers are interested in reducing dependence on chemical inputs, so biological controls (defined in the narrow sense) can be expected to play an important role in Integrated Pest Management (IPM) systems (Jacobsen et al., 2004). Good agricultural practices (GAP) including appropriate site selection, crop rotations, tillage, fertility and water management, provide the foundation for successful pest management by providing a fertile growing environment for the crop. The use of disease-resistant varieties, developed through conventional breeding or genetic engineering, provides the next line of defense. However, such measures are not always sufficient to be productive or economically sustainable. In such cases, the next step would be to deploy biorational controls of diseases. These include BCAs, introduced as inoculants or amendments, as well as active ingredients directly derived from natural origins and having a low impact on the environment and non-target microorganisms (Guetsky et al., 2001; Jacobsen et al., 2004).

If these foundational options are not sufficient to ensure plant health and/ or economically sustainable production, then less specific and less harmful synthetic chemical toxins can be used to ensure productivity and profitability. With the growing interest in reducing chemical inputs, companies involved in the manufacturing and marketing of BCAs should experience continued growth. However, stringent quality control measures must be adopted so that farmers get quality products. New, more effective and stable formulations also will need to be developed.

Most fungal pathogens are susceptible to one or more biocontrol strategies, but practical implementation on a commercial scale has been constrained by a number of factors. Cost, convenience, efficacy and reliability of biological controls are important considerations, but only in relation to the alternative disease control strategies. Cultural practices (e.g., good sanitation, soil preparation and water management) and host resistance can go a long way towards controlling many diseases, so biocontrol should be applied only when such agronomic practices are insufficient for effective disease control. As long as petroleum is cheap and abundant, the cost and convenience of chemical pesticides will be difficult to surpass. However, if the infection court or target pathogen can be effectively colonized using inoculation, the ability of the living organism to reproduce could greatly reduce application costs.

In general, although, regulatory and cultural concerns about the health and safety of specific classes of pesticides are the primary economic drivers promoting the adoption of biological control strategies in urban and rural landscapes (Timms-Wilson et al., 2004). Self-perpetuating biological controls (e.g., hypovirulence of the chestnut blight pathogen) are also needed for control of diseases in forested and rangeland ecosystems where high application rates over larger land areas are not economically feasible. In terms of efficacy and reliability, the greatest successes in biological control have been achieved in situations where environmental conditions are most controlled or predictable and where biocontrol agents can preemptively colonize the infection court (Fravel, 2005). Monocyclic, soilborne and post-harvest fungal diseases have been controlled effectively by biological control agents that act as bioprotectants (i.e., preventing infections). Specific applications for the high value crops targeting specific diseases (e.g., downy mildew, powdery mildew and several other fungal diseases) have also been adopted (Kessel et al., 2005). As research unravels the various conditions needed for successful biocontrol of different fungal diseases, the adoption of BCAs in IPM systems is bound to increase in the years ahead.

CONCLUSIONS

Due to the serious environmental and health problems that wide spread use of chemical pesticides has created in the world, search for alternative safe methods is unavoidable. Biological control of plant diseases has been the subject of numerous research projects in recent years (Bargabus et al., 2004; Benhamou, 2004; Chisholm et al., 2006; Heydari, 2007; Islam et al., 2005). There is a growing demand for biologically based pest management practices. Recent surveys of both conventional and organic growers indicate an interest in using biocontrol products suggesting that the market potential of biocontrol products will increase in the future (Joshi and Gardener, 2006). Application of different biological control strategies has been successful in the greenhouse industry and continues to increase (Jacobsen et al., 2004). An upswing in commercial interests has also developed in the past few years and prospects for increased growth are positive. The Biopesticide Industry Alliance has formed and it is now actively promoting the value and efficacy of biopesticides (including those that control fungal plant pathogens). Clearly, the future success of the biological control industry will depend on innovative business management, product marketing, extension education and research (Timms-Wilson et al., 2004; Joshi and Gardener, 2006).

Increased demand for organic products in home gardening activities by using non-chemical methods has enlarged the market for biocontrol products. The field of plant pathology will contribute substantially to making the 21st century the ages of biotechnology by the development of innovative biocontrol strategies. A variety of research questions remain to be fully answered about the nature of biological control and the means to most effectively manage it under production conditions. Advanced molecular techniques are now being used to characterize the diversity, abundance and activities of microbes that live in and around plants, including those that significantly impact plant health (Joshi and Gardener, 2006). Still, much remains to be learned about the microbial ecology of both plant pathogens and their microbial antagonists in different agricultural systems. Fundamental work remains to be done on characterizing the different mechanisms by which organic amendments reduce plant disease including those caused by fungal pathogens. More studies on the practical aspects of mass production and formulation need to be undertaken to make new biocontrol products stable, effective, safer and more cost-effective.

Fungal pathogens are among the most important factors that cause serious damages and losses to plants. Harmful impacts of the chemical pesticides on the environment and non-target organisms have clearly been documented. The need for the development of non-chemical alternative strategies to protect plants against plant diseases including fungal pathogens is therefore clear. Biological control using fungal and bacterial antagonists to manage plant diseases seems to be a promising alternative strategy and have successfully been applied to control some diseases on different plants and crops. Biocontrol strategies may also be used to manage other plant diseases including foliar ones. Some of the important factors that affect the efficacy of microbial biocontrol agents in controlling plant diseases which should carefully be considered include method of application, formulation of biocontrol microorganisms and timing of application. Various composts and organic amendments as other means of biological control have also been tested on some plants and proven to be promising.

There are many products composed of living organisms, primarily bacteria and fungi, being sold that claim they will increase plant health. However, for any material to be considered a biological fungicide the Environmental Protection Agencies and Organizations must register it (Bloom et al., 2003). This registration indicates that the safety of the product to humans, non-humans (fish for example) and the environment has been determined. Materials that have not been approved should be used with caution.

Complete elimination of chemical pesticides for controlling plant pests and diseases in modern agriculture may be impossible, but a logical reduction in their application is absolutely feasible. To have a sustainable agricultural system with minimum contamination and risks to the environment, a combination of all available methods should be applied to manage pest problems and this can be achieved by Integrated Pest Management (IPM). Implementation of IPM strategies may be the safest solution for management of pest problems including fungal diseases in every cropping system and with no doubt biological control is one of the most important components of Integrated Pest Management which can lead us toward a sustainable agricultural system in the future.

REFERENCES

1.   Agrios, N.A., 1988. Plant Pathology. 3rd Edn., Academic Press, USA., pp: 220-222.

2.   Anderson, A.J., P. Habibzadegah-Tari and C.S. Tepper, 1988. Genetic studies on the role of an agglutinin-in root colonization byPseudomonas putida. Applied Environ. Microbiol., 54: 375-380.

3.   Anderson, L.M., V.O. Stockwell and J.E. Loper, 2004. An extracellular protease of-Pseudomonas fluorescens inactivates antibiotics of Pantoea agglomerans. Phytopathology, 94: 1228-1234.

4.   Ardakani, S., A. Heydari, N. Khorasani, R. Arjmandi and M. Ehteshami, 2009. Preparation of new biofungicides using antagonistic bacteria and mineral compounds for controlling cottn seedling damping-off disease. J. Plant Prot. Res., 49: 49-55.

5.   Audenaert, K., T. Pattery, P. Cornelis and M. Hofte, 2002. Induction of systemic resistance to-Botrytis cinerea in tomato by Pseudomonas aeruginosa 7NSK2: role of salicylic acid, pyochelin and pyocyanin. Mol. Plant-Microbe. Interact., 15: 1147-1156.

6.   Baker, K.F., 1987. Evolving concepts of biological control of plant pathogens. Annu. Rev. Phytopathol., 25: 67-85.

7.   Bankhead, S.B., B.B. Landa, E. Lutton, D.M. Weller and B.B. Gardener, 2004. Minimal changes in rhizosphere population structure following root colonization by wild type and transgenic biocontrol strains. FEMS Microb. Ecol., 49: 307-318.

8.   Bargabus, R.L., N.K. Zidack, J.E. Sherwood and B.J. Jacobsen, 2002. Characterization of systemic resistance in sugar beet elicited by a non-pathogenic, phylosphere-colonizing Bacillus mycoides, biological control agent. Physiol. Mol. Plant Pathol., 61: 289-298.

9.   Bargabus, R.L., N.K. Zidack, J.E. Sherwood and B.J. Jacobsen, 2004. Screening for the identification of potential biological control agents that induce systemic acquired resistance in sugar beet. Biol. Control, 30: 342-350.

10.   Benhamou, N. and I. Chet, 1997. Cellular and molecular mechanisms involved in the interaction between Trichoderma harzianum andPythium ultimum. Applied Environ. Microbiol., 63: 2095-2099.

11. Benhamou, N., 2004. Potential of the mycoparasite, Verticillium lecanii, to protect citrus fruit-against Penicillium digitatum, the causal agent of green mold: A comparison with-the effect of chitosan. Phytopathol., 94: 693-705.

12. Berg, G., A. Krechel, M. Ditz, R.A. Sikora, A. Ulrich and J. Hallmann, 2005. Endophytic and ectophytic potato-associated bacterial communities differ in structure and antagonistic function against plant pathogenic fungi. FEMS Microbiol. Ecol., 51: 215-229.

13. Biermann, B. and R.G. Linderman, 1983. Use of vesicular-arbuscular mycorrhizal roots, intraradical vesicles and extraradical vesicles as inoculum. New Phytol., 95: 97-105.

14. Bloom, B., R. Ehlers, S. Haukeland-Salinas, H. Hoddanen and K. Jung et al., 2003. Biological control agents: Safety and regulatory policy. BioControl, 48: 477-484.

15. Bull, C.T., K.G. Shetty and K.V. Subbarao, 2002. Interactions between Myxobacteria, plant pathogenic fungi and biocontrol agents. Plant Dis., 86: 889-896.

16. Chisholm, S.T., G. Coaker, B. Day and B.J. Staskawicz, 2006. Host-microbe interactions: Shaping the evolution of the plant immune response. Cell, 124: 803-814.

17. Cook, R.J., 1993. Making greater use of introduced microorganisms for biological control of plant pathogens. Annu. Rev. Phytopathol., 31: 53-80.

18. De Capdeville, G., C.L. Wilson, S.V. Beer and J.R. Aist, 2002. Alternative disease control agents induce resistance to blue mold in harvested Red Delicious apple fruit. Phytopathology, 92: 900-908.

19. De Meyer, G. and M. Hofte, 1997. Salicylic acid produced by the rhizobacterium Pseudomonas aeruginosa 7NSK2 induces resistance to leaf infection by Botrytis cinera on bean. Phytopathology, 87: 588-593.

20. Duchesne, L.C., 1994. Role of Ectomycorrhizal Fungi in Biocontrol. In: Mycorrhizae and Plant Health, Pfleger, F.L. and R.G. Linderman (Eds.). APS Press, Paul, MN., pp: 27-45.

21. El-Ghaouth, A., J.L. Smilanick, G.E. Brown, A. Ippolito, M. Wisniewski and C.L. Wilson, 2000. Application of Candida saitoana and glycolchitosan for the control of post harvest diseases of apple and citrus fruit under semi-commercial conditions. Plant Dis., 84: 243-248.

22. Elad, Y. and R. Baker, 1985. Influence of trace amounts of cations and siderophore-producing pseudomonads on chlamydospore germination of

Fusarium oxysporum. Phytopathology, 75: 1047-1052.

23. Fitter, A.H. and J. Grabaye, 1994. Interactions between mycorrhizal fungi and other soil organisms. Plant Soil, 159: 123-132.

24. Fravel, D., 2005. Commercialization and implementation of biocontrol. Annu. Rev. Phytopathol., 43: 337-359.

25. Garcia-Garrido, J.M. and J.A. Ocampo, 1989. Effect of VA mycorrhizal infection of tomato on damage caused by Pseudomonas syringae. Soil Biol. Biochem., 21: 165-167.

26. Guetsky, R., D. Shtienberg, Y. Elad and A. Dinoor, 2001. Combining biocontrol agents to reduce the variability of biological control. Phytopathology, 91: 621-627.

27. Haas, D. and G. Defago, 2005. Biological control of soil-borne pathogens by fluorescent pseudomonads. Nat. Rev. Microbiol., 3: 307-319.

28. Harman, G.E., C.R. Howell, A. Viterbo, I. Chet and M. Lorito, 2004. Trichoderma species-opportunistic, avirulent plant symbionts. Nat. Rev. Microbiol., 2: 43-56.

29. Heydari, A. and I.J. Misaghi, 1998. Biocontrol activity of Burkholderia cepacia against Rhizoctonia solani in herbicide-treated soils. Plant Soil, 202: 109-116.

30. Heydari, A. and I.J. Misaghi, 1999. Herbicide-Mediated Changes in the Populations and Activity of Root Associated Microorganisms: a Potential Cause of Plant Stress. In: Hand Book of Plant and Crop Stress, Pessarakli, M. (Ed.). 2nd Edn., Marcel Dekker Press, New York.

31. Heydari, A. and I.J. Misaghi, 2003. The role of rhizosphere bacteria in herbicide-mediated increase in Rhizoctonia solani-induced cotton seedling damping-off. Plant Soil, 257: 391-396.

32. Heydari, A., 2007. Biological Control of Turfgrass Fungal Diseases. In: Turfgrass Management and Physiology, Pessarakli, M. (Ed.). CRC Press, Florida, USA.

33. Heydari, A., H. Fattahi, H.R. Zamanizadeh, N.H. Zadeh and L. Naraghi, 2004. Investigation on the possibility of using bacterial antagonists for biological control of cotton seedling damping-off in greenhouse. Applied Entomol. Phytopathol., 72: 51-68.

34. Heydari, A., I.J. Misaghi and G.M. Balestra, 2007. Pre-emergence herbicides influence the efficacy of fungicides in controlling cotton seedling damping-off in the field. Int. J. Agric. Res., 2: 1049-1053.

35. Heydari, A., I.J. Misaghi and W.B. McCloskey, 1997. Effects of three soil-applied herbicides on populations of plant disease suppressing

bacteria in the cotton rhizosphere. Plant Soil, 195: 75-81.

36. Hoitink, H.A.J. and M.J. Boehm, 1999. Biocontrol within the context of soil microbial communities: A substrate-dependent phenomenon. Annu. Rev. Phytopathol., 37: 427-446.

37. Howell, C.R. and R.D. Stipanovic, 1980. Suppression of Pythium ultimum induced damping-off of cotton seedlings by Pseudomonas fluorescens and its antibiotic pyoluteorin. Phytopathology, 70: 712-715.

38. Howell, C.R., R.C. Beier and R.D. Stipanovic, 1988. Production of ammonia by Enterobacter cloacae and its possible role in the biological control of Pythium pre-emergence damping-off by the bacterium. Phytopathology, 78: 1075-1078.

39. Islam, M.T., Y. Hashidoko, A. Deora, T. Ito and S. Tahara, 2005. Suppression of damping-off-disease in host plants by the rhizoplane bacterium Lysobacter sp. strain SB-K88 is-linked to plant colonization and antibiosis against soilborne peronosporomycetes. Appl. Environ. Microbiol., 71: 3786-3796.

40. Jacobsen, B.J., N.K. Zidack and B.J. Larson, 2004. The role of Bacillus-based biological control agents in integrated pest management systems: Plant diseases. Phytopathology, 94: 1272-1275.

41. Janisiewicz, W.J. and D.L. Peterson, 2004. Susceptibility of the stem pull area of mechanically harvested apples to blue mold decay and its control with a biocontrol agent. Plant Dis., 88: 662-664.

42. Janisiewicz, W.J. and L. Korsten, 2002. Biological control of post harvest disease of fruits. Ann. Rev. Phytopathol., 40: 411-441.

43. Joshi, R. and B.B.M. Gardener, 2006. Identification and characterization of novel genetic markers associated with biological control activities in Bacillus subtilis. Phytopathology, 96: 145-154.

44. Kageyama, K. and E.B. Nelson, 2003. Differential inactiviation of seed exudates stimulation of-Pythium ultimum sporangium germination byEnterobacter cloacae influences biological control efficacy on different plant species. Applied Environ. Microbiol., 69: 1114-1120.

45. Katska, V., 1994. Interrelationship between vesicular-arbuscular mycorrhiza and rhizosphere microflora in apple replant disease. Biol. Plant, 36: 99-104.

46. Keel, C., C. Voisard, C.H. Berling, G. Kahir and G. Defago, 1989. Iron sufficiency is a prerequisit for suppression of tobacco black root rot byPseudomonas fluorescnes strain CHA0 under gnotobiotic contiditions. Phytopathology, 79: 584-589.

47. Kerry, B.R., 2000. Rhizosphere interactions and the exploitation of microbial agents for the biological control of plant-parasitic nematodes. Ann. Rev. Phytopathol., 38: 423-441.

48. Kessel, G.J.T., J. Kohl, J.A. Powell, R. Rabbinge and W. van der Werf, 2005. Modeling spatial characteristics in the biological control of fungi at the leaf scale: Competitive substrate colonization by Botrytis cinerea and the saprophytic antagonist Ulocladium atrum. Phytopathology, 95: 439-448.

49. Khodakaramian, G., A. Heydari and G.M. Balestra, 2008. Evaluation of pseudomonads bacterial isolates in biological control of citrus bacterial canker disease. Int. J. Agric. Res., 3: 268-272.

50. Kloepper, J.W., C.M. Ryu and S. Zhang, 2004. Induce systemic resistance and promotion of plant growth by Bacillus spp. Phytopathology, 94: 1259-1266.

51. Kloepper, J.W., J. Leong, M. Teintze and M.N. Schroth, 1980. Pseudomonas siderophores: A mechanism explaining disease suppression in soils. Curr. Microbiol., 4: 317-320.

52. Kovach, J., R. Petzoldt and G.E. Harman, 2000. Use of honey and bumble bees to disseminate Trichoderma harzianum 1295-22 to strawberries for Botrytis control. Biol. Control, 18: 235-242.

53. Lafontaine, P.J. and N. Benhamou, 1996. Chitosan treatment: An emerging strategy for enhancing resistance of greenhouse tomato plants to infection by Fusarium oxysporum f.sp. radicis-lycopersici. Biocont. Sci. Technol., 6: 111-124.

54. Leclere, V., M. Bechet, A. Adam, J.S. Guez and B. Wathelet et al., 2005. Mycosubtilin overproduction by Bacillus subtilis BBG100 enhances the organism's antagonistic and biocontrol activities. Applied Environ. Microbiol., 71: 4577-4584.

55. Leeman, M., J.A. van Pelt, F.M. den Ouden, M. Heinsbroek, P.A.H.M. Bakker and B. Schippers, 1995. Induction of systemic resistance byPseudomonas fluorescens in radish cultivars differing in susceptibility to Fusarium wilt, using novel bioassay. Eur. J. Plant Pathol., 101: 655-664.

56. Linderman, R.G., 1994. Role of VAM Fungi in Biocontrol. In: Mycorrhizae and Plant Health. Pfleger, F.L. and R.G. Linderman (Eds.). American Phytological Society, St. Paul. MN, APS Press, USA., ISBN-10: 0890541582.

57. Lo, C.T., E.B. Nelson and G.E. Harman, 1995. Improved biocontrol efficacy of Trichoderma harzianum for foliar phases of turf diseases by

use of spray applications. Plant Dis., 81: 1132-1138.

58. Lo, C.T., E.B. Nelson and G.E. Harman, 1996. Biological control of turfgrass diseases with a rhizosphere component strain of Trichoderma harzianum. Plant Dis., 82: 736-741.

59. Lo, C.T., E.B. Nelson and G.E. Harman, 1997. Biological control of Pythium, Rhizoctonia and Sclerotinia infected diseases of turfgrass withTrichoderma harzianum. Phytopathology, 84: 1372-1379.

60. Loper, J.E. and J.S. Buyer, 1991. Siderophores in microbial interactions of plant surfaces. Mol. Plant-Microbe Interact, 4: 5-13.

61. Maurhoffer, M., C. Hase, J.P. Matraux and G. Defago, 1994. Induction of systemic resistance of tobacco to tobacco necrosis virus by the root-colonizing Pseudomonas fluorescens strains CHAO: Influence of the gacA gene and pyoverdine production. Phytopathology, 84: 139-146.

62. McKellar, M.E. and E.B. Nelson, 2003. Compost-induced suppression of Pythium damping-off is mediated by fatty-acid-metabolizing seed-colonizing microbial communities. Applied Environ. Microbiol., 69: 452-460.

63. McSpadden-Gardener, B.B. and D.M. Weller, 2001. Changes in populations of rhizosphere bacteria associated with take-all disease of wheat. Applied Environ. Microbiol., 67: 4414-4425.

64. Mercier, J. and S.E. Lindow, 2001. Field performance of antagonistic bacteria identified in a novel assay for biological control of firelight. Biol. Control, 22: 66-71.

65. Meziane, H., I. van der Sluis, L.C. van Loon, M. Hofte and P.A.H.M. Bakker, 2005. Determinants of Pseudomonas putida WCS358 involved in inducing systemic resistance in plants. Mol. Plant Pathol., 6: 177-185.

66. Milgroom, M.G. and P. Cortesi, 2004. Biological control of chestnut blight with hypovirulence:a critical analysis. Annu. Rev. Phytopathol., 42: 311-338.

67. Moyne, A.L., R. Shelby, T.E. Cleveland and S. Tuzun, 2001. Bacillomycin D: An iturin with antifungal activity against Aspergillus flavus. J. Applied Microbiol., 90: 622-629.

68. Naraghi, L., A. Heydari, A. Karimi-Roozbehani and D. Ershad, 2004. Isolation of Talaromyces flavus from Golestan cotton fields and its antagonistic effects on Verticillium dahliae the causal agent of cotton verticillium wilt. Iran. J. Plant Pathol., 39: 109-2004.

69. Ordentlich, A., Y. Elad and I. Chet, 1988. The role of chitinase of Serratia marcescens in biocontrol of Sclerotium rolfsii. Phytopathology, 78: 84-

87.

70. Paulitz, T.C. and R.R. Belanger, 2001. Biological control in greenhouse systems. Annu. Rev. Phytopathol., 39: 103-133.

71. Phillips, A.D., T.C. Fox, M.D. King, T.V. Bhuvaneswari and L.R. Teuber, 2004. Microbial products trigger amino acid exudation from plant roots. Plant Physiol., 136: 2887-2894.

72. Press, C.M., J.E. Loper and J.W. Kloepper, 2001. Role of iron in rhizobacteria mediated induced systemic resistance of cucumber. Phytopathology, 91: 593-598.

73. Ramette, A., Y. Moenne-Loccoz and G. Defago, 2003. Prevalence of fluorescent-pseudomonads producing antifungal phloroglucinols and/or hydrogen cyanide in soils naturally suppressive or conducive to tobacco root rot. FEMS Microb. Ecol., 44: 35-43.

74. Ryu, C.M., M.A. Farag, C.H. Hu, M.S. Reddy, J.W. Kloepper and P.W. Pare, 2004. Bacterial volatiles induce systemic resistance in Arabidopsis. Plant Physiol., 134: 1017-1026.

75. Scheuerell, S.J., D.M. Sullivan and W.F. Mahaffee, 2005. Suppression of seedling damping-off caused by Pythium ultimum, P. irregulare andRhizoctonia solani in container media amended with a diverse range of pacific northwest compost sources. Phytopathology, 95: 306-315.

76. Schouten, A, G. van den Berg, V. Edel-Hermann, C. Steinberg and N. Gautheron et al., 2004. Defense responses of Fusarium oxysporum to 2,4-DAPG, a broad spectrum antibiotic produced by Pseudomonas fluorescens. Mol. Plant-Microbe Interact, 17: 1201-1211.

77. Shah-Smith, D.A. and R.G. Burns, 1997. Shelf-life of a biocontrol Pseudomonas putida applied to the sugar beet seeds using commercial coatings. Biocontrol Sci. Technol., 7: 65-74.

78. Shahraki, M., A. Heydari and N. Hassanzadeh, 2009. Investigation of antibiotic, siderophore and volatile metabolites production by Bacillus and Pseudomonas bacteria. Iran. J. Biol., 22: 71-85.

79. Shanahan, P., D.J. O'Sullivan, P. Simpson, J.D. Glennon and F. O`Gara, 1992. Isolation of 2,4-diacetylphloroglucinol from a fluorescent pseudomonad and investigation of physiological parameters influencing its production. Applied Environ. Microbiol., 58: 353-358.

80. Shishido, M., C. Miwa, T. Usami, Y. Amemiya and K.B. Johnson, 2005. Biological control efficiency of fusarium wilt of tomato by nonpathogenic F. oxysporum Fo-B2 in different environments. Phytopathology, 95: 1072-1080.

81.	Silva, H.S.A., R.D.S. Romeiro, D. Macagnan, B.D.A. Halfeld-Vieira, M.C.B. Pereira and A. Mounteer, 2004. Rhizobacterial induction of systemic resistance in tomato plants: Non-specific protection and increase in enzyme activities. Biol. Control, 29: 288-295.

82.	Smith, K.P., M.J. Havey and J. Handelsman, 1993. Suppression of cottony leak of cucumber with Bacillus cereus strain UW85. Plant Dis., 77: 139-142.

83.	Spadaro, D. and M.L. Gullino, 2005. Improving the efficacy of biocontrol agents against soilborne pathogens. Crop Prot., 24: 601-613.

84.	Tari, P.H. and A.J. Anderson, 1988. Fusarium wilt suppression and agglutinability of Pseudomonas putida. Applied Environ. Microbiol., 54: 2037-2041.

85.	Thomashow, L.S. and D.M. Weller, 1988. Role of a phenazine antibiotic from Pseudomonas fluorescens in biological control ofGaeumannomyces graminis var. Tritici. J. Bacteriol., 170: 3499-3508.

86.	Thomashow, L.S., D.M. Weller, R. Bonsall and L.S. Pierson, 1990. Production of de antibiotic phenazine 1-carboxylic acid by fluorescent pseudomonad species in the rhizosphere of wheat. Applied Environ. Microbiol, 56: 908-912.

87.	Thomashow, L.S., R.F. Bonsall and D.M. Weller, 2002. Antibiotic Production by Soil and Rhizosphere Microbes in situ. 2nd Edn., ASM Press, Washington DC., pp: 638-647.

88.	Timms-Wilson, T.M., K. Kerry and J.B. Mark, 2004. Risk assessment for engineered bacteria used in biocontrol of fungal disease in agricultural crops. Plant Soil, 266: 57-60.

89.	Vallad, G.E. and R. Goodman, 2004. Systemic acquired resistance and induced systemic resistance in conventional agriculture. Crop Sci., 44: 1920-1934.

90.	Van Dijk, K. and E.B. Nelson, 2000. Fatty acid competition as a mechanism by which Enterobacter cloacae suppresses Pythium ultimumsporangium germination and damping-off. Applied Environ. Microbiol., 66: 5340-5347.

91.	Van Loon, L.C., P.A.H.M. Bakker and C.M.J. Pieterse, 1998. Systemic resistance induced by rhizosphere bacteria. Annu. Rev. Phytopathol., 36: 453-483.

92.	Van Peer, R. and B. Schippers, 1992. Lipopolysaccharides of plant-growth promoting Pseudomonas spp. strain WCS 417r induce resistance in carnation to Fusarium wilt. Neth. Eur. J. Plant Pathol., 98: 129-139.

93. Van Wees, S.C., C.M. Pieterse, A. Trijssenaar, Y.A. van't Westende, F. Hartog and L.C. van Loon, 1997. Differential induction of systemic resistance in Arabidopsis by biocontrol bacteria. Mol. Plant-Microbe Interact, 10: 716-724.

94. Wang, H., S.F. Hawang, K.F. Chang, G.D. Turnbull and R.J. Howard, 2003. Suppression of important pea diseases by bacterial antagonists. Bio Control, 48: 447-460.

95. Weller, D.M. and R.J. Cook, 1983. Suppression of take-all of wheat by seed treatments with fluorescent pseudomonads. Phytopathology, 73: 463-469.

96. Weller, D.M., J.M. Raaijmakers, B.B.M. Gardener and L.S. Thomashow, 2002. Microbial populations responsible for specific soil suppressiveness to plant pathogens. Annu. Rev. Phytopathol., 40: 309-348.

97. Wilhite, S.E., R.D. Lumsden and D.C. Strancy, 2001. Peptide synthetase gene in Trichoderma virens. Applied Environ. Microbiol., 67: 5055-5062.

98. Yoshihisa, H., S. Zenji, H. Fukushi, K. Katsuhiro, S. Haruhisa and S. Takahito, 1989. Production of antibiotics by Pseudomonas cepacia as an agent for biological control of soilborne Plant pathogens. Soil Biol. Biochem., 21: 723-728.

99. Zaki, K., I.J. Misaghi, A. Heydari and M.N. Shatla, 1998. Control of cotton seedling damping-off in the field by Burkholdria cepacia. Plant Dis., 82: 291-293.

Chapter 6

COMMON PROCESSES IN PATHOGENESIS BY FUNGAL AND OOMYCETE PLANT PATHOGENS, DESCRIBED WITH GENE ONTOLOGY TERMS

Shaowu Meng[1,3], Trudy Torto-Alalibo[2], Marcus C Chibucos[2,4], Brett M Tyler[2] and Ralph A Dean[1]

[1]Fungal Genomics Laboratory, Center for Integrated Fungal Research, North Carolina State University, Raleigh, NC 27695, USA

[2]Virginia Bioinformatics Institute, Virginia Polytechnic and State University, Blacksburg, VA 24061, USA

[3]Hayes Laboratory, Lineberger Comprehensive Cancer Center, School of Medicine, CB# 7295, University of North Carolina at Chapel Hill, Chapel Hill, NC 27599-7295, USA

[4]Institute for Genome Sciences, University of Maryland School of Medicine, Baltimore, MD 21201, USA

ABSTRACT

Plant diseases caused by fungi and oomycetes result in significant economic losses every year. Although phylogenetically distant, the infection processes by these organisms share many common features. These include dispersal of an infectious particle, host adhesion, recognition, penetration, invasive growth, and lesion development. Previously, many of these common processes did not have corresponding Gene Ontology (GO) terms. For example, no GO terms existed to describe processes related to the appressorium, an important structure for infection by many fungi and oomycetes. In this mini-review, we identify common features of the pathogenic processes of fungi and oomycetes and create a pathogenesis model using 256 newly developed and 38 extant GO terms, with an emphasis on the appressorium and signal transduction. This set of standardized GO terms provides a solid base to further compare and contrast the molecular underpinnings of fungal and oomycete pathogenesis.

COMMON PATHOGENESIS PROGRAMS OF FUNGI AND OOMYCETES

Oomycetes, although phylogenetically very distant, share many common morphological and physiological features with the true fungi [1–3]. For example, they have similar filamentous, branching, indeterminate bodies, and they acquire nutrition by secreting digestive enzymes and then absorbing the resultant breakdown products. More importantly, fungi and oomycetes share a unique capability compared with other microbial pathogens, namely that they are able to breach cuticles of host plants and establish infection rapidly [4]. Consequently, both are causal agents of many destructive plant diseases and are responsible for significant economic losses every year.

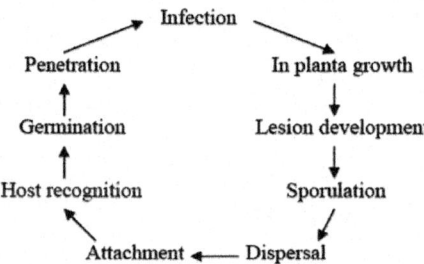

Figure 1: A generalized diagram displaying infection and disease cycle caused by fungi and oomycetes.

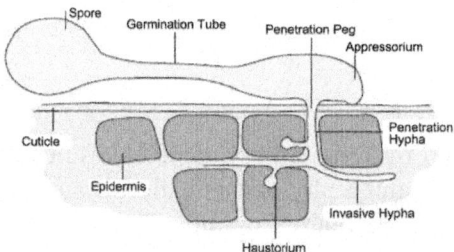

Figure 2: The infection process in fungal and oomycete pathogens. Modified by permission from Schumann, G. L., 1991, Plant diseases: Their biology and social impact, American Phytopathological Society, St. Paul, MN.

In this review, we summarize common mechanisms of pathogenesis displayed by oomycetes and fungi. Pathogenesis by a fungus or oomycete is a complex process. Briefly, it includes the following steps: dispersal and arrival of an infectious particle (usually a spore of some kind) in the vicinity of the host, adhesion to the host, recognition of the host (which may occur prior to adhesion), penetration into the host, invasive growth within the host, lesion development in the host, and finally production of additional infectious particles [5, 6] (see Figures 1, 2).

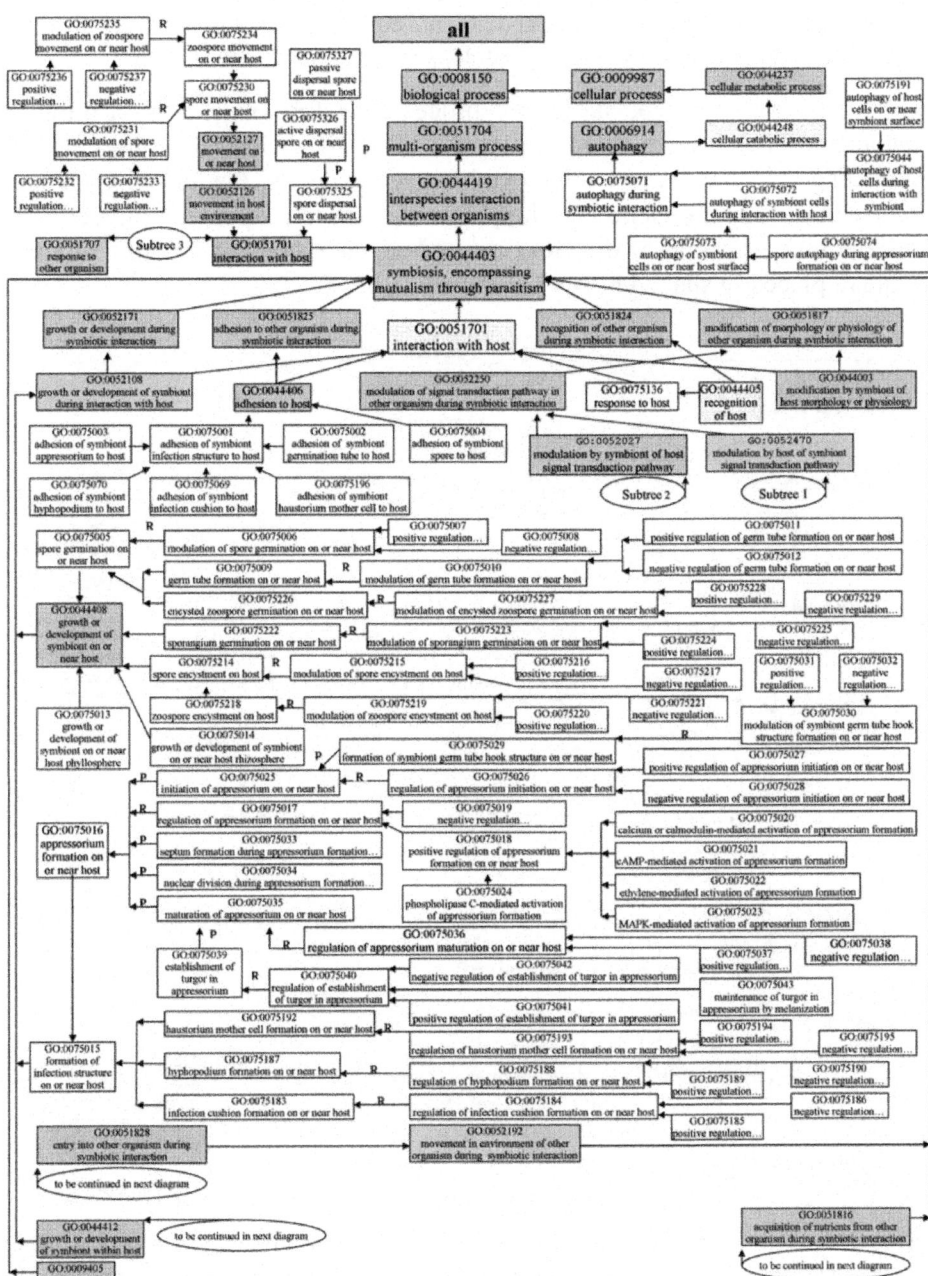

Figure 3: Gene Ontology terms for processes related to infection and disease (Part 1). Subtree 1 and 2 are depictured in Figure 5, and Subtree 3 is depictured in Figure 6. Shaded boxes indicate pre-existing GO terms, and unshaded boxes represent GO terms developed under the PAMGO project. "R" indicates "regulates relationship",

"P" indicates "part of relationship", and null indicates "is a relationship" (see the Gene Ontology website at http://www.geneontology.org for further information).

In order to describe the entire process, we formulate a description of pathogenesis using standardized terms from the Gene Ontology (GO), including 256 new terms developed by members of the PAMGO (Plant-Associated Microbe Gene Ontology) consortium http://pamgo.vbi.vt.edu, an official interest group of the GO Consortium, as well as 38 extant GO terms that are placed in shaded boxes in Figures 3, 4,5, 6.

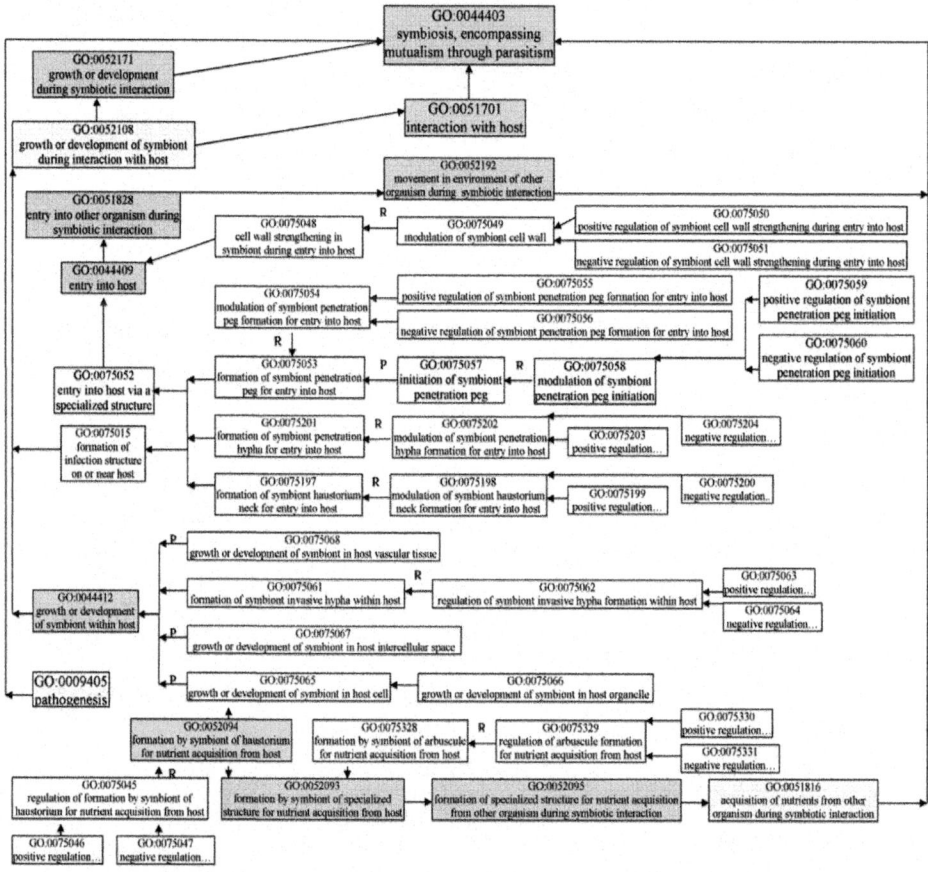

Figure 4: Gene Ontology terms for processes related to infection and disease (Part 2). Shaded boxes indicate pre-existing GO terms, and unshaded boxes represent GO terms developed under the PAMGO project. "R" indicates "regulates relationship", "P" indicates "part of relationship", and null indicates "is a relationship" (see the Gene Ontology website at http://www.geneontology.org for further information).

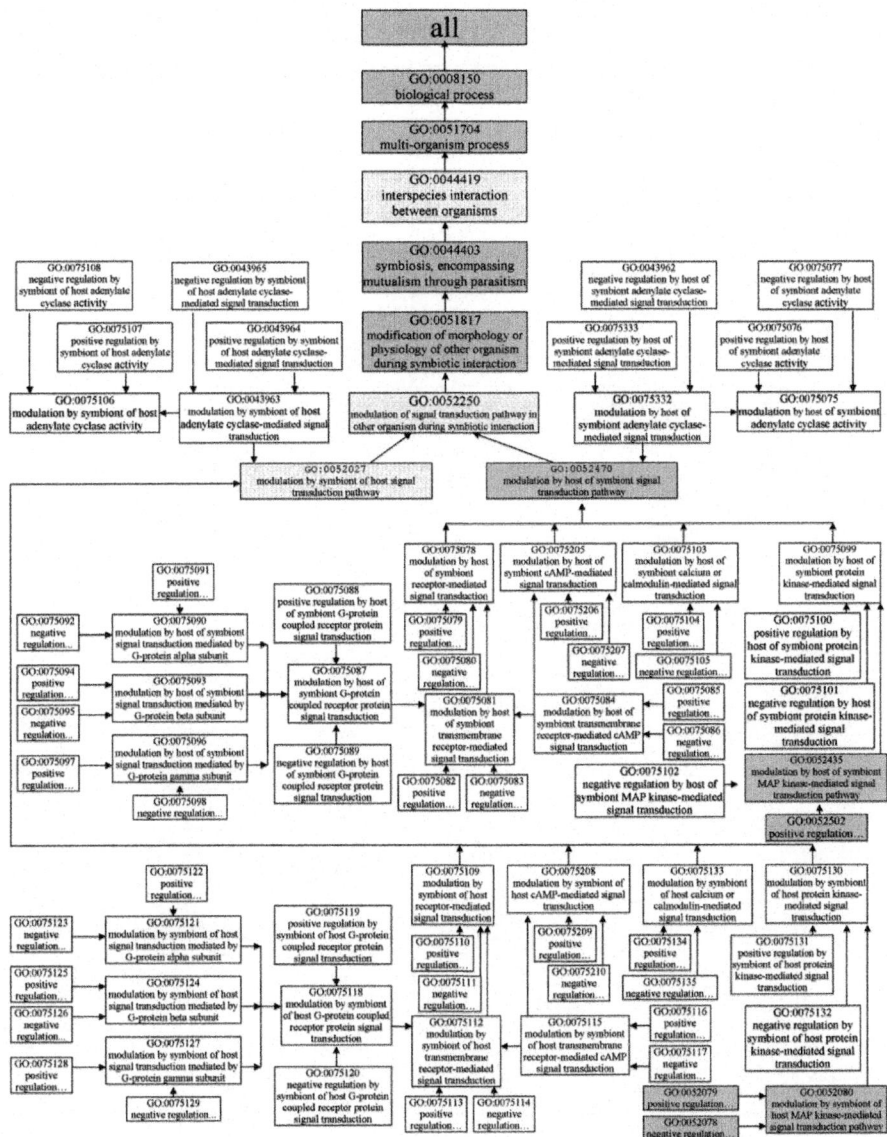

Figure 5: Gene Ontology terms for signal transduction processes related to infection and disease (Part 1). Subtree 1 consists of GO terms intending to annotate host gene products that stimulate signal transduction in symbiont. Subtree 2 represents the opposite perspective of Subtree 1. Shaded boxes indicate pre-existing GO terms, and unshaded boxes represent GO terms developed under the PAMGO project.

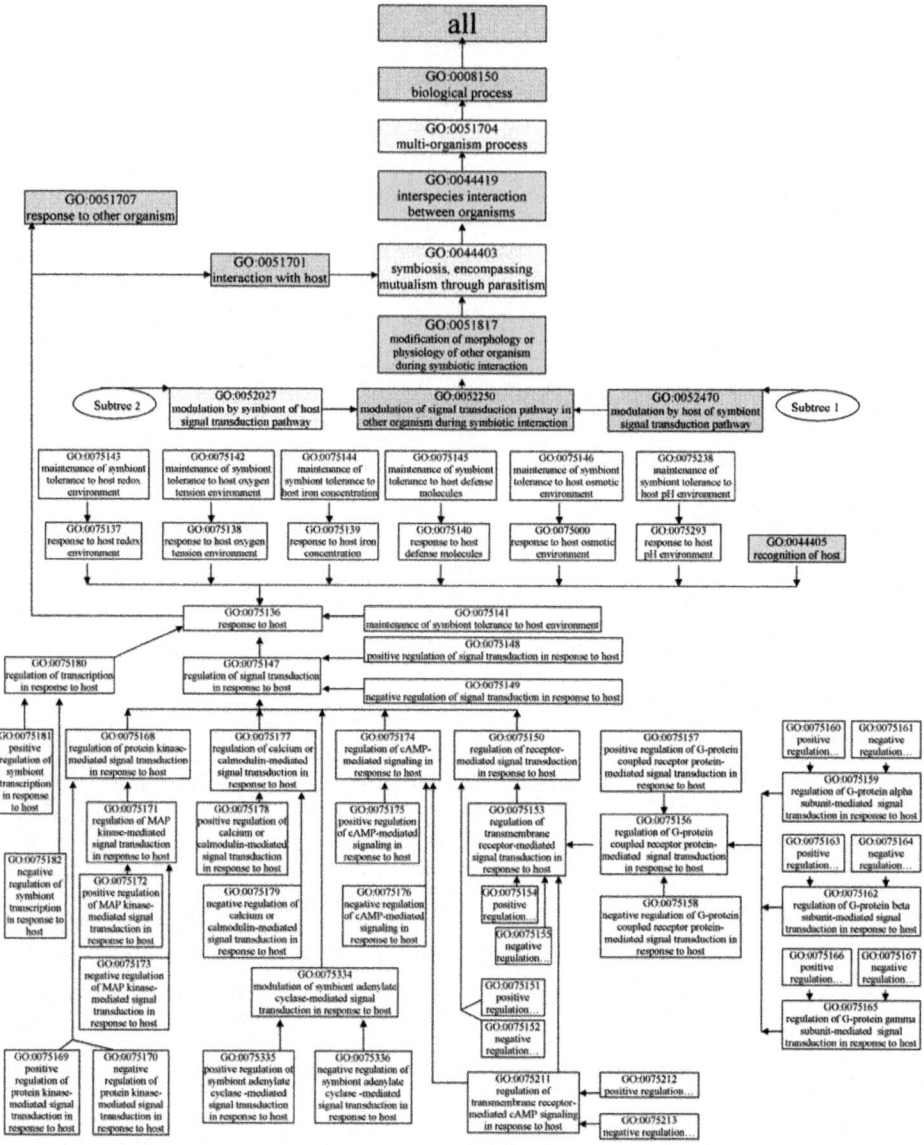

Figure 6: Gene Ontology terms for signal transduction processes related to infection and disease (Part 2). Subtree 3 consists of GO terms intending to annotate symbiont gene products that stimulate signal transduction in symbiont in response to host. Shaded boxes indicate pre-existing GO terms, and unshaded boxes represent GO terms developed under the PAMGO project.

All of the 256 new terms are placed within GO under the node "GO ID 0044403 symbiosis, encompassing mutualism through parasitism" (note the original, broad definition of "symbiosis" used in GO, which specifies use of the words "symbiont" and "host" as the smaller and larger, respectively, of the symbiotic interactants). Among these new terms, every term that starts with "modulation" or "regulation" has two child terms, one is "positive regulation of...", and the other is "negative regulation of..." Note that these child terms are general GO terms; "position regulation," for example, includes induction, upregulation, stimulation, etc. Four diagrams (see Figures 3, 4, 5, 6) encompassing the 256 new and 38 extant GO terms explicitly depict our description of pathogenesis, with an emphasis on appressorium formation and signal transduction. More details about each step are presented in the following sections.

Spore Dispersal

Dispersal of spores is the most common process to initiate new infections [5], though direct infection by hyphae may occur. An example of the latter is the spread of ectomycorrhizal basidiomycetes in forest soils. Dispersal mechanisms can be grouped into two types: one is passive dispersal by wind, water or animal [7], and the other is active dispersal such as shooting ascospores through the boundary layer of air surrounding the fruiting body by forcible discharge [8]. Similarly, spores can be grouped into two types according to their motility. In fungi non-motile spores include sexual spores such as ascospores, rust urediniospores, sclerotia and conidiospores, while non-motile oomycete spores include oospores, sporangiospores and conidia. Motile spores with flagella, called zoospores, are ubiquitous among oomycetes and are also found in chytrid fungi [9]. Additionally, spores vary in requirements for dormancy. Some spores, such as zoospores, must encyst and differentiate to acquire qualities of dormancy before they become true spores [10]. Different spores vary in the length of the dormancy period; for example, some ascospores and oospores show extended dormancy, while others such as zoospores and ascomycete conidia are usually short-lived.

Three new GO terms including the term "GO ID 0075325 spore dispersal on or near host" were developed under the node "GO ID 0051701 ...interaction with host" to describe the mechanisms of spore dispersal. New terms describing active or passive dispersal mechanisms were placed as children of "spore dispersal on or near host" (see Figure 3).

Eight new GO terms describing spore motility were listed under the node "GO ID 0052127 movement on or near host". The term "GO ID 0075230 spore movement on or near host" is central to these eight terms with "GO ID 0075234 zoospore movement on or near host" as the principal child term (see Figure 3).

Similarly, eight new GO terms were added under "GO ID 0044408 growth or development of symbiont on or near host" to describe spore encystment. The term "GO ID 0075214 spore encystment on host" is central to these eight terms, with the term "GO ID 0075218 zoospore encystment on host" as the main child term (see Figure 3).

These 19 new terms are appropriate for annotating gene products known to be involved in spore dispersal. For example, inhibition of the vacuolar H+-ATPase by potassium nitrate causes a reduction in vacuole expulsion in zoospores of the oomycete *Phytophthora nicotianae* and leads to premature encystment [11]. Thus, H+-ATPase negatively regulates zoospore encystment and can be annotated with the new term «GO ID 0075221 negative regulation of zoospore encystment on host».

Adhesion to the Host

Adhesion of spores to the host involves physical and chemical processes [3]. Typically, when spores reach the surface of a host tissue, they attach via adhesion molecules [5]. A germination tube then emerges from the spore or the encysted zoospore (see Figure 2). From the germination tube, a growth hypha or an infection structure such as an appressorium [12–16] develops, which also becomes firmly attached to the host surface via adhesion molecules. A variety of other infection structures such as hyphopodia [17–19], haustorium mother cells [20–23], or infection cushions [24] are generated by fungal pathogens after germinating on the host surface. These all serve a common function of facilitating the pathogen's entry into the host tissue. It should be noted that the sporangia of many oomycetes may germinate directly to form an infection hypha, or else in the presence of abundant water they may differentiate, through specialized cleavage vesicles, into 10–30 zoospores that can individually disperse to initiate sites of infection [25].

Seven new GO terms under the parent, "GO ID 0044406 adhesion to host", were developed to describe in detail the biological process of adhesion to a host. The term "GO ID 0075001 adhesion of symbiont infection structure to host" is central to this section. Among the seven terms, five terms that describe adhesion of a specific infection structure, including appressorium, hyphopodium, haustorium mother cell, infection cushion, or germination tube, are children of "adhesion of symbiont infection structure to host" (see Figure 3).

To describe spore germination on or near host tissue, 16 new terms under the parent, "GO ID 0044408 growth or development of symbiont on or near host", were developed. The 16 terms cover spore germination, sporangium germination, encysted zoospore germination, and germ tube formation. The term "GO ID 0075005 spore germination on or near host" is central to this section. Major relationships among the sixteen terms are shown in Figure 3.

The 23 new GO terms in this section are useful for annotating pathogen gene products involved in adhesion to host tissue. For example, Car (cyst-germination-specific acidic repeat) proteins of the oomycete *Phytophthora infestans* are transiently expressed during germination of cysts (i.e., encysted zoospores) and during formation of appressoria, and they are localized at the surface of germlings. These proteins have considerable sequence homology to mucins and have an internal octapeptide repeat with the mucin consensus sequence TTYAPTEE. Therefore, like mucins, Car proteins should serve as a mucous cover protecting the germling and assisting in adhesion to the leaf surface [26]. Thus, the Car proteins can be annotated with the new terms «GO ID 0075226 encysted zoospore germination on or near host» and «GO ID 0075001 adhesion of symbiont infection structure to host», using the GO evidence code ISS (Inferred from Sequence or Structural Similarity).

Signal Transduction during Recognition of the Host

Signal transduction is an integral component of the host recognition process. Examples include protein kinase-mediated signal transduction [27], receptor-mediated signal transduction [28], G-protein coupled receptor protein signal transduction, G-protein subunit-mediated signal transduction [29], cAMP-mediated signal transduction [30], calcium or calmodulin-mediated signal transduction [31], and adenylate cyclase-mediated signal transduction [12].

In order to adequately describe signal transduction during symbiont interaction with its host, three sets of new terms were developed. Signal transduction pathways involved in the recognition between host and symbiont are generally quite extensively characterized and consequently 127 new terms were developed.

The first set of new terms is intended for annotation of host gene products that stimulate symbiont signal transduction (see Subtree 1, which includes terms under the node "GO ID 0052470 modulation by host of symbiont signal transduction pathway" in Figure 5). This set has 37 new terms. Five of these terms describing different types of signal transduction pathways are children of "GO ID 0052470" (see Subtree 1 in Figure 5).

The second set of new terms is intended for annotation of symbiont gene products that stimulate host signal transduction (see Subtree 2, which includes terms under the node "GO ID 0052027 modulation by symbiont of host signal transduction pathway" in Figure 5). This set has 36 new GO terms and has the same structure as the first set (see Subtree 2 in Figure 5). The terms in the second set are essentially the converse of the terms in the first set. For example, the term "GO ID 0075130 modulation by symbiont of host protein kinase-mediated signal transduction" in the second term set has a complementary term "GO ID 0075099 modulation by host of symbiont protein kinase-mediated signal transduction" in the first term set.

The third set of new terms is intended for annotation of symbiont gene products that stimulate symbiont signal transduction in response to the host (see Subtree 3, which includes terms under the nodes "GO ID 0051701 interaction with host" and "GO ID 0051707 response to other organism" in Figure 6). This set has 56 new GO terms. The new term "GO ID 0075136 response to host" is central to the 56 new terms. This term has 10 child terms that describe 10 symbiont responses to host organisms, such as "GO ID 0075147 regulation of signal transduction in response to host", "GO ID 0075140 response to host defense molecules", and "GO ID 0075180 regulation of transcription in response to host" etc. (see Figure 6). Eight of the 10 terms have their own child and lower level offspring terms, and each of those "response" terms has a child term such as "maintenance of symbiont tolerance to host ..." (see details in Figure 6).

The term "GO ID 0075147 regulation of signal transduction in response to host" has five children to describe different types of signal transduction, similar to the five child terms of "GO ID 0052470 modulation by host of symbiont signal transduction pathway" in the first set. Each of the five terms has child terms for positive regulation and negative regulation.

The three sets of new GO terms can be used to explicitly describe genes of signal transduction pathways involved in host recognition. For instance, the PMK1 gene of the rice blast fungus *Magnaporthe oryzae* encodes a mitogen-activated protein kinase (MAPK), which is a key component in the MAPK signaling cascade and is involved in appressorium formation and infectious growth [32]. Thus, the PMK1 protein can be annotated with the term «GO ID 0075171 regulation of MAP kinase-mediated signal transduction in response to host». Note that this gene product would not be annotated with "GO ID 0052435 modulation by host of symbiont MAP kinase-mediated signal transduction pathway» since this latter GO term is reserved to annotate host gene products. Similarly, this protein should not be annotated with "GO ID 0052080 modulation by symbiont of host MAP kinase-mediated signal transduction pathway" since PMK1 belongs to the symbiont's and not the host's signaling transduction pathway.

In addition, the modulation terms have children that describe more specific kinds of signal transduction. For example, «GO ID 0075168 regulation of protein kinase-mediated signal transduction in response to host» has a child

«GO ID 0075171 regulation of MAP kinase-mediated signal transduction in response to host» (see details in Figure 6).

Penetration into the Host

Pathogens have evolved several mechanisms that include structural and/or enzymatic components in order to enter into their plant hosts [5]. Many fungi, such as *Alternaria alternata, Colletotrichum graminicola, M. oryzae, Pyrenophora teres*, and many oomycetes, such as *P. infestans* and *Phytophthora cinnamomi*, develop appressoria to directly penetrate plant cuticles [13, 33–38]. An appressorium is a highly specialized structure that differentiates from the end of a symbiont germ tube. It is a swollen, dome-shaped or cylindrical organ, from which a narrow penetration peg emerges to rupture the plant cuticle and cell wall [33]. The penetration peg extends and forms a penetration hypha to penetrate through the epidermal cells and emerge into the underlying tissue [34, 35]. In some instances, penetration is driven by astoundingly high turgor pressures within the appressoria [36, 38]. Generation of turgor is due in part to a thick inner cell wall layer of melanin in mature appressoria. In other instances, cell wall degrading enzymes may play a primary role in or may facilitate the penetration process [39–41]. Appressoria produced by some fungi, such as rust fungi, do not penetrate directly through the cuticle, but gain entry through stomata [42].

Sixty-four new GO terms were developed to describe the biological process of penetration into the host, and they form two groups. The first group includes 43 new GO terms related to infection structures established on the outside of the host tissue, such as appressoria, hyphopodia, infection cushions, and haustorium mother cells. The second group has 21 new terms related to specialized structures that directly pierce the surface of the host, for example penetration pegs, penetration hyphae, and haustorium necks.

All of the 43 terms in the first group are children or lower level offspring of "GO ID 0052108 growth or development of symbiont during interaction with host". The core of this group is "GO ID 0075015 formation of infection structure on or near host". Twenty-eight terms in this group are related to appressorium formation. In particular, five of the 28 terms describe in detail the process of appressorium formation, namely "GO ID 0075025 initiation of appressorium on or near host", "GO ID 0075034 nuclear division during appressorium formation on or near host", "GO ID 0075033 septum formation during appressorium formation on or near host", "GO ID 0075035 maturation of appressorium on or near host", and "GO ID 0075017 regulation of appressorium formation on or near host" (see details in Figure 3). Besides the child term "GO ID 0075016 appressorium formation on or near host", the term

"GO ID 0075015 formation of infection structure on or near host" has three more detailed child terms: "GO ID 0075192 haustorium mother cell formation on or near host", "GO ID 0075187 hyphopodium formation on or near host", and "GO ID 0075183 infection cushion formation on or near host" (see details in Figure 3).

All of the 21 terms in the second group are children or lower level offspring of "GO ID 0044409 entry into host". The core of this group is "GO ID 0075052 entry into host via a specialized structure", which has three child terms related to penetration peg, penetration hypha, or haustorium neck for entry into the host (see details in Figure 4).

The 64 new terms can be used to annotate the gene products of penetration-related genes. For example, genes involved in melanin biosynthesis in the rice blast fungus, such as ALB1, RSY1 and BUF1, are required for appressorium function since mutants lacking these genes make appressoria, but are unable to penetrate susceptible rice leaves [43]; these can be annotated with the term "GO ID 0075053 formation of symbiont penetration peg for entry into host".

Invasive Growth within the Host

After successful penetration, invasive hyphae are formed that ramify through the host tissue [44, 45]. In some cases, special structures, such as a haustorium or an arbuscle, are formed in host cells for the symbiont to absorb nutrition [22,23].

To describe invasive growth, 15 new GO terms were developed that are children or lower level offspring of "GO ID 0044412 growth or development of symbiont within host". The term "GO ID 0075065 growth or development of symbiont in host cell" has two children, "GO ID 0052094 formation by symbiont of haustorium for nutrient acquisition from host" and "GO ID 0075066 growth or development of symbiont in host organelle". Additionally, arbuscules produced by mycorrhizal fungi are a type of structure functionally similar to haustoria, and thus "GO ID 0075328 formation by symbiont of arbuscule for nutrient acquisition from host" is a sibling of "GO ID 0052094" (see details in Figure 4).

The 15 new GO terms in this section meet the need to annotate pathogen genes that are involved in invasive growth. For example, the *MST12* gene in the rice blast fungus *M. grisea* was found to regulate infectious growth but not appressorium formation [46]. In particular, no obvious defects in vegetative growth, conidiation, or conidia germination were observed in*MST12* deletion mutants. Also, *MST12* mutants produce typical dome-shaped and melanized appressoria. When inoculated through wound sites, *MST12* mutants fail to cause spreading lesions and appear to be defective in infectious growth. As a result, *MST12* mutants are nonpathogenic [46]. Thus, the *MST12* gene can be annotated with the term "GO ID 0075061 formation of symbiont invasive hypha within host».

Lesion Development in the Host

The eventual result of infection in most cases is lesion development. A lesion can be defined as any abnormality involving any tissue or organ due to any disease or any injury (cited from MedicineNet.com). Not surprisingly, there are many types of lesions including those caused by damage such as cold injury or insects' bites etc. It is difficult to define lesions objectively, as this requires a subjective judgment on what constitutes abnormal or damage and from what perspective, ranging for example from perturbation of a few cells to death of an entire tissue or organ. Similarly, formation of a lesion is not a specific process belonging to either the pathogen or the host and can be highly dependent on the environment. Therefore, at this time only one term, "GO ID 0009405 pathogenesis", is appropriate for genes involved in lesion formation.

Other New GO Terms

Six new terms were placed jointly under the nodes "GO ID 0006914 autophagy" and "GO ID 0044403 symbiosis, encompassing mutualism through parasitism". The term "GO ID 0075071 autophagy during symbiotic interaction" is the core of the six terms, and it has two complementary children, "GO ID 0075044 autophagy of host cells during interaction with symbiont" and "GO ID 0075072 autophagy of symbiont cells during interaction with host". The latter term has a child "GO ID 0075073 autophagy of symbiont cells on or near host surface", which itself has a lower level child "GO ID 0075074 spore autophagy during appressorium formation on or near host" (see details in Figure 3).

The six autophagy-related GO terms are applicable to describe the functions of several genes in fungal pathogens during symbiotic interaction. For example, formation of a functional appressorium in the rice blast fungus requires autophagic cell death of the conidium, which is controlled by the *MgATG8* gene. Deletion of *MgATG8* results in impaired autophagy, arrested conidial cell death, and a nonpathogenic fungus [14]. Thus, *MgATG8* can be annotated with the new term "GO ID 0075074 spore autophagy during appressorium formation on or near host".

CONCLUSION

Two hundred fifty-six new GO terms were developed to annotate genes or gene products involved in common pathogenic processes in fungi and oomycetes, including spore dispersal, host adhesion, recognition, penetration, and invasive growth. These new GO terms provide the opportunity to apply a standard set of terms to annotate gene products of fungi, oomycetes, and their associated hosts, as well as those of other plant-associated pathogens and their hosts. The ability to compare and contrast these annotations for widely different plant-associated microbes and their hosts, using a standardized vocabulary, will greatly facilitate the identification of unique and conserved features of pathogenesis across different kingdoms. In addition, such comparisons should provide insight into the evolution of pathogenic processes.

ACKNOWLEDGEMENTS

All authors read and approved the final manuscript. We thank Candace Collmer, Michelle Gwinn Giglio, and the editor at The Gene Ontology Consortium Jane Lomax for their comments and suggestions in developing these PAMGO terms. This work is a part of PAMGO project, which is supported by the USDA NRI-CSREES (grant number 2005-35600-16370), and the National Science Foundation (grant number EF-0523736).

This article has been published as part of *BMC Microbiology* Volume 9 Supplement 1, 2009: The PAMGO Consortium: Unifying Themes In Microbe-Host Associations Identified Through The Gene Ontology. The full contents of the supplement are available online at http://www.biomedcentral.com/1471-2180/9?issue=S1.

REFERENCES

1.   Money NP: Why oomycetes have not stopped being fungi. Mycology Research. 1998, 102 (6): 767-768. 10.1017/S095375629700556X.

2.   Latijnhouwers M, Wit PJGMD, Govers F: Oomycetes and fungi: similar weaponry to attack plants. Trends in Microbiology. 2003, 11 (10): 462-469. 10.1016/j.tim.2003.08.002.

3.   Epstein L, Nicholson RL: Adhesion and adhesives of fungi and oomycetes. Biological Adhesives. Edited by: Smith AM, Callow JA. 2006, Springer-Verlag Berlin Heidelberg

4.   Soanes DM, Richards TA, Talbot NJ: Insights from sequencing fungal and oomycete genomes: what can we learn about plant disease and the evolution of pathogenicity?. The Plant Cell. 2007, 19: 3318-3326. 10.1105/tpc.107.056663.

5.   Agrios GN: Plant pathology. 2005, Elsevier Academic Press, London, UK, Fifth

6.   Colditz F, Krajinski F, Niehaus K: Plant proteomics upon fungal attack. Plant Proteomics. Edited by: Šamaj J, Thelen J. 2007, Springer

7.   Ingold GT: Dispersal in Fungi. 1953, Clarendon Press, Oxford; Oxford University Press, New York

8.   Trail F: Fungal cannons: explosive spore discharge in the Ascomycota. FEMS Microbiol Lett. 2007, 276: 12-18. 10.1111/j.1574-6968.2007.00900.x.

9.   James TY, Letcher PM, Longcore JE, Mozley-Standridge SE, Porter D, Powell MJ, Griffith GW, Vilgalys R: A molecular phylogeny of the flagellated fungi (Chytridiomycota) and description of a new phylum (Blastocladiomycota). Mycologia. 2006, 98 (6): 860-871. 10.3852/mycologia.98.6.860.

10. Griffin DH: Fungal Physiology. 1996, Published by Wiley_Default

11. Mitchell HJ, Hardham AR: Characterisation of the water expulsion vacuole in *Phytophthora nicotianae* zoospores. Protoplasma. 1999, 206: 118-130. 10.1007/BF01279258.

12. Choi W, Dean RA: The adenylate cyclase gene MAC1 of *Magnaporthe grisea* controls appressorium formation and other aspects of growth and development. Plant Cell. 1997, 9 (11): 1973-1983. 10.1105/tpc.9.11.1973.

13. Hardham AR: The cell biology behind *Phytophthora* pathogenicity. Australasian Plant Pathology. 2001, 30: 91-98. 10.1071/AP01006.

14. Veneault-Fourrey C, Barooah MK, Egan MJ, Talbot NJ: Autophagic fungal cell death is necessary for infection by the rice blast fungus. Science. 2006, 312 (5773): 580-583. 10.1126/science.1124550.

15. Talbot NJ: Fungal genomics goes industrial. Nature Biotechnology. 2007, 25 (5): 542-543. 10.1038/nbt0507-542.

16. Grenville-Briggs LJ, Anderson VL, Fugelstad J, Avrova AO, Bouzenzana J, Williams A, Wawra S, Whisson SC, Birch PRJ, Bulone V, West PV: Cellulose synthesis in *Phytophthora infestans* is required for normal appressorium formation and successful infection of potato. The Plant Cell. 2008, 20: 720-738. 10.1105/tpc.107.052043.

17. Onyile AB, Edwards HH, Gessner RV: Adhesive material of the hyphopodia of *Buergenerula spartinae*. Mycologia. 1982, 74 (5): 777-784. 10.2307/3792864.

18. Du M, Schard CL, Nuckles EM, Vaillancourt LJ: Using mating-type gene sequences for improved phylogenetic resolution of*Collectotrichum* species complexes. Mycologia. 2005, 97 (3): 641-658. 10.3852/mycologia.97.3.641.

19. Sukno SA, García VM, Shaw BD, Thon MR: Root infection and systemic colonization of maize by *Colletotrichum graminicola*. Applied and environmental microbiology. 2008, 74 (3): 823-832. 10.1128/ AEM.01165-07.

20. Heath MC, Heath IB: Ultrastructural changes associated with the haustorial mother cell ceptum during haustorium formation in*Uromyces phaseoli* var. *vignae*. Protoplasma. 1975, 84: 297-314. 10.1007/ BF01279359.

21. Glidewell DC, Mims CW: Ultrastructure of the haustorial apparatus in the rust fungus *Kunkelia nitens*. Botanical Gazette. 1979, 140 (2): 148-152. 10.1086/337071.

22. Mendgen K, Dressler E: Culturing *Puccinia coronata* on a cell monolayer of the *Avena saliva* coleoptile. Phytopath. 1983, 108: 226-234. 10.1111/ j.1439-0434.1983.tb00583.x.

23. Wiethölter N, Horn S, Reisige K, Beike U, Moerschbacher BM: In vitro differentiation of haustorial mother cells of the wheat stem rust fungus, *Puccinia graminis* f. sp. *tritici*, triggered by the synergistic action of chemical and physical signals. Fungal Genetics and Biology. 2003, 38: 320-326. 10.1016/S1087-1845(02)00539-X.

24. Demirci E, Döken MT: Host penetration and infection by the anastomosis groups of *Rhizoctonia solani* Kühn isolated from potatoes. Tr J of Agriculture and Forestry. 1998, 22: 609-613.

25. Birch PRJ, Cooke DEL: Mechanisms of infection: Oomycetes. Encyclopedia of Plant and Crop Science. 2004, 1 (1): 697-700.

26. Görnhardt B, Rouhara I, Schmelzer E: Cyst germination proteins of the potato pathogen *phytophthora infestans* share homology with human mucins. Mol Plant-Micro Interact. 2000, 13 (1): 32-42. 10.1094/ MPMI.2000.13.1.32.

27. Zhao X, Kim Y, Park G, Xu J-R: A mitogen-activated protein kinase cascade regulating infection-related morphogenesis in *Magnaporthe grisea*. The Plant Cell. 2005, 17: 1317-1329. 10.1105/tpc.104.029116.

28. Ligterink W, Kroj T, Nieden UZ, Hirt H, Scheel D: Receptor-mediated activation of a MAP kinase in pathogen defense of plants. Science. 1997, 276 (27): 2054-2057. 10.1126/science.276.5321.2054.

29. Miwa T, Takagi Y, Shinozaki M, Yun C-W, Schell WA, Perfect JR, Kumagai H, Tamaki H: Gpr1, a putative G-protein-coupled receptor, regulates morphogenesis and hypha formation in the pathogenic fungus *Candida albicans*. Eukaryotic Cell. 2004, 3 (4): 919-931. 10.1128/ EC.3.4.919-931.2004.

30. Lafon A, Han K-H, Seo J-A, Yu J-H, d'Enfert C: G-protein and cAMP-mediated signaling in aspergilli: A genomic perspective. Fungal Genetics and Biology. 2006, 43 (7): 490-502. 10.1016/j.fgb.2006.02.001.

31. Praveen RJ, Reena G, Subramanyam C: Calmodulin-dependent protein phosphorylation during conidial germination and growth of *Neurospora crassa*. Mycol Res. 1997, 101: 1484-1488. 10.1017/S0953756297004255.

32. Xu J-R, Hamer JE: MAP kinase and cAMP signaling regulate infection structure formation and pathogenic growth in the rice blast fungus *Magnaporthe grisea*. Genes & Dev. 1996, 10: 2696-2706. 10.1101/gad.10.21.2696.

33. DeZwaan TM, Carroll AM, Valent B, Sweigard JA: *Magnaporthe grisea* Pth11p is a novel plasma membrane protein that mediates appressorium differentiation in response to inductive substrate cues. The Plant Cell. 1999, 11: 2013-2030. 10.2307/3871094.

34. Clergeot P-H, Gourgues M, Cots J, Laurans F, Latorse M-P, Pépin R, Tharreau D, Notteghem J-L, Lebrun M-H: *PLS1*, a gene encoding a tetraspanin-like protein, is required for penetration of rice leaf by the fungal pathogen *Magnaporthe grisea*. PNAS. 2001, 98 (12): 6963-6968. 10.1073/pnas.111132998.

35. Park G, Bruno KS, Staiger CJ, Talbot NJ, Xu J-R: Independent genetic mechanisms mediate turgor generation and penetration peg formation during plant infection in the rice blast fungus. Molecular Microbiology. 2004, 53 (6): 1695-1707. 10.1111/j.1365-2958.2004.04220.x.

36. Wang ZY, Jenkinson JM, Holcombe LJ, Soanes DM, Veneault-Fourrey C, Bhambra GK, Talbot NJ: The molecular biology of appressorium turgor generation by the rice blast fungus *Magnaporthe grisea*. Biochem Soc Trans. 2005, 33 (Pt 2): 384-388.

37. Blanco FA, Judelson HS: A bZIP transcription factor from *Phytophthora* interacts with a protein kinase and is required for zoospore motility and plant infection. Molecular Microbiology. 2005, 56 (3): 638-648. 10.1111/j.1365-2958.2005.04575.x.

38. Liu XH, Lu JP, Zhang L, Dong B, Min H, Lin FC: Involvement of a *Magnaporthe grisea* serine/threonine kinase gene, MgATG1, in appressorium turgor and pathogenesis. Eukaryot Cell. 2007, 6 (6): 997-1005. 10.1128/EC.00011-07.

39. Tonukari NJ, Scott-Craig JS, Walton JD: The *Cochliobolus carbonum* SNF1 gene is required for cell wall-degrading enzyme expression and virulence on maize. Plant Cell. 2000, 12: 237-248. 10.1105/tpc.12.2.237.

40. Li D, Ashby AM, Johnstone K: Molecular evidence that the extracellular

cutinase Pbc1 is required for pathogenicity of *Pyrenopeziza brassicae* on oilseed rape. Mol Plant-Micro Interact. 2003, 16: 545-552. 10.1094/MPMI.2003.16.6.545.

41. Aro N, Pakula T, Penttila M: Transcriptional regulation of plant cell wall degradation by filamentous fungi. FEMS Microbiology Reviews. 2005, 29: 719-739. 10.1016/j.femsre.2004.11.006.

42. Prats E, Llamas MJ, Jorrin J, Rubiales D: Constitutive coumarin accumulation on sunflower leaf surface prevents rust germ tube growth and appressorium differentiation. Crop science. 2007, 47: 1119-1124. 10.2135/cropsci2006.07.0482.

43. Chumley FG, Valent B: Genetic analysis of melanin-deficient, nonpathogenic mutants of *Magnaporthe grisea*. Mol Plant-Micro Interact. 1990, 3: 135-143.

44. Walker SK, Chitcholtan K, Yu YP, Christenhusz GM, Garrill A: Invasive hyphal growth: An F-actin depleted zone is associated with invasive hyphae of the oomycetes *Achlya bisexualis* and *Phytophthora cinnamomi*. Fungal Genetics and Biology. 2006, 43 (5): 357-365. 10.1016/j.fgb.2006.01.004.

45. Bassilana M, Blyth J, Arkowitz RA: Cdc24, the GDP-GTP exchange factor for Cdc42, is required for invasive hyphal growth of *Candida albicans*. Eukaryotic cell. 2003, 2 (1): 9-18. 10.1128/EC.2.1.9-18.2003.

46. Park G, Xue C, Zheng L, Lam S, Xu J-R: *MST12* regulates infectious growth but not appressorium formation in the rice blast fungus *Magnaporthe grisea*. Mol Plant-Micro Interact. 2002, 15 (3): 183-192. 10.1094/MPMI.2002.15.3.183.

Chapter 7

AN UPDATE ON POLYGALACTURONASE-INHIBITING PROTEIN (PGIP), A LEUCINE-RICH REPEAT PROTEIN THAT PROTECTS CROP PLANTS AGAINST PATHOGENS

Raviraj M. Kalunke[1], Silvio Tundo[1], Manuel Benedetti[2], Felice Cervone[2], Giulia De Lorenzo[2*] and Renato D'Ovidio[1]

[1]Dipartimento di Scienze e Tecnologie per l'Agricoltura, le Foreste, la Natura e l'Energia, Università della Tuscia, Viterbo, Italy

[2]Dipartimento di Biologia e Biotecnologie "Charles Darwin", Sapienza Università di Roma, Roma, Italy

Polygalacturonase inhibiting proteins (PGIPs) are cell wall proteins that inhibit the pectin-depolymerizing activity of polygalacturonases secreted by microbial pathogens and insects. These ubiquitous inhibitors have a leucine-rich repeat structure that is strongly conserved in monocot and dicot plants. Previous reviews have summarized the importance of PGIP in plant defense and the structural basis of PG-PGIP interaction; here we update the current knowledge about PGIPs with the recent findings on the composition and evolution of *pgip* gene families, with a special emphasis on legume and cereal crops. We also update the information about the inhibition properties of single *pgip* gene products against microbial PGs and the results, including field tests, showing the capacity of PGIP to protect crop plants against fungal, oomycetes and bacterial pathogens.

INTRODUCTION

Successful colonization of plant tissues by microbial pathogens requires the overcoming of the cell wall. To this end, pathogens produce a wide array of plant cell wall degrading enzymes (CWDEs), among which endo-polygalacturonases (PGs; EC 3.2.1.15) are secreted at very early stages of the infection process (ten Have et al., 1998). PGs cleave the α-(1–4) linkages between the D-galacturonic acid residues of homogalacturonan, the main component of pectin, causing cell separation and maceration of the host tissue. To counteract the activity of PGs, plants deploy the cell wall polygalacturonase inhibiting proteins (PGIPs) that inhibit the pectin-depolymerizing activity of PGs. No plant species or mutants totally lacking PGIP activity have been characterized so far. The structure of PGIPs is typically formed by 10 imperfect leucine-rich repeats (LRRs) of 24 residues each, which are organized to form two β-sheets, one of which (sheet B1) occupies the concave inner side of the molecule and contains residues crucial for the interaction with PGs (Di Matteo et al., 2003). In addition to PG inhibition, the interaction between PGs and PGIPs promotes the formation of oligogalacturonides (OGs), which are elicitors of a variety of defense responses (Cervone et al., 1989; Ridley et al., 2001; Ferrari et al., 2013). Since many aspects of the PGIP biology have been already summarized in previous reviews (De Lorenzo et al., 2001; De Lorenzo and Ferrari, 2002; D'Ovidio et al., 2004a; Gomathi and Gnanamanickam, 2004; Shanmugam, 2005; Di Matteo et al., 2006; Federici et al., 2006; Cantu et al., 2008;Misas-Villamil and van der Hoorn, 2008; Protsenko et al., 2008; Reignault et al., 2008; Lagaert et al., 2009), here we present an overview of the recent findings on genome composition and evolution of *pgip* gene families and on the efficacy of PGIP to limit the development of diseases caused by microbial pathogens in crop plants.

PGIP Genes and their Genomic Organization

Early characterization of a polygalacturonase-inhibiting activity was reported in 1970s (Albersheim and Anderson, 1971) and the first *pgip* gene was isolated 20 years later in French bean (Toubart et al., 1992). Since then, several PGIPs and a large number of *pgip* genes have been characterized. Up to now more than 170 complete or partial *pgip* genes from dicot and monocot plants have been deposited in nucleotide databases (e.g., http://www.ncbi.nlm.nih.gov/). Most of these genes have been identified as *pgip* genes on the basis of sequence identity but only a few of them have been shown to encode proteins with PG-inhibitory activity.

Genome analysis has shown that *pgip* genes did not undergo a large expansion and may exist as single genes, as in diploid wheat species (Di Giovanni et al., 2008), or organized into gene families, the members of which are organized in tandem and can vary from two, as in *Arabidopsis thaliana* (Ferrari et al., 2003), to sixteen, as in *Brassica napus* (Hegedus et al., 2008). The majority of *pgip* genes are intronless, however, some of them can contain a short intron as in *Atpgip1* and *Atpgip2* (Ferrari et al., 2003). Moreover, *pgip* genes can be inactivated by transposon elements as in cultivated and wild wheat where the occurrence of *Copia*-retrotransposon and *Vacuna* transposons has been reported (Di Giovanni et al., 2008). Characterized *pgip* loci are shown in Figure 1. Like other families of defense-related genes, *pgip* families show variation in the expression pattern of the different members, some of which are constitutive, others are tissue-specific and, in most cases, up-regulated following stress stimuli (see reviews indicated above; Table 1).

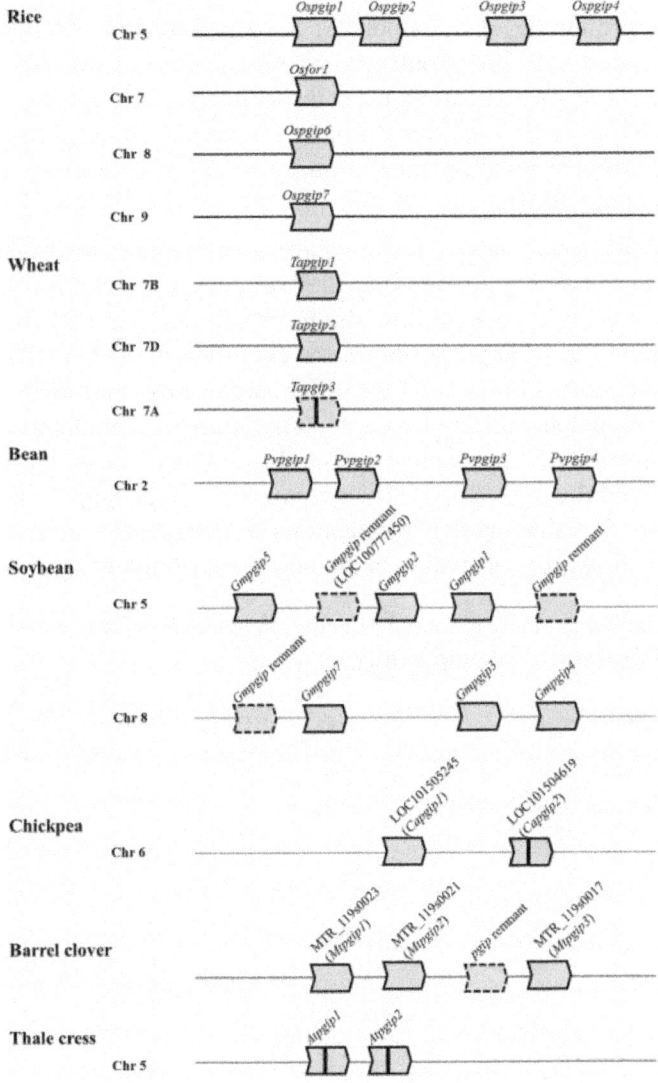

Figure 1. Schematic representation of the genomic organization *pgip* **families in rice, wheat, bean, soybean, chickpea, barrel clover and thale cress**. Each block-arrow with compound-type lines represents a predicted *pgip* gene and a block-arrow with

dash type lines represents a predicted pseudo-gene or remnant gene. Vertical line within block-arrow indicates introns (*Capgip2*, *Atpgip1*, and *Atpgip2*) or a Copia retrotransposon (*Tapgip3*). The direction of the arrow indicates ATG to stop codon. The location of *pgip* genes of legume species are based on Kalunke et al. (2014), those of rice and wheat on Janni et al. (2006) and Di Giovanni et al. (2008), and those of thale crest on Ferrari et al. (2003). Chr, chromosome.

At the protein level, members of a *pgip* family show both functional redundancy and sub-functionalization (De Lorenzo et al., 2001;Federici et al., 2006). As suggested previously, these features likely have an adaptive significance for combating more efficiently a broad array of pathogens (Ferrari et al., 2003) or responding more rapidly to diverse environmental stimuli (D'Ovidio et al., 2004b).

In support of this view, a recent analysis of the genomic organization and composition of the legume *pgip* families suggested that the forces driving the evolution of the *pgip* genes follow the birth-and-death model (Kalunke et al., 2014), similarly to what proposed for the evolution of NBS-LRR-type *R* genes (Michelmore and Meyers, 1998). This possibility is based on genomic features that include inferred recent duplications, diversification as well as pseudogenization of *pgip* copies, as found in soybean, bean, barrel clover and chickpea (Kalunke et al., 2014). The organization of the *pgip* families therefore supports the view that tandem duplications are frequent in stress-related genes and are beneficial for survival in challenging environments (Oh et al., 2012).

Table 1. Treatments or stress stimuli affecting *pgip* **expression in some plant species with a well characterized pgip family**

Pgip family	Treatments or stress stimuli	References
Rice	Abscisic acid (ABA), brassinosteroid, gibberellic acid (GA), 3-indole acetic acid (IAA), jasmonic acid (JA), kinetin, naphthalene acetic acid (NAA), salicylic acid (SA); *Rhizoctonia solani* (necrotrophic fungus)	Janni et al., 2006; Lu et al., 2012
Wheat	*Bipolaris sorokiniana* (necrotrophic fungus) and mechanical wounding	Janni et al., 2013
Bean	Oligogalacturonides (OGs); mechanical wounding; *Botrytis cinerea*, *Sclerotinia sclerotiorum* (necrotrophic fungi); *Colletotrichum lindemuthianum* (hemibiotrophic fungus)	Bergmann et al., 1994; Nuss et al., 1996; Devoto et al., 1997; D'Ovidio et al., 2004b; Oliveira et al., 2010; Kalunke et al., 2011
Soybean	Mechanical wounding; *S. sclerotiorum* (necrotrophic fungus)	D'Ovidio et al., 2006; Kalunke et al., 2014
M. truncatula	JA, SA, ABA; *Colletotrichum trifolii* (hemibiotrophic fungus)	Song and Nam, 2005
Rapeseed	JA, SA, mechanical wounding; *S. sclerotiorum*	Hegedus et al., 2008
Pepper	SA, Methyl jasmonate (Me-JA), ABA, wounding, cold treatment	Wang et al., 2013
Arabidopsis	OGs; JA; *B. cinerea*; *Stemphylium solani* (necrotrophic fungus); aluminum, low-pH, cold; geminivirus	Ferrari et al., 2003; Ascencio-Ibañez et al., 2008; Sawaki et al., 2009; Di et al., 2012; Kobayashi et al., 2014

Inhibition Activity of PGIPs

A number of papers deals with the inhibition activity of PGIPs purified from several plant tissues. This aspect has been reviewed several years ago (De Lorenzo et al., 2001); here, we present an update of this information (Table 2).

Because purified PGIPs may contain a mix of highly similar PGIP isoforms, the activity detected in a tissue may result from the contribution of the activities of different PGIPs expressed in that tissue. An appropriate approach to study the inhibition activity of individual PGIP isoforms is their expression in a heterologous system. However, only a few of the more than 170 *pgip* genes isolated so far from different plant species have been investigated. As reported in Table 3, individual heterologous expression and analysis of all members of a *pgip* family has been performed only for *Arabidopsis* (Ferrari et al., 2003), common bean (D›Ovidio et al., 2004b), soybean (D›Ovidio et al., 2006; Kalunke et al., 2014) and wheat (Janni et al., 2013). PGIPs have been expressed in prokaryotic systems, as a fusion with the maltose-binding protein (MBP) (Jang et al., 2003; Table 3) or using lower temperature for bacterial growth (Chen et al., 2011), in *Pichia pastoris* and in plants by stable transformation or, transiently, by virus-mediated expression (Table 3). In some cases, the proteins were successfully expressed, but did not show any inhibitory activity *in vitro*, as, for example, in the case of some GmPGIPs (D'Ovidio et al., 2006). GmPGIP3, but not GmPGIP1, GmPGIP2, and GmPGIP7 showed inhibitory activity, whereas no expression of GmPGIP5 was obtained (D'Ovidio et al., 2006; Kalunke et al., 2014). Similarly, TaPGIP1 and TaPGIP2, encoded by the two members of the wheat *pgip* family, were successfully expressed but showed no inhibition activity (Janni et al., 2013).

Table 2. Bulk PGIP purified from plants and tested against microbial PGs. These data update those reported in De Lorenzo et al. (2001)

Plant	Tissue	PGIP preparation	Polygalacturonases		References
			Inhibited	Not inhibited	
Tomato (*Solanum lycopersici* L.)	Stem	Crude extract	*Ralstonia solanacearum*		Schacht et al., 2011
Tobacco (*Nicotiana tabacum* L.)	Nectar		*Botrytis cinerea*		Thornburg et al., 2003
Potato (*Solanum tuberosum* L.)		Gel chromatography	*Aspergillus niger* *Fusarium moniliforme*[§] *Fusarium solani* isolate 1402	*Fusarium solani* (isolate 3122)	Machinandiarena et al., 2001
Common Bean (*Phaseolus vulgaris* L.)	Leaves	PG-Sepharose chromatography	*Fusarium anthophilum* *Fusarium circinatum* *Fusarium subglutinans.* *Fusarium proliferatum* isolate 1152 *Fusarium proliferatum* PVS-Fu 64 *Fusarium sacchari* *Fusarium fujikuroi* *F. thapsinum* *Fusarium moniliforme*[§] FC-10	*Fusarium verticillioides* *Fusarium proliferatum* ISPAVEmc 1189 *Fusarium nygamai* *Fusarium moniliforme*[§] PD	Raiola et al., 2008 Sella et al., 2004

Leek (*Allium ampeloprasum* L.)	Basal leaves	Mono-S chromatography		*Fusarium anthophilum* *Fusarium circinatum* *Fusarium subglutinans* *Fusarium proliferatum* *Fusarium sacchari* *Fusarium fujikuroi* *Fusarium verticillioides* *Fusarium proliferatum* ISPAVEmc 1189 *Fusarium nygamai*	Raiola et al., 2008
Asparagus (*Asparagus officinalis* L.)	White spear	Mono-S chromatography		*Fusarium anthophilum* *Fusarium circinatum* *Fusarium subglutinans.* *Fusarium proliferatum* *Fusarium sacchari* *Fusarium fujikuroi* *Fusarium verticillioides* *Fusarium proliferatum* ISPAVEmc 1189 *Fusarium nygamai*	Raiola et al., 2008
Pepper (*Capsicum annuum* L.)	Fruit	Ion-exchange chromatography	*Colletotrichum gleosporoides*, *Colletotrichum capsici*, *Colletotrichum lindemuthianum* *Sclerotium rolfsi* *Fusarium moniliforme*[§]		Shivashankar et al., 2010

Plant	Tissue	PGIP preparation	Polygalacturonases		References
			Inhibited	Not inhibited	
Guava (*Psidium guajava* L.)	Fruit	Purified using a Sephadex G-100	*Aspergillus niger*		Deo and Shastri, 2003
"Oroblanco" grapefruit hybrid (*Citrus grandis* × *C. paradisi* Macf.)	Fruit	Anion exchange chromatography	*Penicillium italicum* *Botrytis cinerea*		D'hallewin et al., 2004
Apple (*Malus domestica* L.)	Fruit Fruit skin Parenchymal tissues	Partial purified Partial purified	*Colletotrichum acutatum* *Botryosphaeria dothidea* *Monilia fructigena*	*Glomerella cingulata*	Gregori et al., 2008 Lee et al., 2006 Buza et al., 2004
Cantaloupe (*Cucumis melo* L.)	Fruit	Cation exchange chromatography	*Phomopsis cucurbitae* *Aspergillus niger* *Fusarium solani*	*Didymella bryoniae* *Rhizopus* PG *Fusarium verticillioides*	Fish and Davis, 2004
Cotton (*Gossypium hirsutum* L.)	Stem	PG-affinity chromatography	*Aspergilus niger*		James and Dubery, 2001
Pear (*Pyrus communis* L.)	Fruit	Partial purified	*Verticillium dahliae* *Botrytis cinerea* *Venturia nashicola*		Ladu et al., 2012; Faize et al., 2003
Pearl millets (*Pennisetum glaucum* (L) R. Br.)	Seedlings	Crude extract	*Aspergilus niger*		Prabhu et al., 2012

Grass pea (*Lathyrus sativus* L.)	Seeds	Gel-filtration chromatography	*Aspergilus niger*		Tamburino et al., 2012
			Rhizopus spp		
Orange (*Citrus reticulate* L.)	Fruit	Partial purified	*Diaprepes abbreviatus*		Doostdar et al., 1997
Blue mustard (*Chorispora bungeana*)	Leaves, stem, root	Partial purified	*Aspergillus niger*		Di et al., 2009
			Stemphylium solani		
Ginseng (*Panax ginseng* L.)		Crude extract	*Colletotrichum gloeosporioides*		Sathiyaraj et al., 2010
			Phythium ultimum		
			Fusarium oxysporum		
			Rhizoctonia solani		
Bread wheat (*Triticum aestivum* L.)	Leaves	Cation exchange chromatography	*Cochliobolus sativus*	*Aspergillus niger* (EPG I and II)	Kemp et al., 2003
				Cryphonectria parasitica	
				Postia placenta	
				Fusarium moniliforme[§]	
				Colletotrichum lindemuthianum	
				Aspergillus niger exopolygalacturonase	
				Ralstonia solanacearum	

Plant	Tissue	PGIP preparation	Polygalacturonases		References
			Inhibited	Not inhibited	
Durum wheat (*Triticum turgidum ssp. dicoccoides*)	Leaves	Crude extract	*Fusarium graminearum* *Bipolaris sorokiniana* *Stenocarpella maydis*	*Fusarium phyllophylum*	Janni et al., 2013

[§] *Reclassified as Fusarium phyllophilum (Mariotti et al., 2008).*

Table 3. *Pgip* genes individually expressed in plants or in heterologous systems and tested for inhibition activity against microbial PGs

Species	Gene	Heterologous systems	Origin of purified PG		References
			Inhibited	Not inhibited	
Common bean (*Phaseolus vulgaris* L.)	PvPGIP1	Transgenic tomato		*Fusarium oxysporum* *Botrytis cinerea* *Alternaria solani*	Desiderio et al., 1997
			Stenocarpella maydis *Aspergillus niger*		Berger et al., 2000
	PvPGIP1 PvPGIP2 PvPGIP3 PvPGIP4	PVX/*Nicotiana benthamiana*	*Aspergillus niger* *Fusarium moniliforme[§]* *Stenocarpella maydis* *Colletotrichum acutatum* *Botrytis cinerea*	*Lygus rugulipennis* *Adelphocoris lineolatus* *Orthops kalmi* *Closterotomus norwegicus*	D'Ovidio et al., 2006; Frati et al., 2006
	PvPGIP2	Transgenic wheat	*Bipolaris sorokiniana* *F. graminearum*	*Claviceps purpurea*	Janni et al., 2008; Volpi et al., 2013
		Transgenic *Brassica napus*	*Rhizoctonia solani*		Akhgari et al., 2012
		Transgenic sugarbeet	*Fusarium phyllophilum* FC10		Mohammadzadeh et al., 2012
		PVX/*Nicotiana benthamiana*	*Fusarium phyllophilum* FC-10 *Fusarium phyllophilum* 10241 *Fusarium phyllophilum* 25219 *Fusarium phyllophilum* 25218	*Fusarium phyllophilum* 25305 *Fusarium verticillioides* 62264 *Fusarium verticillioides* PD	Mariotti et al., 2008

Species	Gene	Heterologous systems	Origin of purified PG		References
			Inhibited	Not inhibited	
Runner bean (*Phaseolus coccineus* L.)	PcPGIP2	PVX/*Nicotiana benthamiana*	*Fusarium moniliforme*[§] *Aspergillus niger* *Colletotrichum lupini* *Botrytis cinerea*		Farina et al., 2009
Tepary bean (*Phaseolus acutifolius* L.)	PaPGIP2	PVX/*Nicotiana benthamiana*	*Fusarium moniliforme*[§] *Aspergillus niger* *Colletotrichum lupini* *Botrytis cinerea*		Farina et al., 2009
Lima bean (*Phaseolus lunatus* L.)	PlPGIP2	PVX/*Nicotiana benthamiana*	*Fusarium moniliforme*[§] *Aspergillus niger* *Colletotrichum lupini* *Botrytis cinerea*		Farina et al., 2009
Soybean (*Glycine max* L.)	GmPGIP1 GmPGIP2	PVX/*Nicotiana benthamiana*		*Sclerotinia sclerotiorum* PGb *Sclerotinia sclerotiorum* PGa *Fusarium moniliforme*[§] *Botrytis aclada* *Aspergillus niger* *Botrytis cinerea* *Colletotrichum acutatum* *Fusarium graminearum* *Lygus rugulipennis* *Adelphocoris lineolatus* *Orthops kalmi* *Closterotomus norwegicus*	D'Ovidio et al., 2006; Frati et al., 2006

Species	Gene	Heterologous systems	Origin of purified PG		References
			Inhibited	Not inhibited	
	GmPGIP3	PVX/*Nicotiana benthamiana*	*Sclerotinia sclerotiorum* PGb *Sclerotinia sclerotiorum* PGa *Fusarium moniliforme*[§] *Botrytis aclada* *Aspergillus niger* *Botrytis cinerea* *Colletotrichum acutatum* *Fusarium graminearum*		D'Ovidio et al., 2006; Frati et al., 2006
	GmPGIP4	PVX/*Nicotiana benthamiana*		*Sclerotinia sclerotiorum* PGb *Sclerotinia sclerotiorum* PGa *Fusarium moniliforme*[§] *Botrytis aclada* *Aspergillus niger* *Botrytis cinerea* *Colletotrichum acutatum* *Fusarium graminearum*	D'Ovidio et al., 2006; Frati et al., 2006
	GmPGIP7	PVX/*Nicotiana benthamiana*		*Sclerotinia sclerotiorum* *Fusarium graminearum* *Colletotrichum acutatum* *Aspergillus niger*	Kalunke et al., 2014
Pepper (*Capsicum annum* L.)	CaPGIP1 CaPGIP2	*Escherichia coli*	*Alternaria alternata* *Colletotrichum nicotianae*		Wang et al., 2013

Species	Gene	Heterologous systems	Origin of purified PG Inhibited	Not inhibited	References
Rapeseed (Brassica napus L.)	BnPGIP1	Pichia pastoris	Sclerotinia sclerotiorum PG6		Bashi et al., 2013
Chinese cabbage (Brassica rapa L.)	BrPGIP2	Transgenic Brassica rape	Pectobacterium carotovorum Botryosphaeria dothidea		Hwang et al., 2010
	BrPGIP2	Escherichia coli	Sclerotinia sclerotiorum		HuangFu et al., 2014
Grapevine (Vitis vinifera L.)	VvPGIP1	Transgenic tobacco	Botrytis cinerea PGI Botrytis cinerea PG4 Botrytis cinerea PG6 Aspergillus. niger PGA Aspergillus niger PGB Aspergillus niger PGI	Botrytis cinerea PG3 Aspergillus niger PGII Botrytis cinerea PG2	Joubert et al., 2006 Joubert et al., 2007
Apple (Malus domestica Borkh.)	MdPGIP1	Transgenic tobacco	Colletotrichum lupini Botryosphaeria obtusa Diaporthe ambigua	Aspergillus niger	Oelofse et al., 2006
		Transgenic potato	Verticillium dahliae		Gazendam et al., 2004
Pear (Pyrus communis L.)	PpPGIP	Transgenic grape	Botrytis cinerea		Agüero et al., 2005
		Transgenic tomato	Botrytis cinerea		Powell et al., 2000
		Transgenic persimmon	Botrytis cinerea		Tamura et al., 2004
Raspberry (Rubus idaeus L.)	RiPGIP	Transgenic pea	Stenocarpella maydis Colletotrichum lupini		Richter et al., 2006
Wheat (Triticum aestivum L.)	TaPGIP1 TaPGIP2	PVX/Nicotiana benthamiana		Fusarium phyllophylu Stenocarpella maydis Bipolaris sorokiniana Fusarium graminearum	Janni et al., 2013
Rice (Oryza sativa L.)	OsPGIP1	PVX/Nicotiana benthamiana	Sclerotinia sclerotiorum Fusarium moniliforme§ Fusarium graminearum Aspergillus niger Botrytis cinerea		Janni et al., 2006
	OsFOR1	Escherichia coli BL21	Aspergillus niger PG		Jang et al., 2003
Pearl millet [Pennisetum glaucum (L.) R. Br.]	PglPGIP1	Escherichia coli SHuffle® T7 Express	Aspergillus niger, AnPGII	Fusarium moniliforme, FmPGIII	Prabhu et al., 2014
Arabidopsis thaliana	AtPGIP1 AtPGIP2	Transgenic Arabidopsis	Colletotrichum gloeosporioides Stenocarpella maydis Botrytis cinerea Fusarium graminearum	Aspergillus niger Fusarium moniliforme§ Lygus rugulipennis Adelphocoris lineolatus Orthops kalmi Closterotomus norwegicus	Frati et al., 2006; Ferrari et al., 2012, 2003

§Reclassified as Fusarium phyllophilum FC10 (Mariotti et al., 2008).

The absence of inhibition activity *in vitro* may also reflect the possibility that some PGIPs are active only in the *in planta* environment, as suggested by Joubert et al. (2006) in the case of the *Botrytis cinerea* BcPG2 and VvPGIP1 from grapevine (*Vitis vinifera L.*). These proteins do not interact *in vitro*, although VvPGIP1 reduces symptoms caused by BcPG2 upon co-infiltration in leaves. The number and sources of PGs tested is also limited; only a few studies have been carried out against PGs of bacteria and insects (Doostdar et al., 1997; D'Ovidio et al., 2004b; Frati et al., 2006; Hwang et al., 2010; Schacht et al., 2011; Kirsch et al., 2012). The limitations of data prevents to draw conclusions about correlations between PGIPs of specific plant families and specific pathogens. Notably, PG produced by a highly detrimental pathogen, *Fusarium verticillioides*, is not inhibited by any known PGIP (see Table 2). This PG has been a target of an unsuccessful attempt to render PvPGIP2 an efficient inhibitor against this PG (see below, Benedetti et al., 2011a).

The utilization of *pgip* genes for crop protection relies on the identification of inhibitors with broad specificities against the many PGs produced by phytopathogens and/or the construction of novel PGIPs with stronger and broader inhibitor activity. Many more PGIPs than those reported in Tables 2, 3 exist in nature and are likely to have different specificities against microbial PGs, considering that single amino acid changes are able to change specificity of the inhibitors (Leckie et al., 1999). Searching for PGIPs with novel specificities may allow to count on a much larger reservoir of possible genes for crop protection. A direct and simple strategy to isolate PGIPs with recognition capability against a given PG may be based on affinity chromathography methods, similar to that originally used to purify PGIP from *P. vulgaris* (Cervone et al., 1987), and mass spectrometry. Attempts to drive *in vitro* evolution of PGIPs to generate proteins with improved inhibition properties have not been successful yet (Benedetti et al., 2011a).

The occurrence of PG-inhibiting activity in crude leaf protein extracts of tetraploid wild wheat (*T. dicoccoides*) possessing nonfunctional *pgip* genes (Di Giovanni et al., 2008) suggested the existance of *pgip* genes with a sequence divergent from the classical one. This possibility, which deserves further investigation, is also supported by the finding that the wheat tissue contains PG-inhibiting proteins with N-terminal sequences (Lin and Li, 2002; Kemp et al., 2003) different from TaPGIP1 and TaPGIP2 (Janni et al., 2013) and from the *pgip*sequences reported so far (http://www.ncbi.nlm.nih.gov/nucleotide/). Recently, a wheat gene with some sequence similarity to *pgip* genes has been reported and was shown to be involved in the defense response against *Fusarium graminearum* (Hou et al., 2014).

Structural Studies on the PG-PGIP Interaction

Thus, the possibility of engineering new forms of PGIPs depends on the detailed structural knowledge of the PG-PGIP interaction. Several structural studies have been performed (Mattei et al., 2001; King et al., 2002;Benedetti et al., 2011b, 2013; Gutierrez-Sanchez et al., 2012), but a high resolution 3D-structure of the PG-PGIP complex is still missing. The enzyme-inhibitor combinations that have been more extensively investigated, are those that PGIP2 from *Phaseolus vulgaris* (PvPGIP2) forms with PG from *A. niger* (AnPGII),*F. phyllophilum* (FpPG) and *C. lupini* (ClPG). Site-directed mutagenesis has shown that the residues involved in the interaction are located in the concave surface of the inhibitor (Leckie et al., 1999; Federici et al., 2001;Spinelli et al., 2009; Benedetti et al., 2011b, 2013). Computational methods such as the Codon Substitution Model in combination with the Desolvation Energy Calculation and the Repeat Conservation Mapping (RCM;Helft et al., 2011) have pinpointed several residues of PvPGIP2 responsible for the PG-inhibiting activity (Casasoli et al., 2009).

On the other hand, residues of PG that are critical for the interaction with PGIP have been also studied. FvPG is 92.5% identical to FpPG, but is inhibited by neither PvPGIP2 nor other known PGIPs. By both loss- and gain-of-function site-directed mutations, a single amino acid at position 274 of both FvPG and FpPG was demonstrated to act as a switch for recognition by PvPGIP2 (Raiola et al., 2008; Benedetti et al., 2013). Unfortunately, the lack of high-resolution structural information on the PG-PGIP complex does not allow to precisely identify the contacting residue in PGIP. Moreover, both PGs and PGIPs are glycosylated proteins (Caprari et al., 1993; Lim et al., 2009); however, whether glycosylation plays a role in the PGIP-PG interaction requires further investigation. For example, glycosylation in pearl millet PGIP was found to affect pH and temperature stability of the protein but not its capability of inhibiting AnPGII (Prabhu et al., 2015).

A single PGIP may display different mechanisms of PG inhibition (competitive, non competitive and mixed) suggesting that the protein is highly versatile in recognizing different epitopes of various PGs (Federici et al., 2001; King et al., 2002; Sicilia et al., 2005; Bonivento et al., 2008). Consequently, many 3D-models based on docking predictions have been proposed so far (Sicilia et al., 2005; Maulik et al., 2009; Prabhu et al., 2014). Techniques such as the mass amide exchange mass spectrometry in the case of AnPGII and FpPG and the Small Angle X-ray Scattering (SAXS) in the case of FpPG and ClPG have produced models that, in some cases, are discordant. For example, while the mass amide exchange mass spectrometry predicts that the area of FpPG in contact with PvPGIP2 is located at the N-terminus and predominantly on the underside of the enzyme beta-barrel structures (Gutierrez-Sanchez et al., 2012), the SAXS analysis indicates that the protein region in contact with PvPGIP2 is located at the C-terminus of the enzyme and includes the loops surrounding the active site cleft. A site-directed mutagenesis analysis has been used to validate this second view (Benedetti et al., 2013). In general, low resolution techniques such as SAXS analysis or mass amide exchange mass spectrometry require validation by site-directed mutagenesis to locate the contacting residues in a protein complex. The X-ray crystallography, successfully used to solve several high-resolution structures of PGs (van Santen et al., 1999; Federici et al., 2001; Bonivento et al., 2008) and that of PvPGIP2 (Di Matteo et al., 2003), was so far unsuccessful in the case of the PG-PGIP complex. This is probably due to the intrinsic instability of the PG-PGIP interaction, which only occurs, under apoplastic conditions of pH and ionic strength, through the contact of only a few, sometimes only one, residues (Leckie et al., 1999).

The use of a cross-linker for stabilizing the PG-PGIP complex coupled to techniques that allow the protein analysis directly in solution, such as SAXS and NMR spectroscopy (Wand and Englander, 1996; Nietlispach et al., 2004), may be a valid alternative in order to obtain a detailed map of the contacting residues but this requires a subsequent validation by site-directed mutagenesis.

PGIPs Engineered in Dicot Crops

The important role of PGIP in plant defense has been demonstrated by overexpressing *pgip* genes in several plant species. In these experiments, the source of the used genes was either the same plant species utilized for transformation or a different one (Table 4). The transformation of the model plant *A. thaliana* has been particularly useful to highlight the potentiality of several *pgip* genes, namely the endogenous *Atpgip1* and*Atpgip2*, the bean *Pvpgip2* and the rapeseed (*Brassica napus*) *Bnpgip1* or *Bnpgip2*. Arabidopsis plants overexpressing *Atpgip1* or *Atpgip2* showed a significant reduction of disease symptoms caused by *B. cinerea*(Ferrari et al., 2003) and were less susceptible against the hemibiotrophic fungal pathogen *F. graminearum*(Ferrari et al., 2012), the major causal agent of Fusarium head blight (FHB).

Conversely, silencing of their expression using an antisense *Atpgip*, led to enhanced susceptibility (Ferrari et al., 2006). Arabidopsis plants expressing *Pvpgip2*, encoding an efficient inhibitor of the *B. cinerea* PG (ten Have et al., 1998), showed reduction of disease symptoms caused by *B. cinerea* and those expressing the rapeseed genes *Bnpgip1* and*Bnpgip2* delayed the symptoms caused by *S. sclerotiorum* (Bashi et al., 2013).

The protective potential of *pgip* genes has also been demonstrated in transgenic crops. The first transgenic crop plant obtained by using a*pgip* gene and tested against pathogenic microorganisms were tomatos expressing PvPGIP1 from *P. vulgaris*. These plants, however, did not show any increased resistance against *Fusarium oxysporum* f. sp. *lycopersici*, *B. cinerea*, and *Alternaria solani*. The negative result was due to the inability of PvPGIP1 to inhibit the PGs secreted by these fungi, as shown by *in vitro* inhibition assays and led to discovery of other forms of PGIPs and eventually to the existence of a complex PGIP family in French bean (Desiderio et al., 1997).

Table 4. List of transgenic crops produced using the gene coding for PGIP and their response to fungal, oomycetes or bacterial phytopathogens

Transgenic crops	PGIP gene[c]	Tested against fungal, oomycetes or bacterial phytopathogens	References
Tomato[a] (Solanum lycopersicum L.)	PcPGIP	Botrytis cinerea[*]	Powell et al., 2000, 1994
	PvPGIP1	Fusarium oxysporum f.sp. lycopersici[†]	Desiderio et al., 1997
		Botrytis cinerea[†]	
		Alternaria solani[†]	
Tobacco[a] (Nicotiana tabacum L.)	PvPGIP2	Botrytis cinerea[*]	Manfredini et al., 2005
		Rhizoctonia solani[*]	Borras-Hidalgo et al., 2012
		Phytophthora parasitica[*]	
		Peronospora hyoscyami[*]	
	CaPGIP1	Alternaria alternata[*]	Wang et al., 2013
		Colletotrichum nicotianae[*]	
	VvPGIP1	Botrytis cinerea[*]	Joubert et al., 2006
	BrPGIP2	Pectobacterium carotovorum[*]	Hwang et al., 2010
Potato[a] (Solanum tuberosum L.)	MdPGIP1 StPGIP	Verticillium dahliae[†]	Gazendam et al., 2004; Guo
		Verticillium dahliae[*]	et al., 2014
Brassica rapa[a]	BrPGIP2	Pectobacterium carotovorum[*]	Hwang et al., 2010
Rapeseed[a] (Brassica napus L.)	BnPGIP2	Sclerotinia sclerotiorum[*]	HuangFu et al., 2014
Pea[a] (Pisum sativum L.)	RiPGIP	Glomus intraradices[Ψ]	Hassan et al., 2012
Grapevine[a] (Vitis vinifera L.) Rice[a] (Oriza sativa L.)	PcPGIP OsPGIP1	Botrytis cinerea[*]	Agüero et al., 2005; Wang et al.,
		Xylella fastidiosa[*]	2014b
		Rhizoctonia solani	
Wheat[b] (Triticum aestivum L., Triticum durum Desf.)	PvPGIP2GmPGIP3	Bipolaris sorokiniana[*]	Janni et al., 2008
		Fusarium graminearum[*]	Ferrari et al., 2012
		Claviceps purpurea[†]	Volpi et al., 2013; Wang et al.,
		Bipolaris sorokiniana[*]	2014a
		Gaeumannomyces graminis var. tritici[*]	
Arabidopsis thaliana L.[a]	PvPGIP2	Botrytis cinerea[*]	Manfredini et al., 2005
	AtPGIP1 AtPGIP2	Fusarium graminearum[*]	Ferrari et al., 2012
	BnPGIP1 BnPGIP2	Sclerotinia sclerotiorum[*]	Bashi et al., 2013

[a] The transgenic gene was under control of CaMV 35S promoter.
[b] The transgenic gene was under control of Ubiquitin promoter.
[c] Pc, Pyrus communis; Pv, Phaseolus vulgaris; Ca, Capsicum annum; Vv, Vitis vinifera; Br, Brassica rapa; Md, Malus domestica; St, Solanum torvum; Ri, Rubus idaeus; Ac, Actinidia deliciosa; At, Arabidopsis thaliana; Bn, Brassica napa.
[*] Showed enhanced resistance.
[†] No evidence of enhanced resistance.
[Ψ] No effect on mycorrhization.

A few years later, transgenic tomato plants expressing a pear (*Pyrus communis* L.) PGIP (PcPGIP) capable of inhibiting the PGs secreted by *B. cinerea*, showed a reduction of disease lesions caused by this fungus both on ripening fruit (15% reduction) and leaves (about 25% reduction). The initial establishment of infection was not affected in the transgenic plants but the later colonization of the host tissue was significantly reduced (Powell et al., 2000).

Tobacco has been the most used crop plant for testing the effect of PGIP expression on resistance to pathogens. Constitutive and high-level expression of *Pvpgip2* (from *P. vulgaris*), *Vvpgip1* (from *V. vinifera*), *Capgip1*[from pepper (*Capsicum annum*)] and *Brpgip2* (from *B. rapa*) have been obtained in transgenic tobacco. Plants expressing PvPGIP2 showed about 35% reduction of symptoms caused by *B. cinerea* (Manfredini et al., 2005) and, more recently, were shown to display reduced disease symptoms against *Rhizoctonia solani* and two oomycete pathogens, *Phytophthora parasitica* var. *nicotianae* and the blue mold-causing agent*Peronospora hyoscyami* f. sp. *tabacina* (Borras-Hidalgo et al., 2012). Notably, the experiments against *P. hyoscyami* f. sp. *tabacina* were performed in the field during seasonal conditions that favor the pathogen spreading. In agreement with what observed under controlled conditions, resistance of transgenic plants was comparable to that exhibited by *Nicotiana* species (*N. rustica*, *N. debneyi* and *N. megalosiphon*) that are highly resistant to blue mold disease. These transgenic plants expressing PvPGIP2 represented the first example of PGIP-expressing plants subjected to field trails. Recently, transgenic rice expressing OsPGIP1 showed also improved resistance against *Rhizoctonia solani* in field experiments (Wang et al., 2014b).

Transgenic tobacco plants expressing the grapevine *pgip* gene *Vvpgip1* (Joubert et al., 2006) also showed a reduced (from 47 to 69%) disease susceptibility to *B. cinerea* infection. As for plants expressing PvPGIP2, the resistance phenotype correlated with the accumulation of VvPGIP1 as well as with its capability of inhibiting the activity of PG secreted by *B. cinerea*, namely BcPG1, BcPG3, and BcPG6. Several observations, however, suggest that PGIP may improve resistance by mechanisms other than classical PGIP-PG inhibition. For example, non-infected transgenic tobacco plants expressing *Vvpgip1* show modified expression patterns of genes involved in various metabolic pathways (Alexandersson et al., 2011) and an altered cell wall structure (Nguema-Ona et al., 2013). In these plants, lignin accumulation and arabinoxyloglucan-cellulose re-organization leads to a general strengthening/reinforcing of the cell wall that may contribute to an improved resistance against *B. cinerea*.

A reduction of disease symptoms (about 50%) caused by *Alternaria alternata* and *Colletotrichum nicotianae*was also observed in transgenic tobacco lines expressing the pepper CaPGIP1 and, once again, resistance correlated with the inhibition capacity of purified CaPGIP1 against PG activity of both fungal pathogens (Wang et al., 2013).

Within the Solanaceae family, transgenic potato (*Solanum tuberosum*) plants expressing the gene *StPGIP1*from *S. torvum* showed a 50% reduction of wilt disease symptoms caused by *Verticillium dahliae* and a normal plant growth (Guo et al., 2014). Transgenic potato plants overexpressing the apple *pgip1* gene showed protection against the same fungal pathogen but displayed an extended juvenile phase (Gazendam et al., 2004). Transgenic grapevine (*V. vinifera*) plants constitutively expressing the pear PcPGIP gene represent an interesting example of the potential of PGIP for protection against pathogens other than fungi and oomycetes. These plants show a delayed development of the Pierce's disease (PD) caused by bacterial pathogen *Xylella fastidiosa* (Agüero et al., 2005). Not only leaf scorching and *Xylella* titre were reduced but also plants showed a better re-growth after pruning compared to infected untransformed controls. Moreover, an inverse dose-effect relationship was shown between development of PD and levels of PcPGIP activity in the tissues. The improved resistance of the grapevine plants expressing PcPGIP against a bacterial pathogen was unexpected, because until then the PGIP inhibition activity was thought to be limited to fungal and insect PGs (Cervone et al., 1990;Johnston et al., 1993; D›Ovidio et al., 2004b). It was later shown that pear PcPGIP inhibits the PG encoded by*X. fastidiosa* and that PG activity is a virulence factor of this pathogen (Roper et al., 2007; Pérez-Donoso et al., 2010). The observation that PcPGIP is present in xylem exudates of non-transgenic scions grafted on transgenic rootstocks expressing PcPGIP suggests that grafting of non transgenic varieties on transgenic rootstocks represents, in this case, a useful agronomical practice for plant protection (Agüero et al., 2005).

The results obtained with *X. fastidiosa* prompted further investigations on the capability of PGIP of controlling bacterial diseases (summarized in Table 4). Transgenic tobacco plants expressing *B. rapa* BrPGIP2 were resistant against *Pectobacterium carotovorum*, the causal agent of the soft rot disease, with a strong reduction (66–88%) of the symptoms as compared to wild-type plants (Hwang et al., 2010). The resistance correlated with the inhibitory activity against *P. carotovorum* PG activity found in the total protein extracts of the transgenic plants (Hwang et al., 2010). Also chinese cabbage (*B. rapa* ssp. *pekinensis*) plants overexpressing BrPGIP2 showed higher resistance against *P. carotovorum* and produced normal looking pods-like structures with no viable seeds. Combination of crossing with non-transgenic plants did not restore fertility of the transgenic plants, suggesting that mechanisms such as ploidy changes occurring during the tissue culture stage or changes in cell-wall architecture of sexual organs are responsible for the abnormality (Hwang et al., 2010).

No phenotypic abnormalities were, instead, found in transgenic tobacco plants expressing BrPGIP2 (Hwang et al., 2010), nor in rapeseed plants overexpressing the *B. napus Bnpgip2*. The latter plants displayed a significant reduction of rot caused by the necrotrophic fungal pathogen *S. sclerotiorum* (HuangFu et al., 2014).

Additional PGIP-transgenic crops include pea (*Pisum sativum* L.), transformed with *Ripgip* from raspberry (*Rubus idaeus* L.) (Richter et al., 2006), persimmon (*Diospyros kaki* L.) and apple (*Malus domestica* Borkh.) transformed with pear PcPGIP (Szankowski et al., 2003; Tamura et al., 2004), sugarbeet (*Beta vulgaris* L.) transformed with bean *Pvpgip2* (Mohammadzadeh et al., 2012), chickpea transformed with either *Ripgip* or a*pgip* gene from kiwi fruit (Senthil et al., 2004), tobacco transformed with PpPGIP gene from *Pyrus pyrifolia*Nakai (Liu et al., 2013) and maize (*Zea mays* L.) transformed with bean *Pvpgip1* (O›Kennedy et al., 2001). The response of these plants to pathogens has not been reported yet. Transgenic pea plants expressing RiPGIP were instead evaluated for their response to beneficial microorganisms. *Glomus intraradices*, an arbuscular mycorrhizal fungus, colonized roots of transgenic plants at an extend comparable to that observed in control non transgenic plants, indicating that the expression of RiPGIP does not affect mycorrhization (Hassan et al., 2012).

PGIPs Engineered in Monocot Crops

Although the low pectin content of cereal species like wheat and rice indicates that this cell wall component may have a marginal role during infection, results show that the expression of PGIP in transgenic plants limits some diseases caused by fungal pathogens (Janni et al., 2008; Ferrari et al., 2012; Wang et al., 2014a,b). In our labs, the bean *Pvpgip2* gene was used under the constitutive promoter of the maize unbiquitin gene (*Ubi-1*) to transform both durum and bread wheat by particle bombardment. PvPGIP2 was correctly targeted to the apoplast and the transgenic plants did not show any major morphological and growth defects. Transgenic wheat showed a significant reduction (46–50%) of foliar spot blotch symptoms caused by the hemibiotrophic fungal pathogen *Bipolaris sorokiniana* and improved resistance (25–30%) against the hemibiotrophic fungal pathogen *F. graminearum* (Ferrari et al., 2012), the major causal agent of FHB in wheat. A reduced degradability of the transgenic tissue by PG treatments correlated with the capacity of PvPGIP2 to inhibit PG activity of *B. sorokiniana* and less strongly PG of *F. graminearum* (Janni et al., 2008; Ferrari et al., 2012). An interesting aspect of the wheat plants expressing PvPGIP2 is that, under moderate infection with *F. graminearum*, the reduced FHB symptoms are concomitant with a greater amount of total starch in the grains as compared to control plants (D'Ovidio et al., 2012). On the other hand, wheat plants expressing PvPGIP2 were susceptible to the biotrophic fungal pathogen *Claviceps purpurea*, the causal agent of ergot disease probably because PvPGIP2 is not able to inhibit the activity of *C. purpurea* CpPG1 and CpPG2 (Volpi et al., 2013). Recently, transgenic wheat expressing the soybean GmPGIP3 was shown to be resistant to both take-all and common root rot diseases caused by the fungal pathogen *Gaeumannomyces graminis* var. *tritici* and *B. sorokiniana*, respectively; symptoms were reduced of about 47–83% and 42–60%, respectively (Wang et al., 2014a). Similarly, the expression of OsPGIP1 in transgenic rice enhanced resistence against *Rhizoctonia solani*in field tests and resistance was related with the expression levels of OsPGIP1 (Wang et al., 2014b).

CONCLUDING REMARKS AND FUTURE CHALLENGES

The results reported in this review clearly indicate that PGIP is useful to improve resistance in different crop species. High-level expression of PGIP does not prevent infection but limits significantly the colonization of the host tissue with a consequent positive impact on crop yield and product quality. The efficacy of PGIP to control diseases has been demonstrated against fungi, oomycetes and bacteria and is equally efficient against necrorophic and hemibiotrophic pathogens. The experiments performed with biotrophs do not allow to draw any clear conclusion since the only fungal biotrophic pathogen analyzed, *C. purpures*, produced PG activity that was not inhibited by the PGIP expressed in the transgenic plants (Volpi et al., 2013). The identification and development of PGIPs with stronger and broader inhibitory capacities may be useful to utilize these proteins in crop protection. Germplasm analysis to identify novel PGIPs is still limited (Farina et al., 2009) and the initial attempts to drive *in vitro* evolution of PGIP to generate proteins with improved inhibition properties have not been particularly successful (Benedetti et al., 2011a). Structural studies should be implemented in order to obtain a detailed map of the contacts between various PGs and PGIPs. This is necessary not only for constructing novel inhibitors with stronger activities but also for future programs of genome editing in which the existing genes of a plant species may be ameliorated to better adapt to new virulent strains of microorganisms evolving in nature.

The available results support the notion that inhibition of the microbial PG by PGIP is a prerequisite of the inhibitors to confer resistance to transgenic plants against microbes. The delay of symptoms is often related to the capacity of PGIP to inhibit the PG activity secreted by the pathogens and, consequently, to reduce both tissue maceration and favor the release of OGs, as summarized in Figure 2. However, this aspect of the PGIP›s biology needs further investigation. In some cases PGIP has been reported to confer resistance without any evidence of PG-inhibition *in vitro* (Joubert et al., 2006). Moreover, some evidence suggests that the capability of reducing tissue maceration is associated with the property of PGIP to bind pectin, likely shielding this component of the cell wall from PG activity (Spadoni et al., 2006). In this regard the observation that transgenic plants expressing PGIPs exhibit an altered gene expression and cell wall composition is also intriguing. It is not yet clear the mechanism that links the ectopic expression of PGIP to alteration of gene expression and whether this contributes to disease resistance (Alexandersson et al., 2011; Nguema-Ona et al., 2013).

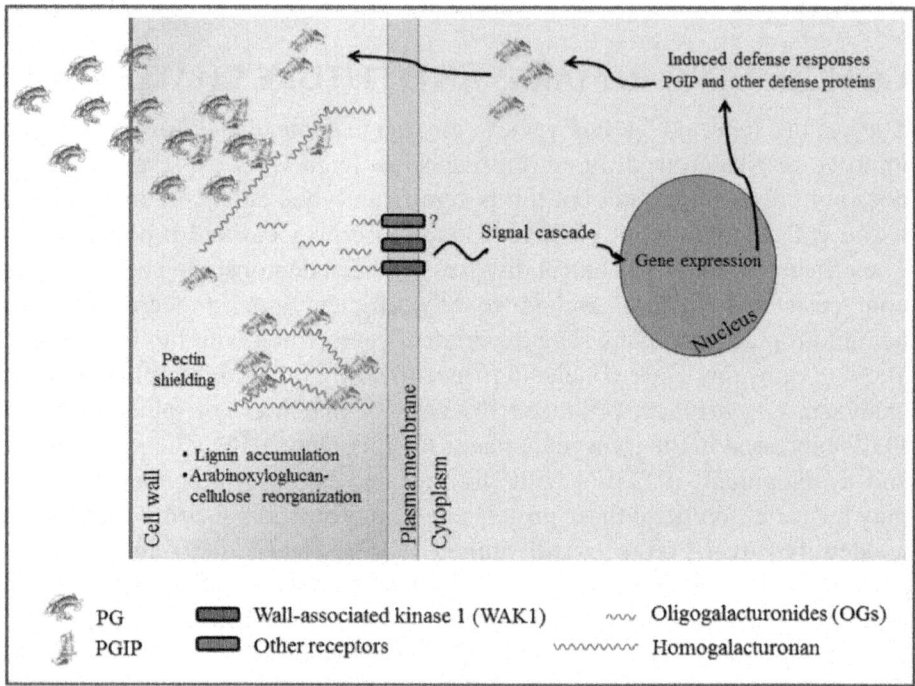

Figure 2. A model for the role of PGIP in the defense response against pathogens. Delay of symptoms is related to the inhibitory activity of PGIP toward PGs secreted by the pathogens and likely to the accumulation of oligogalacturonide (OG) elicitors, which are recognized by WAK1 and likely other receptors not yet characterized. Cell wall modification and pectin shielding could also play a role. Signaling cascades activated by OGs are described in Ferrari et al. (2013).

An important but very little explored aspect of the PGIP biology is its possible role in processes of growth and development. Although plants overexpressing PGIPs do not show obvious morphological alterations, indeed several reports point to PGIP as a player in development. PGIP are induced, not only by phosphate deficiency, but also by auxin treatment and in mutants defective in SIZ1, a SUMO (small ubiquitin-related modifier) E3 ligase that is involved in several stress responses, including Pi starvation, and flowering (Sato and Miura, 2011). Suppression of PGIPs under the control ABA insensitive 5

(ABI5) transcription factor accompanies promotion of seed germination by the peroxisomal ABC transporter PED3 (Kanai et al., 2010). Upregulation of *PGIP2*correlates with the acquisition of competence to form green callus in an auxin-rich callus induction medium (Che et al., 2007) and occurs in Arabidopsis tissue culture lines in which the expression of the peroxidases PRX33 and PRX34 is knocked down by antisense expression (O'Brien et al., 2012), whereas PGIP1 was identified in a proteomic study performed on Arabidopsis etiolated hypocotyls used as a model of cells undergoing elongation followed by growth arrest within a short time (Irshad et al., 2008). Finally, both PGIP1 and PGIP2 are associated with cell wall stabilization at low pH under the control of the zinc-finger protein STOP1 (Sensitive to Proton Rhizotoxicity 1) and STOP2 (Kobayashi et al., 2014). A role of PGIP not only in defense but also in growth and development implies that the inhibitor may affect one or more of the many endogenous PGs expressed by plants. This is also an unexplored aspect of the PGIP biology and, at the moment, only one very old evidence is available showing that PGIP may have a plant-derived PG partner (Cervone et al., 1990).

ACKNOWLEDGMENTS

This research was supported by the Italian Ministry of University and Research (PRIN 2010–2011) to RD.

REFERENCES

1. Agüero, C. B., Uratsu, S. L., Greve, C., Powell, A. L. T., Labavitch, J. M., Meredith, C. P., et al. (2005). Evaluation of tolerance to Pierce's disease and Botrytis in transgenic plants of *Vitis vinifera* L. expressing the pear PGIP gene. *Mol. Plant Pathol.* 6, 43–51. doi: 10.1111/j.1364-3703.2004.00262.x

2. Akhgari, A., Motallebi, B., and Zamani, M. (2012). Bean polygalacturonase-inhibiting protein expressed in transgenic *Brassica napus* inhibits polygalacturonase from its fungal pathogen *Rhizoctonia solani*. *Plant Prot. Sci.* 48, 1–9.

3. Albersheim, P., and Anderson, A. J. (1971). Proteins from plant cell walls inhibit polygalacturonases secreted by plant pathogens. *Proc. Natl. Acad. Sci. U.S.A.* 68, 1815–1819.

4. Alexandersson, E., Becker, J. V. W., Jacobson, D., Nguema-Ona, E., Steyn, C., Denby, K. J., et al. (2011). Constitutive expression of a grapevine polygalacturonase-inhibiting protein affects gene expression and cell wall properties in uninfected tobacco. *BMC Res. Notes* 4:493.

doi: 10.1186/1756-0500-4-493

5.   Ascencio-Ibanez, J. T., Sozzani, R., Lee, T.-J., Chu, T.-M., Wolfinger, R. D., Cella, R., et al. (2008). Global analysis of Arabidopsis gene expression uncovers a complex array of changes impacting pathogen response and cell cycle during geminivirus infection. *Plant Physiol.* 148, 436–454. doi: 10.1104/pp.108.121038

6.   Bashi, Z. D., Rimmer, S. R., Khachatourians, G. G., and Hegedus, D. D. (2013). *Brassica napus* polygalacturonase inhibitor proteins inhibit *Sclerotinia sclerotiorum* polygalacturonase enzymatic and necrotizing activities and delay symptoms in transgenic plants. *Can. J. Microbiol.* 59, 79–86. doi: 10.1139/cjm-2012-0352

7.   Benedetti, M., Andreani, F., Leggio, C., Galantini, L., Di Matteo, A., Pavel, N. V., et al. (2013). A single amino-acid substitution allows endo-polygalacturonase of *Fusarium verticillioides* to acquire recognition by PGIP2 from *Phaseolus vulgaris*. *PLoS ONE* 8:e80610. doi: 10.1371/journal.pone.0080610

8.   Benedetti, M., Bastianelli, E., Salvi, G., De Lorenzo, G., and Caprari, C. (2011a). Artificial evolution corrects a repulsive amino acid in polygalacturonase inhibiting proteins (PGIPs). *J. Plant Pathol.* 93, 89–95. doi: 10.4454/jpp.v93i1.277

9.   Benedetti, M., Leggio, C., Federici, L., De Lorenzo, G., Pavel, N. V., and Cervone, F. (2011b). Structural resolution of the complex between a fungal polygalacturonase and a plant polygalacturonase-inhibiting protein by small-angle X-ray scattering. *Plant Physiol.* 157, 599–607. doi: 10.1104/pp.111.181057

10.   Berger, D. K., Oelofse, D., Arendse, M. S., Du Plessis, E., and Dubery, I. A. (2000). Bean polygalacturonase inhibitor protein-1 (PGIP-1) inhibits polygalacturonases from *Stenocarpella maydis*. *Physiol. Mol. Plant Pathol.* 57, 5–14. doi: 10.1006/pmpp.2000.0274

11.   Bergmann, C. W., Ito, Y., Singer, D., Albersheim, P., Darvill, A. G., Benhamou, N., et al. (1994). Polygalacturonase-inhibiting protein accumulates in *Phaseolus vulgaris* L. *in response to wounding, elicitors and fungal infection. Plant J.* 5, 625–634. doi: 10.1111/j.1365-313X.1994.00625.x

12.   Bonivento, D., Pontiggia, D., Di Matteo, A., Fernandez-Recio, J., Salvi, G., Tsernoglou, D., et al. (2008). Crystal structure of the endopolygalacturonase from the phytopathogenic fungus *Colletotrichum lupini* and its interaction with polygalacturonase inhibiting proteins. *Proteins* 70, 294–299. doi: 10.1002/prot.21610

13. Borras-Hidalgo, O., Caprari, C., Hernandez-Estevez, I., De Lorenzo, G., and Cervone, F. (2012). A gene for plant protection: expression of a bean polygalacturonase inhibitor in tobacco confers a strong resistance against *Rhizoctonia solani* and two oomycetes. *Front. Plant Sci.* 3:268. doi: 10.3389/fpls.2012.00268

14. Buza, N. L., Krinitsyna, A. A., Protsenko, M. A., and Vartapetyan, V. V. (2004). Role of the polygalacturonidase inhibitor Protein in the ripening of apples and their resistance to *Monilia fructigena*, a causative agent of fruit rot. *Appl. Biochem. Microbiol.* 40, 89–92. doi: 10.1023/B:ABIM.0 000010361.48129.6e

15. Cantu, D., Vicente, A. R., Labavitch, J. M., Bennett, A. B., and Powell, A. L. (2008). Strangers in the matrix: plant cell walls and pathogen susceptibility. *Trends Plant Sci.* 13, 610–617. doi: 10.1016/j. tplants.2008.09.002

16. Caprari, C., Bergmann, C., Migheli, Q., Salvi, G., Albersheim, P., Darvill, A., et al. (1993). *Fusarium moniliforme* secretes four endopolygalacturonases derived from a single gene product. *Physiol. Mol. Plant Pathol.* 43, 453–462. doi: 10.1006/pmpp.1993.1073

17. Casasoli, M., Federici, L., Spinelli, F., Di Matteo, A., Vella, N., Scaloni, F., et al. (2009). Integration of evolutionary and desolvation energy analysis identifies functional sites in a plant immunity protein. *Proc. Natl. Acad. Sci. U.S.A.* 106, 7666–7671. doi: 10.1073/pnas.0812625106

18. Cervone, F., De Lorenzo, G., Degrà, L., Salvi, G., and Bergami, M. (1987). Purification and characterization of a polygalacturonase-inhibiting protein from *Phaseolus vulgaris* L. *Plant Physiol.* 85, 631–637. doi: 10.1104/pp.85.3.631

19. Cervone, F., De Lorenzo, G., Pressey, R., Darvill, A. G., and Albersheim, P. (1990). Can *Phaseolus* PGIP inhibit pectic enzymes from microbes and plants? *Phytochemistry* 29, 447–449. doi: 10.1016/0031-9422(90)85094-V

20. Cervone, F., Hahn, M. G., De Lorenzo, G., Darvill, A., and Albersheim, P. (1989). Host-pathogen interactions. XXXIII. A plant protein converts a fungal pathogenesis factor into an elicitor of plant defense responses. *Plant Physiol.* 90, 542–548. doi: 10.1104/pp.90.2.542

21. Che, P., Lall, S., and Howell, S. H. (2007). Developmental steps in acquiring competence for shoot development in Arabidopsis tissue culture.*Planta* 226, 1183–1194. doi: 10.1007/s00425-007-0565-4

22. Chen, X., Liu, X., Zuo, S., Ma, Y., Tong, Y., Pan, X., et al. (2011). Prokaryotic expression of rice *Ospgip1* gene and bioinformatic analysis

of encoded product. *Rice Sci.* 18, 250–256. doi: 10.1016/S1672-6308(12)60002-X

23. D'hallewin, G., Schirra, M., Powell, A. L., Greve, L. C., and Labavitch, J. M. (2004). Properties of a polygalacturonase-inhibiting protein isolated from 'Oroblanco' grapefruit. *Physiol. Plant.* 120, 395–404. doi: 10.1111/j.0031-9317.2004.00264.x

24. D'Ovidio, R., Laino, P., Janni, M., Botticella, E., Di Carli, M. S., Benvenuto, E., et al. (2012). *"Proteomic analysis of mature kernels of Fusarium graminearum-infected transgenic bread wheat expressing PGIP,"* in *Proceedings of 11th International Gluten Workshop*, eds Z. He and D. Wang (Beijing), 18.

25. D'Ovidio, R., Mattei, B., Roberti, S., and Bellincampi, D. (2004a). Polygalacturonases, polygalacturonase-inhibiting proteins and pectic oligomers in plant-pathogen interactions. *Biochim. Biophys. Acta* 1696, 237–244. doi: 10.1016/j.bbapap.2003.08.012

26. D›Ovidio, R., Raiola, A., Capodicasa, C., Devoto, A., Pontiggia, D., Roberti, S., et al. (2004b). Characterization of the complex locus of bean encoding polygalacturonase-inhibiting proteins reveals subfunctionalization for defense against fungi and insects. *Plant Physiol.* 135, 2424–2435. doi: 10.1104/pp.104.044644

27. D'Ovidio, R., Roberti, S., Giovanni, M. D., Capodicasa, C., Melaragni, M., Sella, L., et al. (2006). The characterization of the soybean polygalacturonase-inhibiting proteins (Pgip) gene family reveals that a single member is responsible for the activity detected in soybean tissues. *Planta* 224, 633–645. doi: 10.1007/s00425-006-0235-y

28. De Lorenzo, G., D›Ovidio, R., and Cervone, F. (2001). The role of polygalacturonase-inhibiting proteins (PGIPs) in defense against pathogenic fungi. *Annu. Rev. Phytopathol.* 39, 313–335. doi: 10.1146/annurev.phyto.39.1.313

29. De Lorenzo, G., and Ferrari, S. (2002). Polygalacturonase-inhibiting proteins in defense against phytopathogenic fungi. *Curr. Opin. Plant Biol.* 5, 295–299. doi: 10.1016/S1369-5266(02)00271-6

30. Deo, A., and Shastri, N. V. (2003). Purification and characterization of polygalacturonase-inhibitory proteins from *Psidium guajava* Linn. (guava) fruit. *Plant Sci.* 164, 147–156. doi: 10.1016/S0168-9452(02)00337-0

31. Desiderio, A., Aracri, B., Leckie, F., Mattei, B., Salvi, G., Tigelaar, H., et al. (1997). Polygalacturonase-inhibiting proteins (PGIPs) with different specificities are expressed in *Phaseolus vulgaris. Mol. Plant Microbe Interact.* 10, 852–860. doi: 10.1094/MPMI.1997.10.7.852

32. Devoto, A., Clark, A. J., Nuss, L., Cervone, F., and De Lorenzo, G. (1997). Developmental and pathogen-induced accumulation of transcripts of polygalacturonase-inhibiting protein in *Phaseolus vulgaris* L. *Planta* 202, 284–292. doi: 10.1007/s004250050130

33. Di, C., Li, M., Long, F., Bai, M., Zheng, X., Xu, S., et al. (2009). Molecular cloning, functional analysis and localization of a novel gene encoding polygalacturonase-inhibiting protein in *Chorispora bungeana*. *Planta* 231, 169–178. doi: 10.1007/s00425-009-1039-7

34. Di, C. X., Zhang, H., Sun, Z. L., Jia, H. L., Yang, L. N., Si, J., et al. (2012). Spatial distribution of polygalacturonase-inhibiting proteins in*Arabidopsis* and their expression induced by *Stemphylium solani* infection. *Gene* 506, 150–155. doi: 10.1016/j.gene.2012.06.085

35. Di Giovanni, M., Cenci, A., Janni, M., and D›Ovidio, R. (2008). A LTR copia retrotransposon and Mutator transposons interrupt *pgip* genes in cultivated and wild wheats. *Theor. Appl. Genet.* 116, 859–867. doi: 10.1007/s00122-008-0719-1

36. Di Matteo, A., Bonivento, D., Tsernoglou, D., Federici, L., and Cervone, F. (2006). Polygalacturonase-inhibiting protein (PGIP) in plant defence: a structural view. *Phytochemistry* 67, 528–533. doi: 10.1016/j.phytochem.2005.12.025

37. Di Matteo, A., Federici, L., Mattei, B., Salvi, G., Johnson, K. A., Savino, C., et al. (2003). The crystal structure of polygalacturonase-inhibiting protein (PGIP), a leucine-rich repeat protein involved in plant defense. *Proc. Natl. Acad. Sci. U.S.A.* 100, 10124–10128. doi: 10.1073/pnas.1733690100

38. Doostdar, H., McCollum, T. G., and Mayer, R. T. (1997). Purification and characterization of an endo-polygalacturonase from the gut of West Indies sugarcane rootstalk borer weevil (*Diaprepes abbreviatus* L.) larvae. *Comp. Biochem. Physiol. Part B Biochem. Mol. Bio.* 118, 861–867.

39. Faize, M., Sugiyama, T., Faize, L., and Ishii, H. (2003). Polygalacturonase-inhibiting protein (PGIP) from Japanese pear: possible involvement in resistance against scab. *Physiol. Mol. Plant Pathol.* 63, 319–327. doi: 10.1016/j.pmpp.2004.03.006

40. Farina, A., Rocchi, V., Janni, M., Benedettelli, S., De Lorenzo, G., and D'Ovidio, R. (2009). The bean polygalacturonase-inhibiting protein 2 (PvPGIP2) is highly conserved in common bean (*Phaseolus vulgaris* L.) germplasm and related species. *Theor. Appl. Genet.* 118, 1371–1379. doi: 10.1007/s00122-009-0987-4

41. Federici, L., Caprari, C., Mattei, B., Savino, C., Di Matteo, A., De Lorenzo, G., et al. (2001). Structural requirements of endopolygalacturonase for the interaction with PGIP (polygalacturonase-inhibiting protein). *Proc. Natl. Acad. Sci. U.S.A.* 98, 13425–13430. doi: 10.1073/pnas.231473698

42. Federici, L., Di Matteo, A., Fernandez-Recio, J., Tsernoglou, D., and Cervone, F. (2006). Polygalacturonase inhibiting proteins: players in plant innate immunity? *Trends Plant Sci.* 11, 65–70. doi: 10.1016/j.tplants.2005.12.005

43. Ferrari, S., Galletti, R., Vairo, D., Cervone, F., and De Lorenzo, G. (2006). Antisense expression of the *Arabidopsis thaliana AtPGIP1* gene reduces polygalacturonase-inhibiting protein accumulation and enhances susceptibility to *Botrytis cinerea*. *Mol. Plant Microbe Interact.* 19, 931–936. doi: 10.1094/MPMI-19-0931

44. Ferrari, S., Savatin, D. V., Sicilia, F., Gramegna, G., Cervone, F., and De Lorenzo, G. (2013). Oligogalacturonides: plant damage-associated molecular patterns and regulators of growth and development. *Front. Plant Sci.* 4:49. doi: 10.3389/fpls.2013.00049

45. Ferrari, S., Sella, L., Janni, M., De Lorenzo, G., Favaron, F., and D'Ovidio, R. (2012). Transgenic expression of polygalacturonase-inhibiting proteins in *Arabidopsis* and wheat increases resistance to the flower pathogen *Fusarium graminearum*. *Plant Biol.* 14, 31–38. doi: 10.1111/j.1438-8677.2011.00449.x

46. Ferrari, S., Vairo, D., Ausubel, F. M., Cervone, F., and De Lorenzo, G. (2003). Tandemly duplicated *Arabidopsis* genes that encode polygalacturonase-inhibiting proteins are regulated coordinately by different signal transduction pathways in response to fungal infection. *Plant Cell Online* 15, 93–106. doi: 10.1105/tpc.005165

47. Fish, W. W., and Davis, A. R. (2004). The purification, physical/chemical characterization, and cDNA sequence of cantaloupe fruit polygalacturonase-inhibiting protein. *Phytopathology* 94, 337–344. doi: 10.1094/PHYTO.2004.94.4.337

48. Frati, F., Galletti, R., De Lorenzo, G., Salarino, G., and Conti, E. (2006). Activity of endo-polygalacturonases in mirid bugs (Heteroptera: Miridae) and their inhibition by plant cell wall proteins (PGIPs). *Eur. J. Entomol.* 103, 515–522. doi: 10.14411/eje.2006.067

49. Gazendam, I., Oelofse, D., and Berger, D. K. (2004). High-level expression of apple PGIP1 is not sufficient to protect transgenic potato against *Verticillium dahliae*. *Physiol. Mol. Plant Pathol.* 65, 145–155. doi: 10.1016/j.pmpp.2005.01.002

50. Gomathi, V., and Gnanamanickam, S. S. (2004). Polygalacturonase-inhibiting proteins in plant defence. *Curr. Sci.* 87, 1211–1217.

51. Gregori, R., Mari, M., Bertolini, P., Barajas, J. A., Tian, J. B., and Labavitch, J. M. (2008). Reduction of *Colletotrichum acutatum* infection by a polygalacturonase inhibitor protein extracted from apple. *Postharvest. Biol. Technol.* 48, 309–313. doi: 10.1016/j.postharvbio.2007.10.006

52. Guo, J.-L., Zhu, Y.-P., Shi, K., Jue, D.-W., Liu, S.-P., Hong, Y.-B., et al. (2014). Over-expression of *Solanum torvum* PGIP enhances resistance to*Verticillium dahliae* in transgenic potato plants. *Bothalia Pretoria.* 44, 392–404.

53. Gutierrez-Sanchez, G., King, D., Kemp, G., and Bergmann, C. (2012). SPR and differential proteolysis/MS provide further insight into the interaction between PGIP2 and EPGs. *Fungal Biol.* 116, 737–746. doi: 10.1016/j.funbio.2012.04.010

54. Hassan, F., Noorian, M. S., and Jacobsen, H. J. (2012). Effect of antifungal genes expressed in transgenic pea (*Pisum sativum* L.) on root colonization with *Glomus intraradices*. *GM Crops Food* 3, 301–309. doi: 10.4161/gmcr.21897

55. Hegedus, D. D., Li, R., Buchwaldt, L., Parkin, I., Whitwill, S., Coutu, C., et al. (2008). *Brassica napus* possesses an expanded set of polygalacturonase inhibitor protein genes that are differentially regulated in response to *Sclerotinia sclerotiorum* infection, wounding and defense hormone treatment. *Planta* 228, 241–253. doi: 10.1007/s00425-008-0733-1

56. Helft, L., Reddy, V., Chen, X., Koller, T., Federici, L., Recio, J. F., et al. (2011). LRR Conservation mapping to predict functional sites within protein leucine-rich repeat domains. *PLoS ONE* 6:e21614. doi: 10.1371/journal.pone.0021614

57. Hou, W., Mu, J., Li, A., Wang, H., and Kong, L. (2014). Identification of a wheat polygalacturonase-inhibiting protein involved in Fusarium head blight resistance. *Eur. J. Plant Pathol.* 141, 731–745. doi: 10.1007/s10658-014-0574-7

58. HuangFu, H., Guan, C., Jin, F., and Yin, C. (2014). Prokaryotic expression and protein function of *Brassica napus* PGIP2 and its genetic transformation. *Plant Biotechnol. Rep.* 8, 171–181. doi: 10.1007/s11816-013-0307-y

59. Hwang, B. H., Bae, H., Lim, H. S., Kim, K. B., Kim, S. J., Im, et al. (2010). Overexpression of polygalacturonase-inhibiting protein 2 (PGIP2) of Chinese cabbage (*Brassica rapa* ssp. pekinensis) increased

resistance to the bacterial pathogen *Pectobacterium carotovorum* ssp. carotovorum.*Plant Cell Tiss. Organ Cult.* 103, 293–305. doi: 10.1007/s11240-010-9779-4

60. Irshad, M., Canut, H., Borderies, G., Pont-Lezica, R., and Jamet, E. (2008). A new picture of cell wall protein dynamics in elongating cells of*Arabidopsis thaliana*: Confirmed actors and newcomers. *BMC Plant Biol.* 8:94. doi: 10.1186/1471-2229-8-94

61. James, J. T., and Dubery, I. A. (2001). Inhibition of polygalacturonase from *Verticillium dahliae* by a polygalacturonase inhibiting protein from cotton. *Phytochemistry* 57, 149–156. doi: 10.1016/S0031-9422(01)00024-3

62. Jang, S., Lee, B., Kim, C., Kim, S. J., Yim, J., Han, J. J., et al. (2003). The *OsFOR1* gene encodes a polygalacturonase-inhibiting protein (PGIP) that regulates floral organ number in rice. *Plant Mol. Biol.* 53, 357–372. doi: 10.1023/B:PLAN.0000006940.89955.f1

63. Janni, M., Bozzini, T., Moscetti, I., Volpi, C., and D'Ovidio, R. (2013). Functional characterisation of wheat Pgip genes reveals their involvement in the local response to wounding. *Plant Biol. Stuttg. Ger.* 15, 1019–1024. doi: 10.1111/plb.12002

64. Janni, M., Giovanni, M., Roberti, S., Capodicasa, C., and D'Ovidio, R. (2006). Characterization of expressed Pgip genes in rice and wheat reveals similar extent of sequence variation to dicot PGIPs and identifies an active PGIP lacking an entire LRR repeat. *Theor. Appl. Genet.* 113, 1233–1245. doi: 10.1007/s00122-006-0378-z

65. Janni, M., Sella, L., Favaron, F., Blechl, A. E., De Lorenzo, G., and D'Ovidio, R. (2008). The expression of a bean PGIP in transgenic wheat confers increased resistance to the fungal pathogen *Bipolaris sorokiniana*. *Mol. Plant Microbe Interact.* 21, 171–177. doi: 10.1094/MPMI-21-2-0171

66. Johnston, D. J., Ramanathan, V., and Williamson, B. (1993). A protein from immature raspberry fruits which inhibits endopolygalacturonases from *Botrytis cinerea* and other micro-organisms. *J. Exp. Bot.* 44, 971–976. doi: 10.1093/jxb/44.5.971

67. Joubert, D. A., Kars, I., Wagemakers, L., Bergmann, C., Kemp, G., Vivier, M. A., et al. (2007). A polygalacturonase-inhibiting protein from grapevine reduces the symptoms of the endopolygalacturonase BcPG2 from *Botrytis cinerea* in *Nicotiana benthamiana* leaves without any evidence for *in vitro* interaction. *Mol. Plant Microbe Interact.* 20, 392–402. doi: 10.1094/MPMI-20-4-0392

68. Joubert, D. A., Slaughter, A. R., Kemp, G., Becker, J. V. W., Krooshof, G. H., Bergmann, C., et al. (2006). The grapevine polygalacturonase-inhibiting protein (*VvPGIP1*) reduces *Botrytis cinerea* susceptibility in transgenic tobacco and differentially inhibits fungal polygalacturonases. *Transgenic Res.* 15, 687–702. doi: 10.1007/s11248-006-9019-1

69. Kalunke, R. M., Cenci, A., Volpi, C., O'Sullivan, D. M., Sella, L., Favaron, F., et al. (2014). The pgip family in soybean and three other legume species: evidence for a birth-and-death model of evolution. *BMC Plant Biol.* 14:189. doi: 10.1186/s12870-014-0189-3

70. Kalunke, R. M., Janni, M., Sella, L., David, P., Geffroy, V., Favaron, F., et al. (2011). Transcript analysis of the bean polygalacturonase inhibiting protein gene family reveals that Pvpgip2 is expressed in the whole plant and is strongly induced by pathogen infection. *J. Plant Pathol.* 93, 141–148. doi: 10.4454/jpp.v93i1.284

71. Kanai, M., Nishimura, M., and Hayashi, M. (2010). A peroxisomal ABC transporter promotes seed germination by inducing pectin degradation under the control of ABI5. *Plant J.* 62, 936–947. doi: 10.1111/j.1365-313X.2010.04205.x

72. Kemp, G., Bergmann, C. W., Clay, R., Van der Westhuizen, A. J., and Pretorius, Z. A. (2003). Isolation of a polygalacturonase-inhibiting protein (PGIP) from wheat. *Mol. Plant Microbe Interact.* 16, 955–961. doi: 10.1094/MPMI.2003.16.11.955

73. King, D., Bergmann, C., Orlando, R., Benen, J. A., Kester, H. C., and Visser, J. (2002). Use of amide exchange mass spectrometry to study conformational changes within the endopolygalacturonase II-homogalacturonan-polygalacturonase inhibiting protein system. *Biochemistry* 41, 10225–10233. doi: 10.1021/bi020119f

74. Kirsch, R., Wielsch, N., Vogel, H., Svatoš, A., Heckel, D. G., and Pauchet, Y. (2012). Combining proteomics and transcriptome sequencing to identify active plant-cell-wall-degrading enzymes in a leaf beetle. *BMC Genomics* 13:587. doi: 10.1186/1471-2164-13-587

75. Kobayashi, Y., Ohyama, Y., Kobayashi, Y., Ito, H., Iuchi, S., Fujita, M., et al. (2014). STOP2 activates transcription of several genes for Al- and low pH-tolerance that are regulated by STOP1 in Arabidopsis. *Mol. Plant* 7, 311–322. doi: 10.1093/mp/sst116

76. Ladu, G., Pani, G., Venditti, T., Dore, A., Molinu, M. G., and D›Hallewin, G. (2012). Natural resistance against pre- and post-harvest pathogens in Sardinian pears germoplasm. *Commun. Agric. Appl. Biol. Sci.* 77, 163–171.

77. Lagaert, S., Beliën, T., and Volckaert, G. (2009). "Plant cell walls: protecting the barrier from degradation by microbial enzymes" *Semin. Cell Dev. Biol.* 20, 1064–1073. doi: 10.1016/j.semcdb.2009.05.008

78. Leckie, F., Mattei, B., Capodicasa, C., Hemmings, A., Nuss, L., Aracri, B., et al. (1999). The specificity of polygalacturonaseinhibiting protein (PGIP): a single amino acid substitution in the solvent-exposed b-strand/ b-turn region of the leucine-rich repeats (LRRs) confers a new recognition capability. *EMBO J.* 18: 2352–2363 doi: 10.1093/emboj/18.9.2352

79. Lee, H. D., Bae, H., Kang, I. K., Byun, J. K., and Kang, S. G. (2006). Characterization of an apple polygalacturonase-inhibiting protein (PGIP) that specifically inhibits an endopolygalacturonase (PG) purified from apple fruits infected with *Botryosphaeria dothidea*. *J. Microbio. Biotech.* 16, 1192–1200.

80. Lim, J. M., Aoki, K., Angel, P., Garrison, D., King, D., Tiemeyer, M., et al. (2009). Mapping glycans onto specific N-linked glycosylation sites of*Pyrus communis* PGIP redefines the interface for EPG-PGIP interactions. *J. Proteome Res.* 8, 673–680. doi: 10.1021/pr800855f

81. Lin, W., and Li, Z. (2002). The partial structure of wheat polygalacturonase inhibiting protein. *Chin. J. Biochem. Mol. Bio.* 18, 197–201.

82. Liu, D., Li, W., He, X., Ding, Y., Chen, C., and Ge, F. (2013). Characterization and functional analysis of a novel PGIP gene from *Pyrus pyrifolia*Nakai cv Huobali. *Acta Physiol. Plant.* 35, 1247–1256. doi: 10.1007/s11738-012-1164-y

83. Lu, L., Zhou, F., Zhou, Y., Fan, X., Ye, S., Wang, L., et al. (2012). Expression profile analysis of the polygalacturonase-inhibiting protein genes in rice and their responses to phytohormones and fungal infection. *Plant Cell Rep.* 31, 1173–1187. doi: 10.1007/s00299-012-1239-7

84. Machinandiarena, M. F., Olivieri, F. P., Daleo, G. R., and Oliva, C. R. (2001). Isolation and characterization of a polygalacturonase-inhibiting protein from potato leaves. Accumulation in response tosalicylic acid, wounding and infection. *Plant Physiol. Biochem.* 39, 129–136. doi: 10.1016/S0981-9428(00)01228-6

85. Manfredini, C., Sicilia, F., Ferrari, S., Pontiggia, D., Salvi, G., Caprari, C., et al. (2005). Polygalacturonase-inhibiting protein 2 of *Phaseolus vulgaris* inhibits BcPG1, a polygalacturonase of*Botrytis cinerea* important for pathogenicity, and protects transgenic plants from infection.*Physiol. Mol. Plant Pathol.* 67, 108–115. doi: 10.1016/j.pmpp.2005.10.002

86. Mariotti, L., Casasoli, M., Migheli, Q., Balmas, V., Caprari, C., and De Lorenzo, G. (2008). Reclassification of *Fusarium verticillioides* (syn. *F.*

moniliforme) strain FC-10 as *F. phyllophilum. Mycol. Res.* 112, 1010–1011. doi: 10.1016/j.mycres.2008.07.004

87. Mattei, B., Bernalda, M. S., Federici, L., Roepstorff, P., Cervone, F., and Boffi, A. (2001). Secondary structure and post-translational modifications of the leucine-rich repeat protein PGIP (polygalacturonase-inhibiting protein) from *Phaseolus vulgaris. Biochemistry* 40, 569–576. doi: 10.1021/bi0017632

88. Maulik, A., Ghosh, H., and Basu, S. (2009). Comparative study of protein-protein interaction observed in Polygalacturonase-inhibiting proteins from *Phaseolus vulgaris* and *Glycine max* and Polygalacturonase from *Fusarium moniliforme. BMC Genomics* 10:S19. doi: 10.1186/1471-2164-10-S3-S19

89. Michelmore, R. W., and Meyers, B. C. (1998). Clusters of resistance genes in plants evolve by divergent selection and a birth-and-death process.*Genome Res.* 8, 1113–1130.

90. Misas-Villamil, J. C., and van der Hoorn, R. A. (2008). Enzyme-inhibitor interactions at the plant-pathogen interface. *Curr. Opin. Plant Biol.* 11, 380–388. doi: 10.1016/j.pbi.2008.04.007

91. Mohammadzadeh, R., Zamani, M., Motallebi, M., Norouzi, P., Jourabchi, E., Benedetti, M., et al. (2012). Agrobacterium tumefaciens-mediated introduction of polygalacturonase inhibiting protein 2 gene (PvPGIP2) from Phaseolus vulgaris into sugar beet (Beta vulgaris L.). *Aust. J. Crop Sci.* 6, 1290–1297.

92. Nguema-Ona, E., Moore, J. P., Fagerström, A. D., Fangel, J. U., Willats, W. G., Hugo, A., et al. (2013). Overexpression of the grapevine PGIP1 in tobacco results in compositional changes in the leaf arabinoxyloglucan network in the absence of fungal infection. *BMC Plant Biol.* 13:46. doi: 10.1186/1471-2229-13-46

93. Nietlispach, D., Mott, H. R., Stott, K. M., Nielsen, P. R., Thiru, A., and Laue, E. D. (2004). Structure determination of protein complexes by NMR.*Methods Mol. Biol.* 278, 255–288. doi: 10.1385/1-59259-809-9:255

94. Nuss, L., Mahe, A., Clark, A. J., Grisvard, J., Dron, M., Cervone, F., et al. (1996). Differential accumulation of PGIP (polygalacturonase inhibiting protein) mRNA in two near-isogenic lines of *Phaseolus vulgaris* L. upon infection with *Colletotrichum lindemuthianum. Physiol. Mol. Plant Pathol.* 48, 83–89.

95. O'Kennedy, M., Burger, J., and Berger, D. (2001). Transformation of elite white maize using the particle inflow gun and detailed analysis of a

low-copy integration event. *Plant Cell Rep.* 20, 721–730. doi: 10.1007/s002990100383

96. O'Brien, J. A., Daudi, A., Finch, P., Butt, V. S., Whitelegge, J. P., Souda, P., et al. (2012). A peroxidase-dependent apoplastic oxidative burst in cultured arabidopsis cells functions in MAMP-elicited defence. *Plant Physiol.* 158, 2013–2027. doi: 10.1104/pp.111.190140

97. Oelofse, D., Dubery, I. A., Meyer, R., Arendse, M. S., Gazendam, I., and Berger, D. K. (2006). Apple polygalacturonase inhibiting protein1 expressed in transgenic tobacco inhibits polygalacturonases from fungal pathogens of apple and the anthracnose pathogen of lupins. *Phytochemistry* 67, 255–263. doi: 10.1016/j.phytochem.2005.10.029

98. Oh, D.-H., Dassanayake, M., Bohnert, H. J., and Cheeseman, J. M. (2012). Life at the extreme: lessons from the genome. *Genome Biol.* 13, 241. doi: 10.1186/gb-2012-13-3-241

99. Oliveira, M. B., Nascimento, L. B., Junior, M. L., and Petrofeza, S. (2010). Characterization of the dry bean polygalacturonase-inhibiting protein (PGIP) gene family during *Sclerotinia sclerotiorum* (*Sclerotiniaceae*) infection. *Genet. Mol. Res.* 9, 994–1004. doi: 10.4238/vol9-2gmr776

100. Pérez-Donoso, A. G., Sun, Q., Roper, M. C., Greve, L. C., Kirkpatrick, B., and Labavitch, J. M. (2010). Cell wall-degrading enzymes enlarge the pore size of intervessel pit membranes in healthy and *Xylella fastidiosa*-infected grapevines. *Plant Physiol.* 152, 1748–1759. doi: 10.1104/pp.109.148791

101. Powell, A. L. T., Stotz, H. U., Labavitch, J. M., and Bennett, A. B. (1994). *"Glycoprotein inhibitors of fungal polygalacturonases,"* in *Advances in Molecular Genetics of Plant-Microbe Interactions, Current Plant Science and Biotechnology in Agriculture*, eds M. J. Daniels, J. A. Downie, and A. E. Osbourn (Dordrecht: Springer), 399–402.

102. Powell, A. L. T., van Kan, J., ten Have, A., Visser, J., Greve, L. C., Bennett, A. B., et al. (2000). Transgenic expression of pear PGIP in tomato limits fungal colonization. *Mol. Plant Microbe Interact.* 13, 942–950. doi: 10.1094/MPMI.2000.13.9.942

103. Prabhu, S. A., Kini, K. R., Raj, S. N., Moerschbacher, B. M., and Shetty, H. S. (2012). Polygalacturonase-inhibitor proteins in pearl millet: possible involvement in resistance against downy mildew. *Acta Biochim. Biophys. Sinica.* 44, 415–423. doi: 10.1093/abbs/gms015

104. Prabhu, S. A., Singh, R., Kolkenbrock, S., Sujeeth, N., El Gueddari, N. E., Moerschbacher, B. M., et al. (2014). Experimental and bioinformatic

characterization of recombinant polygalacturonase-inhibitor protein from pearl millet and its interaction with fungal polygalacturonases. *J. Exp. Bot.* 65, 5033–5047. doi: 10.1093/jxb/eru266

105. Prabhu, S. A., Wagenknecht, M., Melvin, P., Gnanesh Kumar, B. S., Veena, M., Shailasree, S., et al. (2015). Immuno-affinity purification of PglPGIP1, a polygalacturonase-inhibitor protein from pearl millet: studies on its inhibition of fungal polygalacturonases and role in resistance against the downy mildew pathogen. *Mol. Biol. Rep.* doi: 10.1007/s11033-015-3850-5. [Epub ahead of print].

106. Protsenko, M. A., Buza, N. L., Krinitsyna, A. A., Bulantseva, E. A., and Korableva, N. P. (2008). Polygalacturonase-inhibiting protein is a structural component of plant cell wall. *Biochem. (Moscow)* 73, 1053–1062. doi: 10.1134/S0006297908100015

107. Raiola, A., Sella, L., Castiglioni, C., Balmas, V., and Favaron, F. (2008). A single amino acid substitution in highly similar endo-PGs from*Fusarium verticillioides* and related *Fusarium* species affects PGIP inhibition. *Fungal Genet. Biol.* 45, 776–789. doi: 10.1016/j.fgb.2007.11.003

108. Reignault, P., Valette-Collet, O., and Boccara, M. (2008). The importance of fungal pectinolytic enzymes in plant invasion, host adaptability and symptom type. *Eur. J. Plant Pathol.* 120, 1–11. doi: 10.1007/s10658-007-9184-y

109. Richter, A., Jacobsen, H. J., De Kathen, A., De Lorenzo, G., Briviba, K., Hain, R., et al. (2006). Transgenic peas (*Pisum sativum*) expressing polygalacturonase inhibiting protein from raspberry (*Rubus idaeus*) and stilbene synthase from grape (*Vitis vinifera*). *Plant Cell Rep.* 25, 1166–1173. doi: 10.1007/s00299-006-0172-z

110. Ridley, B. L., O'Neill, M. A., and Mohnen, D. (2001). Pectins: structure, biosynthesis, and oligogalacturonide-related signaling.*Phytochemistry*57, 929–967. doi: 10.1016/S0031-9422(01)00113-3

111. Roper, M. C., Greve, L. C., Warren, J. G., Labavitch, J. M., and Kirkpatrick, B. C. (2007). *Xylella fastidiosa* requires polygalacturonase for colonization and pathogenicity in *Vitis vinifera* grapevines. *Mol. Plant Microbe Interact.* 20, 411–419. doi: 10.1094/MPMI-20-4-0411

112. Sathiyaraj, G., Srinivasan, S., Subramanium, S., Kim, Y. J., Kim, Y. J., Kwon, W. S., et al. (2010). Polygalacturonase inhibiting protein: isolation, developmental regulation and pathogen related expression in *Panax ginseng* C.A. Meyer. *Mol. Biol. Rep.* 37, 3445–3454. doi: 10.1007/s11033-009-9936-1

113. Sato, A., and Miura, K. (2011). Root architecture remodeling induced by phosphate starvation. *Plant Signal. Behav.* 6, 1122–1126. doi: 10.4161/psb.6.8.15752

114. Sawaki, Y., Iuchi, S., Kobayashi, Y., Kobayashi, Y., Ikka, T., Sakurai, N., et al. (2009). STOP1 regulates multiple genes that protect Arabidopsis from proton and aluminum toxicities. *Plant Physiol.* 150, 281–294. doi: 10.1104/pp.108.134700

115. Schacht, T., Unger, C., Pich, A., and Wydra, K. (2011). Endo- and exopolygalacturonases of *Ralstonia solanacearum* are inhibited by polygalacturonase-inhibiting protein (PGIP) activity in tomato stem extracts. *Plant Physiol. Biochem.* 49, 377–387. doi: 10.1016/j.plaphy.2011.02.001

116. Sella, L., Castiglioni, C., Roberti, S., D'Ovidio, R., and Favaron, F. (2004). An endo-polygalacturonase (PG) of *Fusarium moniliforme* escaping inhibition by plant polygalacturonase-inhibiting proteins (PGIPs) provides new insights into the PG-PGIP interaction. *FEMS Microbiol. Lett.* 240, 117–124. doi: 10.1016/j.femsle.2004.09.019

117. Senthil, G., Williamson, B., Dinkins, R. D., and Ramsay, G. (2004). An efficient transformation system for chickpea (*Cicer arietinum* L.). *Plant Cell Rep.* 23, 297–303. doi: 10.1007/s00299-004-0854-3

118. Shanmugam, V. (2005). Role of extracytoplasmic leucine rich repeat proteins in plant defence mechanisms. *Microbiol. Res.* 160, 83–94. doi: 10.1016/j.micres.2004.09.014

119. Shivashankar, S., Thimmareddy, C., and Roy, T. K. (2010). Polygalacturonase inhibitor protein from fruits of anthracnose resistant and susceptible varieties of Chilli (*Capsicum annuum* L). *Indian J. Biochem. Biophys.* 47, 243–248.

120. Sicilia, F., Fernandez-Recio, J., Caprari, C., De Lorenzo, G., Tsernoglou, D., Cervone, F., et al. (2005). The polygalacturonase-inhibiting protein PGIP2 of *Phaseolus vulgaris* has evolved a mixed mode of inhibition of endopolygalacturonase PG1 of *Botrytis cinerea*. *Plant Physiol.* 139, 1380–1388. doi: 10.1104/pp.105.067546

121. Song, K. H., and Nam, Y. W. (2005). Genomic organization and differential expression of two polygalacturonase-inhibiting protein genes from*Medicago truncatula*. *J. Plant Biol.* 48, 467–478. doi: 10.1007/BF03030589

122. Spadoni, S., Zabotina, O., Di Matteo, A., Mikkelsen, J. D., Cervone, F., De Lorenzo, G., et al. (2006). Polygalacturonase-inhibiting protein interacts with pectin through a binding site formed by four clustered

residues of arginine and lysine. *Plant Physiol.* 141, 557–564. doi: 10.1104/pp.106.076950

123. Spinelli, F., Mariotti, L., Mattei, B., Salvi, G., Cervone, F., and Caprari, C. (2009). Three aspartic acid residues of polygalacturonase-inhibiting protein (PGIP) from *Phaseolus vulgaris* are critical for inhibition of *Fusarium phyllophilum* PG. *Plant Biol.* 11, 738–743. doi: 10.1111/j.1438-8677.2008.00175.x

124. Szankowski, I., Briviba, K., Fleschhut, J., Schönherr, J., Jacobsen, H. J., and Kiesecker, H. (2003). Transformation of apple (*Malus domestica*Borkh.) with the stilbene synthase gene from grapevine (*Vitis vinifera* L.) and a PGIP gene from kiwi (*Actinidia deliciosa*). *Plant Cell Rep.* 22, 141–149. doi: 10.1007/s00299-003-0668-8

125. Tamburino, R., Chambery, A., Parente, A., and Di Maro, A. (2012). A Novel Polygalacturonase-Inhibiting Protein (PGIP) from *Lathyrus sativus*L. seeds. *Protein Pept. Lett.* 19, 820–825. doi: 10.2174/092986612801619561

126. Tamura, M., Gao, M., Tao, R., Labavitch, J. M., and Dandekar, A. M. (2004). Transformation of persimmon with a pear fruit polygalacturonase inhibiting protein (PGIP) gene. *Sci. Hortic.* 103, 19–30. doi: 10.1016/j. scienta.2004.04.006

127. ten Have, A., Mulder, W., Visser, J., and van Kan, J. A. (1998). The endopolygalacturonase gene *Bcpg1* is required for full virulence of *Botrytis cinerea*. *Mol. Plant Microbe Interact.* 11, 1009–1016. doi: 10.1094/MPMI.1998.11.10.1009

128. Thornburg, R. W., Carter, C., Powell, A. L., Mittler, R., Rizhsky, R., and Horner, H. T. (2003). A major function of the tobacco floral nectary is defense against microbial attack. *Plant Syst. Evol.* 238, 211–218. doi: 10.1007/s00606-003-0282-9

129. Toubart, P., Desiderio, A., Salvi, G., Cervone, F., Daroda, L., De Lorenzo, G., et al. (1992). Cloning and characterization of the gene encoding the endopolygalacturonase-inhibiting protein (PGIP) of *Phaseolus vulgaris* L. *Plant J.* 2, 367–373. doi: 10.1046/j.1365-313X.1992.t01-35-00999.x

130. van Santen, Y., Benen, J. A., Schroter, K. H., Kalk, K. H., Armand, S., and Visser, J. (1999). 1.68-Å crystal structure of endopolygalacturonase II from *Aspergillus niger* and identification of active site residues by site-directed mutagenesis. *J. Biol. Chem.* 274, 30474–30480. doi: 10.1074/jbc.274.43.30474

131. Volpi, C., Raiola, A., Janni, M., Gordon, A., O'Sullivan, D. M., Favaron,

F., et al. (2013). *Claviceps purpurea* expressing polygalacturonases escaping PGIP inhibition fully infects PvPGIP2 wheat transgenic plants but its infection is delayed in wheat transgenic plants with increased level of pectin methyl esterification. *Plant Physiol. Biochem.* 73, 294–301. doi: 10.1016/j.plaphy.2013.10.011

132. Wand, A. J., and Englander, S. W. (1996). Protein complexes studied by NMR spectroscopy. *Curr. Opin. Biotechnol.* 7, 403–408.

133. Wang, A., Wei, X., Rong, W., Dang, L., Du, L. P., Qi, L., et al. (2014a). GmPGIP3 enhanced resistance to both take-all and common root rot diseases in transgenic wheat. *Funct. Integr. Genomics.* doi: 10.1007/s10142-014-0428-6. [Epub ahead of print].

134. Wang, R., Lu, L., Pan, X., Hu, Z., Ling, F., Yan, Y., et al. (2014b). Functional analysis of *OsPGIP1* in rice sheath blight resistance. *Plant Mol. Biol.* 87, 181–191. doi: 10.1007/s11103-014-0269-7

135. Wang, X., Zhu, X., Tooley, P., and Zhang, X. (2013). Cloning and functional analysis of three genes encoding polygalacturonase-inhibiting proteins from *Capsicum annuum* and transgenic CaPGIP1 in tobacco in relation to increased resistance to two fungal pathogens. *Plant Mol. Biol.* 81, 379–400. doi: 10.1007/s11103-013-0007-6

Chapter 8

UNRAVELING INCOMPATIBILITY BETWEEN WHEAT AND THE FUNGAL PATHOGEN ZYMOSEPTORIA TRITICI THROUGH APOPLASTIC PROTEOMICS

Fen Yang[1], Wanshun Li[2], Mark Derbyshire[3], Martin R Larsen[4], Jason J Rudd[3] and Giuseppe Palmisano[4,5]

[1]Department of Plant and Environmental Sciences, University of Copenhagen, 1871 Frederiksberg C, Denmark

[2] BGI-tech, BGI, 518083 Shenzhen, China

[3] Department of Plant Biology and Crop Science, Rothamsted Research, Harpenden, Hertfordshire AL5 2JQ, United Kingdom

[4] Department of Biochemistry and Molecular Biology, University of Southern Denmark, 5230 Odense M, Denmark

[5] Present address: Institute of Biomedical Science, Dep

ABSTRACT

Background

Hemibiotrophic fungal pathogen *Zymoseptoria tritici* causes severe foliar disease in wheat. However, current knowledge of molecular mechanisms involved in plant resistance to *Z. tritici* and *Z. tritici* virulence factors is far from being complete. The present work investigated the proteome of leaf apoplastic fluid with emphasis on both host wheat and *Z. tritici* during the compatible and incompatible interactions.

Results

The proteomics analysis revealed rapid host responses to the biotrophic growth, including enhanced carbohydrate metabolism, apoplastic defenses and stress, and cell wall reinforcement, might contribute to resistance. Compatibility between the host and the pathogen was associated with inactivated plant apoplastic responses as well as fungal defenses to oxidative stress and perturbation of plant cell wall during the initial biotrophic stage, followed by the strong induction of plant defenses during the necrotrophic stage. To study the role of anti-oxidative stress in Z. *tritici* pathogenicity in depth, a YAP1 transcription factor regulating antioxidant expression was deleted and showed the contribution to anti-oxidative stress in Z. *tritici*, but was not required for pathogenicity. This result suggests the functional redundancy of antioxidants in the fungus.

Conclusions

The data demonstrate that incompatibility is probably resulted from the proteome-level activation of host apoplastic defenses as well as fungal incapability to adapt to stress and interfere with host cell at the biotrophic stage of the interaction.

BACKGROUND

The ascomycete fungus *Zymoseptoria tritici* causes Septoria tritici blotch (STB) in wheat, a foliar disease that poses a significant threat to global food production. Leaf penetration occurs by means of fungal hyphae emerging from geminating, surface-attached spores that enter via stomata. The fungus has a slow intercellular biotrophic symptomless growth, for typically up to 10 days, as hyphae extend in close contact with mesophyll cells, probably utilizing lipid and fatty acid stores for growth [1,2]. Subsequently, the fungus suddenly switches to the necrotrophic growth associated with leakage of nutrients from dying plant cells into the apoplastic spaces, an increase in fungal biomass, enhanced signaling, metabolism and defense responses in host, the appearance of lesions on the leaf surface, and the collapse of the plant tissue [3,4]. Disease transition and appearance of symptoms have been suggested to be triggered by fungal small protein effectors secreted into apoplast [2-4]. Differing from many other phytopathogenic fungi, Z. *tritici* does not form any specialized penetration or feeding structures and remains strictly apoplastic throughout the entire infection cycle.

The plant apoplast is potentially important as a bridge that perceives and transduces signals from the environment to the symplast. Under stress conditions, complex mechanisms, including accumulation of reactive oxygen species (ROS) and changes in the synthesis of extracellular proteins, are activated in the apoplast as a first line of defenses. The secreted plant apoplastic proteins predominantly represent functional categories associated with carbohydrate metabolism, cell wall metabolism, defense, and programmed cell death [5]. As an apoplast-inhabiting fungus, Z. *tritici* need to acquire apoplastic nutrients, shape the plant cell structures, and overcome the activated apoplastic defenses to survive, possibly via secretion of effectors involved in detoxification of defense-related molecules as well as protection against recognition by the plant. This highly dynamic compartment serves as the molecular battlefield that contributes to the success of infection or plant resistance.

Given the crucial role of leaf apoplast in wheat-*Z. tritici* interaction, it is of particular interest to investigate the molecular basis underlying plant apoplastic immunity and the counter defenses that *Z. tritici* evolves at different growth stages. Although there have been advances in understanding mechanisms of wheat responses to *Z. tritici*, involving programming cell death, ROS accumulation, activation of signal transduction, transport and energy metabolism, expression of a broad spectrum of pathogenesis-related (PR) proteins, antioxidants and jasmonic acid biosynthesis genes, and production of small signaling and defense compounds [1,2,4,6-10], well-characterized systematic apoplastic responses to *Z. tritici* in wheat, which are essential for determining the plant fate, are currently lacking.

Considerable studies have been performed to understand *Z. tritici* gene functions, mainly focusing on the necrotrophic growth due to low fungal biomass hardly detectable at the biotrophic stage. Until recently, the emerging high throughput 'omics' and sequencing technologies partly address the issue and enable the discovery of several fungal genes and proteins expressed at different growth stages [2,4,10]. Compared to uncovering the expression of *Z. tritici* genes including the genes encoding cell-wall-degrading enzymes (CWDEs), ROS-scavenging enzymes and putative effector proteins such as LysM as well as the production of the secondary metabolites during the compatible interaction [1-4,11], *Z. tritici in planta* proteins, particularly secreted protein effectors, have not been fully explored at a systematic level. The only proteomic report identified thirty-one proteins and five phosphoproteins of *Z. tritici* mainly involved in basic cellular machinery and signaling at the biotrophic stage of the compatible and incompatible interactions [10]. This study revealed a similarity in fungal protein profiles between two interactions, possibly due to the fact that analysis of whole inoculated leaves resulted in the dominance of most abundant plant and fungal proteins, which largely diluted the information about low abundant fungal proteins likely essential for pathogenicity. A deeper insight into *Z. tritici*-specific strategies of colonization can be achieved by a detailed analysis of host apoplast for the enrichment of fungal identifications, in particular secreted molecules.

The expression of genes encoding ROS-scavenging enzymes is one of the multiple defense systems evolved by *Z. tritici* to thrive in the oxidative apoplastic environment. The YAP1 transcription factor is one of the most important determinants of oxidative stress responses, responsible for transcriptional activation of oxidative stress-associated genes in fungi [12]. YAP1 undergoes a conformational change due to the formation of disulfide bonds upon exposure to oxidative stress and is thus transported from the cytoplasm into the nucleaus to activate gene transcription. YAP1-mediated detoxification of ROS is essential in the virulence of many pathogenic fungi including biotroph *Ustilago maydis*, necrotroph *Alternaria alternate* and human pathogen *Candida albcans* [13]. However, the role of YAP1 in oxidative stress responses during *Z. tritici* colonization is poorly defined.

In order to gain further insights into host resistance and *Z. tritici* pathogenicity, we explored the proteome from apoplastic washing fluid (AWF) with emphasis on the host and the pathogen during the compatible and incompatible interactions. This is a perspective focusing on the direct battle ground and differing from the previous genome-wide studies on plant-*Z. tritici*interaction. The analysis uncovers apoplastic regulatory networks that shape the aspects of the plant physiology in response to the fungus and the principles of fungal modulation of apoplastic immunity. The combat between the host and*Z. tritici* through the activation of their defense mechanisms at the biotrophic stage likely determines their fates. Our data have also expanded on the current models of fungal apoplastic proteome.

RESULTS

Wheat Proteins from AWF in Response to *Z. tritici*

It was difficult to avoid the leakage of symplastic compounds in leaf apoplast during the extraction, particularly from the dying leaf cells in the necrotrophic phase. In order to obtain the important apoplastic components at the necrotrophic stage but avoid severe symplastic contamination, we sampled the leaves at the early time point of the necrotrophic growth (14 days after inoculation, 14 dai) when immature asexual sporulation structure started to form. Malate dehydrogenase (MDH) was used as a cytosolic enzyme marker to evaluate the level of contamination. MDH activities in AWF contained max. 1% of those from the whole leaf extracts in all samples (Additional file 1). The ratio below 1% is considered as little damage to the cells and is acceptable for plant apoplast studies [14]. Approx. 1% MDH activity related to the total leaf extracts was also observed in the apoplastic tissue of ryegrass leaves during senescence [15], and of *V. longisporum*-infected oilseed rape leaves [16] and *A. thaliana* [17]. This recurrent and unavoidable cytoplasmic contamination has already been highlighted by several previous studies of plant apoplast [18-22]. Therefore, we considered apoplastic proteins were enriched in the AWF in the present study.

We applied a mass spectrometry (MS)-based shotgun quantitative proteomics using isobaric tags for relative and absolute quantitation (iTRAQ) labeling. This approach has been developed to analyze the proteome and phosphoproteome of Z. tritici-inoculated wheat leaves [10]. Missing reporter ions of the peptide from some of the samples in the MS/MS spectrum can occur in the iTRAQ-based proteomics due to the incomplete labeling and inefficient fragmentation of the tags, which results in no quantitative ratio. Therefore, the criterion defined for the reliable quantified proteins was that proteins had to be identified in at least two biological replicates with quantitative ratio. This resulted in the identification of 2122 and 2071 wheat proteins and quantification of 607 and 575 proteins in Stakado (resistant cultivar) and Sevin (susceptible cultivar), respectively (Additional file 2). Not surprisingly, cytosolic proteins such as Rubisco and ribosomal proteins were identified in all the samples by this highly sensitive MS approach, which were eliminated for the further analysis.

Based on the selection criteria, 45 proteins in Stakado and 100 proteins in Sevin were found to change in abundance in response to the fungus (Additional file 3). Eleven were detected in both cultivars. Of the regulated proteins, 42% and 60% from Stakado and Sevin were predicted with signal peptides using SignalP program, respectively. Different percentages of proteins with predicted signal sequences were found in the secretome of soybean (65%), grapevine (66%), Arabidopsis(47%) and rice (37%) [18]. There were 43 and 96 differentially expressed proteins identified from Stakado and Sevin whose corresponding transcripts were identified from the previous transcriptome dataset, respectively [4].

Expression cluster analysis revealed that a substantial number of proteins were up-regulated only at the necrotrophic stage (14 dai) in contrast to minor changes at the biotrophic stage (5 dai) in Sevin, whereas proteins could be up- or down- regulated at both time points in Stakado (Figure 1A). The major proteins in each cluster were annotated in carbohydrate metabolism, stress, defense, cell wall metabolism, protein process, and other metabolic processes (Figure 1B). In Stakado, at either time point, the fungus caused a significantly increased expression of several carbohydrate metabolic proteins involved in glycolysis (i.e., 2,3-bisphosphoglycerate-independent phosphoglycerate mutase and phosphoglycerate kinase), pentose phosphate pathway (i.e., 6-phosphogluconate dehydrogenase) and TCA (i.e., aconitate hydratase), as well as stress/defense-related proteins including PR-1, PR-2, PR-3, peroxidase 3, heat shock protein and glycolate oxidase (clusters 1, 3 and 4), in contrast to a decreased expression of cell-wall metabolic proteins including α-L-arabinofuranosidase, β-D-glucan exohydrolase,

1,3(4)-β-glucanase, and β-D-xylosidase (cluster 2 and 5). The homologs of aconitate hydratase, 2,3-bisphosphoglycerate-independent phosphoglycerate mutase and phosphoglycerate kinase were also identified in the study of poplar apoplastic proteome [5]. Additionally, a 1-deoxy-D-xylulose 5-phosphate reductoisomerase responsible for terpenoid biosynthesis displayed a strong accumulation (11-fold change) at 14 dai, indicating the possible involvement of terpenoid in wheat resistance to *Z. tritici*. The pathogen-induced expression of terpenoid synthase gene and production of diterpenoid phytoalexins have been seen in maize and rice [23,24]. Moreover, integration of proteome and previous transcriptome datasets revealed that the majority of regulated proteins during the incompatible interaction did not change at transcription level in the compatible interaction. This result further emphasizes an important role of these proteins in plant resistance to *Z. tritici*. In Sevin, up-regulated proteins at 14 dai consisted of proteins involved in protein process, cell wall metabolism, stress, and defense, including a cold acclimation induced protein, peroxidases and numerous PR-proteins belonging to PR families 1, 2, 3, 4, 5, and 17 (clusters 1, 3, 4 and 5). A good correlation between protein and transcript regulation at the necrotrophic stage was observed, indicating the great amplitude of activation of host defense responses. On the other hand, most proteins that did not change in abundance at the biotrophic stage of the compatible interaction were transcriptionally suppressed, suggesting a complexity of molecular mechanisms in host plant triggered by *Z. tritici*biotrophic growth.

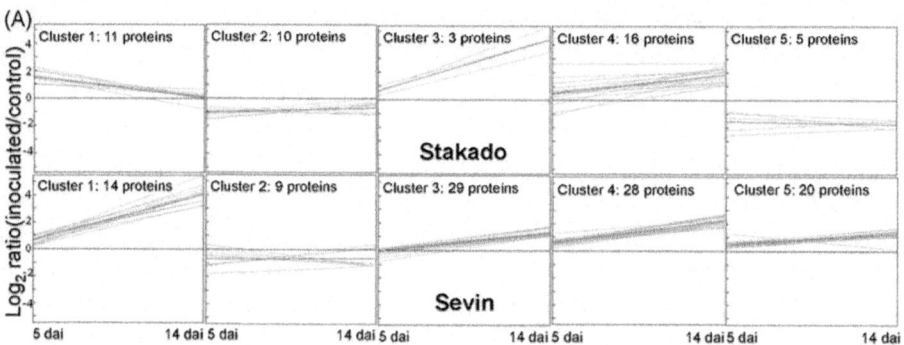

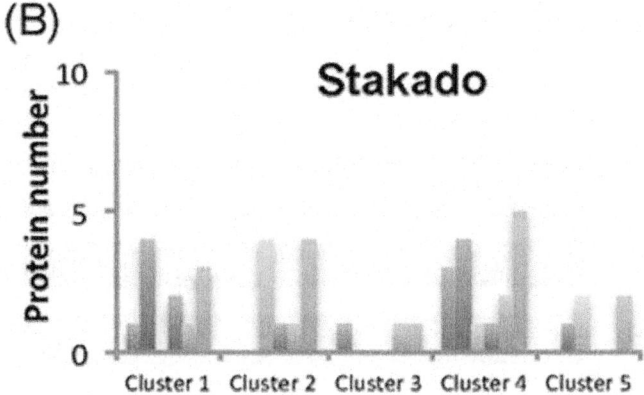

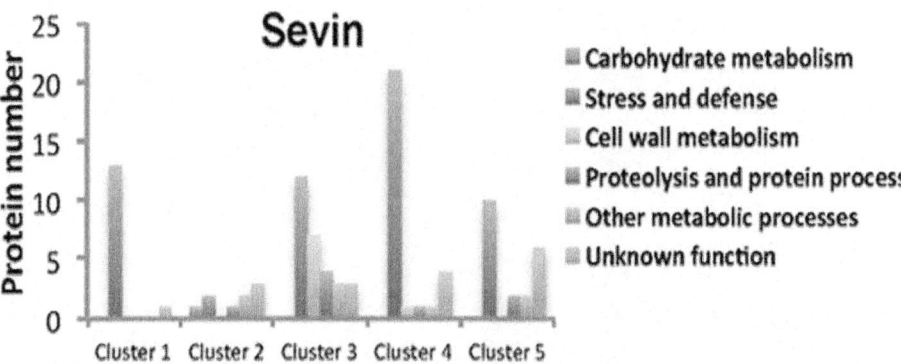

Figure 1: Differentially expressed plant proteins identified from AWF. AWF was isolated from the leaves of wheat cvs. Stakado (resistant) and Sevin (susceptible) inoculated with *Z. tritici* or water (control) at 5 and 14 dai. Three biological samples were prepared. The proteins from AWF were labeled with iTRAQ and subjected to LC-MS/MS analysis. Differentially expressed plant proteins were analyzed by expression clustering **(A)** and functional classification **(B)**.

In order to validate our proteomics data, western blot analysis of PR-1, PR-2, and PR-3 proteins was conducted with three biological replicates of AWF samples. In agreement with the proteomics analysis, the abundance of PR-1, −2, and −3 increased slightly at the early biotrophic stage of the incompatible interaction and markedly at the late stage of the compatible interaction (Figure 2).

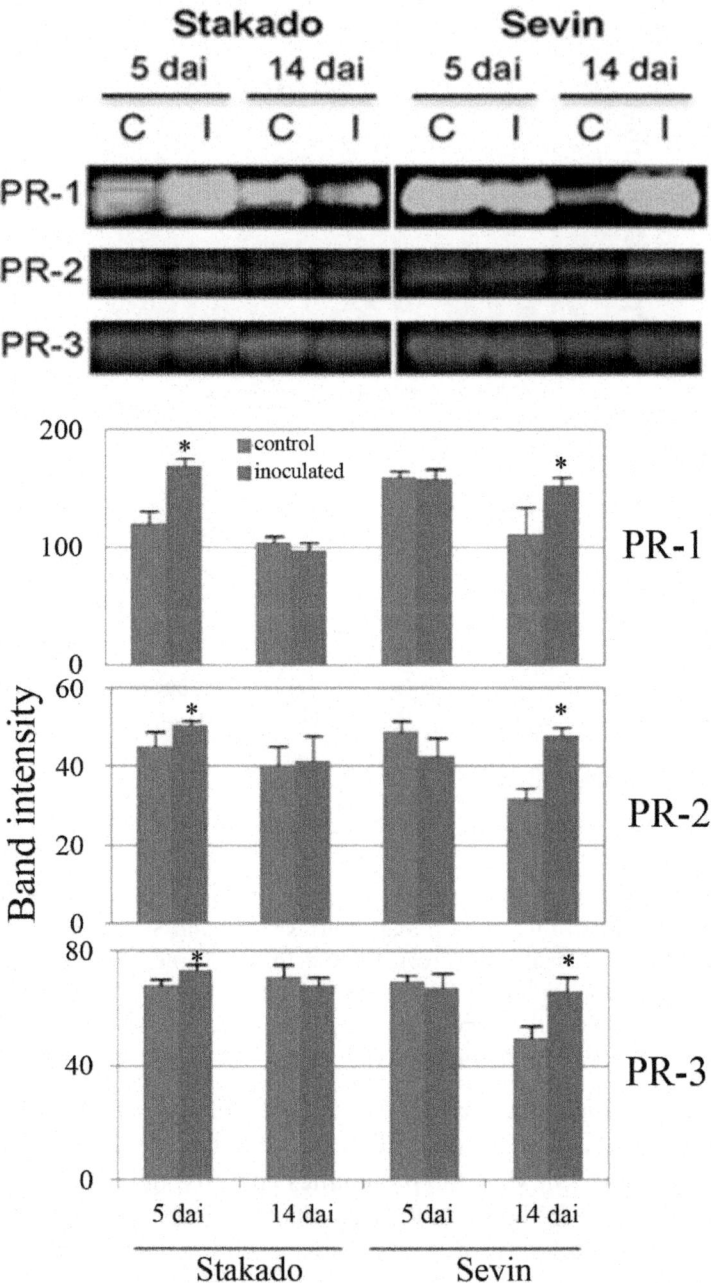

Figure 2: Western blotting validation of proteome data. Protein was extracted from three biological replicates of control (C) and *Z. tritici*-inoculated (I) wheat cvs. Stakado (resistant) and Sevin (susceptible) at 5 and 14 dai and used in western blot analysis.

The representative membranes of PR-1, PR-2 and PR-3 protein expression are shown. Quantification of signals on the membranes was performed by using ImageJ program based on three biological samples. The asterisks indicate significant differences in signal intensity (P 0.05) between control and inoculated samples.

Fungal Proteins Identified from Inoculated Leaf AWF

It is difficult to distinguish whether the protein is produced by the plant, the fungus or both in a plant-fungus interaction system, when the peptide matches to the protein sequences from both host and pathogen databases. Therefore, we have showed the proteins with peptides confidently identified from fungal database only, since the proteins identified from both databases are more likely of plant origin due to high biomass ratio of plant to the fungus. Twenty-four and thirty-one fungal proteins were identified from the inoculated Stakado and Sevin, respectively (Table 1). There were eighteen proteins whose corresponding transcripts were identified from the previous RNA-seq-based transcriptome dataset (Table 1, Additional file 4). Twelve proteins had an N-terminal signal peptide sequence predicted by SignalP, indicating they could be secreted proteins. The identified fungal proteins with known functions could be assigned to biological processes, encompassing metabolism, degradation of host cell wall, signaling, stress, defense, transport, cell mobility, cell component organization, and cell wall degradation and remodeling. Despite similar functional categories of fungal proteins and nine proteins identified from both inoculated cultivars, the overviews of fungal protein profiles were distinct between the compatible and incompatible interactions (Table 1). Apart from the categories of primary carbohydrate, amino acid and protein metabolisms regarded as no significant changes between two interactions, the fungus seemed to preferentially express the proteins involved in signal transduction, transport, and cell wall remodeling during the incompatible interaction, whereas the proteins involved in nucleic acid metabolism, transcription, perturbation of plant cell wall, anti-oxidative stress, and cellular component organization could only be identified in the compatible interaction.

Table 1: *Z. tritici* proteins identified from *Z. tritici* -inoculated wheat leaves

		Ratio (14 d/ 5 d)			
ID	Annotation	S	SK	SV	Biological process
91239	Myosin			1.36	Cell mobility
105948	Actin		1.15	1.11	Cell mobility

110409	Myosin	Y	1.17	1.33	Cell mobility
45905	Chitinase		0.52		Cell wall remodeling
86354	SWI/SNF chromatin re-modeling complex protein			N/A	Cellular component organization
87313	Rrp15p domain-containing protein			1.04	Cellular component organization
68922	Glycoside hydrolase family 62	Y		4.18	Degradation of plant cell wall
70396	α-L-arabinofuranosidase B	Y		5.67	Degradation of plant cell wall
99970	β-glucosidase	Y		2.70	Degradation of plant cell wall
44922	N6 adenine-specific DNA methylase			N/A	Metabolism
46697	ATPase		0.81		Metabolism
52059	Formate-tetrahydrofolate ligase		N/A	1.05	Metabolism
55968	Triosephosphate isomerase		1.63	1.38	Metabolism
64923	Glucose/ribitol dehydro-genase			7.74	Metabolism
68338	Isopenicillin N synthase	Y	1.23		Metabolism
69333	Glycine dehydrogenase		0.67		Metabolism
73114	Hydantoinase/oxoprolinase			2.31	Metabolism
74730	ATP-citrate lyase/succinyl-CoA ligase		1.81		Metabolism
76530	D-isomer specific 2-hy-droxyacid dehydrogenase	Y		N/A	Metabolism
77172	Aldo-keto reductase		1.37	1.70	Metabolism
96092	Trypsin	Y	1.80	0.78	Metabolism
102177	20S proteasome		1.58	0.99	Metabolism
106055	Aconitate hydratase		1.05		Metabolism
107787	Mediator16 domain-con-taining protein			1.86	Metabolism
110230	α -isopropylmalate/homoci-trate synthase			1.35	Metabolism
65824	Serine/threonine-specific protein phosphatase			1.07	Signaling

98343	Ras GTPase		1.64	1.22	Signaling
99493	Ras GTPase		2.33		Signaling
99564	Calmodulin		1.02		Signaling
110139	14-3-3 protein		N/A		Signaling
67250	Catalase/peroxidase	Y		36.9	Stress and defense
75170	Chaperonin 60		0.81		Stress and defense
99959	Heat shock protein 90	Y		1.04	Stress and defense
103593	Copper/Zinc superoxide dismutase			4.33	Stress and defense
105409	Catalase/peroxidase			9.33	Stress and defense
105895	Heat shock protein 70			N/A	Stress and defense
77089	Adaptin			1.05	Transport
83550	Golgi transport complex		1.07		Transport
94552	Ca^{2+}-modulated nonselective polycystin	Y	1.30		Transport
102764	Ankyrin		0.63		Transport
67764	Protein of unknown function DUF185			0.42	--
69789	Hypothetical protein	Y	0.64		--
90006	Hypothetical protein		0.94	1.33	--
92804	Hypothetical protein			0.75	--
103686	Hypothetical protein		0.45		--
109652	Membrane protein containing DUF221	Y		0.45	--

Fungal proteins were identified from wheat leaves of resistant cultivar Stakado (SK) and susceptible cultivar Sevin (SV) inoculated with *Z. tritici* at 5 and 14 days by LC-MS/MS. ID is protein accession number from Joint Genome Institute gene index for *Zymoseptoria tritici*. Proteins whose corresponding transcripts were identified from RNA-seq-based transcriptome dataset [4] are indicated in bold. The expression of these transcripts marked in bold is shown in Additional file 4. Y indicates protein identification containing a signal peptide examined by SignalP (S). The present ratios of protein abundance between 14 days and 5 days were calculated from at least two biological replicates by iTRAQ-117/iTRAQ-115 in labeling-based quantitative proteomics analysis.

The ratios ≥ 2 or ≤ 0.5 were defined as significant change. N/A indicates the identified protein without quantitative data. No ratio value shown indicates the protein was not identified in the cultivar.

Furthermore, the changes in fungal protein expression at two time points were measured. A signaling-related protein Ras GTPase was found to increase the expression during the incompatible interaction. The predominant fungal proteins identified from the inoculated Stakado, did not significantly change in abundance, indicating the restraint fungal growth during the incompatible interaction. A hydantoinase, a glucose/ribitol dehydrogenase, three CWDEs and three ROS-scavenging enzymes strongly accumulated during the compatible interaction, suggesting an increase in carbohydrate metabolism, interfering with host cell and oxidative stress. By comparing the changes in fungal proteins and the corresponding transcripts from the previous study during the compatible interaction, it clearly illustrates a strong multilevel induction of the mechanisms of anti-oxidative stress and host cell wall perturbation in *Z. tritici*, the greater depth and sensitivity of detection of molecules by RNA-seq approach, and the tendency for proteomics methodologies to preferentially detect proteins of higher abundance (Additional file 4).

ZtYAP1 Contributes to Resistance to Oxidative Stress but is not Required for Fungal Pathogenicity

Since the fungus distinctively secreted antioxidants during the compatible interaction, a gene encoding transcription factor YAP1 activating the expression of genes encoding antioxidants was targeted for deletion to examine the role of anti-oxidative stress mechanisms in fungal pathogenicity. The *ZtYAP1* mutant slightly reduced growth on regular Potato Dextrose Agar (PDA) compared to wild type (Figure 3A). Growth of the mutant was strongly inhibited in the presence of H_2O_2, oxidizing compound cumene hydroperoxide, Rose Bengal diacetate generating singlet oxygen or SDS (Figure 3A). Additionally, the mutant retained normal formation and germination of conidia at a rate and magnitude similar to the wild type *in vitro* (data not shown).

To determine whether *ZtYAP1* gene is required for fungal pathogenicity and lesion development, conidial suspensions prepared from the wild type and the mutant were spray-inoculated onto detached susceptible wheat leaves. Disease symptoms on the leaves inoculated with the wild type or the mutant appeared at 10 dai and developed rapidly afterwards (Figure 3B). No clear difference in leaf lesion caused by the wild type and the mutant was observed throughout the entire infection process. The spore production of the mutant in 21-d-infected wheat leaves was approximately three times as high as that of the wild type.

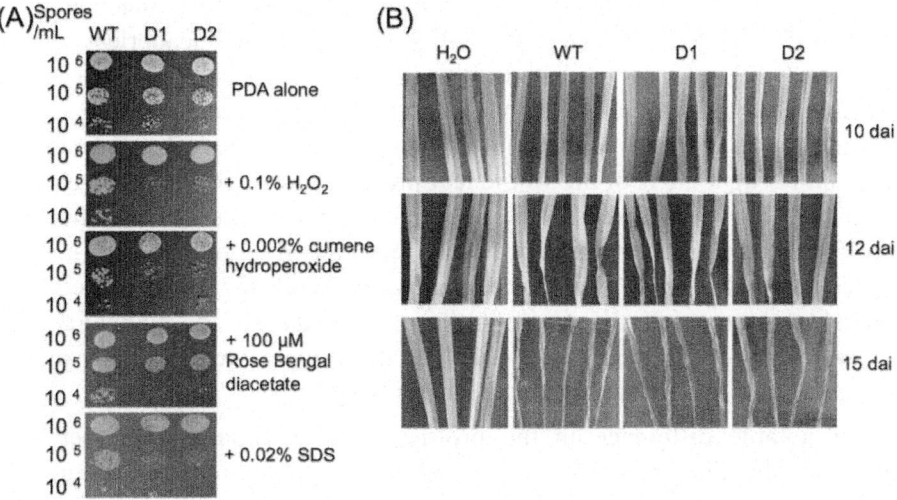

Figure 3: Characterization of *ΔZtYAP1* mutant. **(A)** *ΔZtYAP1* mutant is hypersensitive to oxidants *in vitro*. Sensitivity of *Z. tritici* wild type (WT) and *ΔZtYAP1* deletion strains D1 and D2 was determined by radial growth on PDA supplemented with oxidants or compounds as indicated. Results from one representative of two technical replicates are shown. **(B)** *ΔZtYAP1* mutant is as virulent as WT *in planta*. Fungal pathogenicity was assayed on wheat leaves spray-inoculated with spore suspension (1 x 10^6 spores/mL) prepared from WT, D1, and D2 strains. Photos were taken at 10, 12 and 15 dai.

Discussion

Because of the economic importance as a major fungal pathogen of wheat, *Z. tritici* has emerged as model for studying plant-pathogen interaction. Understanding the molecular mechanisms underlying plant defenses and fungal pathogenicity is a prerequisite for understanding the host-pathogen interaction and can contribute to the development of new strategies of crop protection against pathogen. Thus, we have presented a detailed analysis of proteome from the battle ground apoplast between host wheat and *Z. tritici* during the compatible and incompatible interactions. A comparison of the apoplastic protein profiles of both host and the pathogen between two interactions allows us to gain further insights into such mechanisms.

The analysis of cellular location of protein identifications in the present study shows that approx. half of the differentially expressed plant proteins and fungal proteins from the AWF do not contain signal peptides. This could a consequence of cell lysis leading to the release of cytosolic proteins in the apoplast. Alternatively, evidences are emerging to support that proteins identified in the AWF without symplastic contamination and lacking signal peptides could be secreted by a non-classical secretory mechanism as described in bacteria and fungi [5,18,25]. They are named leaderless secreted proteins (LSPs). Some predicted intracellular proteins such as carbohydrate metabolic proteins found in the AWF can be in fact actively translocated into the extracellular space. Inventories of plant secretome reveal that LSPs may account for up to 50% of the whole leaf proteins identified in the extracellular fluid [25]. The secretome of plants submitted to stresses usually contains more LSPs than unstressed plants, which has been shown in the pathogen-infected maize [26] and salicylic acid-treated *Arabidopsis* cells [27]. With respect to the fungal secretome, investigation of culture supernatant of the fungus*Fusarium graminearum* has indicated that 30% of the protein identifications may be secreted but do not contain signal peptides [28].

A notable difference in the profiles of differentially expressed plant proteins identified from the AWF between compatible and incompatible interactions is the occurrence of regulation of the proteins implicated in carbohydrate metabolism, defense, and stress. It was very likely that the carbohydrate metabolic proteins identified in the AWF were translocated from intracellular space into the apoplast [25]. Plant defense against pathogens is known to be costly in terms of energy, assimilates, reducing equivalents, and carbon skeleton components that are provided by the primary metabolism [29]. Rapid mobilization and metabolism of the carbohydrates are important factors determining the outcome of plant-pathogen interactions. In the incompatible interaction between wheat and *Z. tritici*, it has been shown that accelerated sugar production associated with signal transduction cascades and expression of defense responses, occurs rapidly and intensively, compared to the compatible interaction [10]. Consistent with this, enhanced carbohydrate metabolism occurring in the intracellular space was accompanied by the activation of defense and anti-oxidative stress responses in the apoplast during the biotrophic stage of the incompatible interaction observed in the present study. By contrast, in the compatible interaction, apoplastic defenses were impeded or not activated at protein level and suppressed at transcription level during the establishment of biotrophic growth, but strongly triggered at both levels in the necrotrophic phase [2,4]. These results collectively suggest a contribution of rapid apoplastic defenses and energy metabolism to host resistance to*Z. tritici*.

Host plant cell wall metabolism and remodeling were also distinct in response to the fungus between two interactions. The plant cell wall represents a first line of defense against microbial pathogens since it is a preformed, passive physical barrier limiting access of pathogens and is actively remodeled and reinforced specifically at discrete sites of interaction with the pathogens [30]. During the incompatible interaction, the abundance of the plant proteins that hydrolyze the carbohydrate moieties of arabinogalactan, arabinoxylan, and other polysaccharides diminished, possibly preventing the degradation of cell wall polysaccharides, and thus stabilizing the plant cell wall. Similar results were seen in a study of rice root apoplastic proteome in response to salt stress [31]. Conversely, these proteins largely increased the abundance at the necrotrophic stage of the compatible interaction. On one hand, this phenomenon can result in the release of plant oligosaccharides triggering a set of basal defenses. On the other hand, in agreement with the previous transcriptome analysis [4], the necrotrophic growth of Z. *tritici* induces the high metabolic activity of host cell wall, giving rise to the loss of control of host cell permeability and host necrosis, which allows the fungus thrives on the dead plant.

Moreover, a large group of responsive proteins manifesting increased abundance in both interactions were peroxidases. In addition to catalyzing formation and the consumption of ROS, peroxidases have been implicated in the modification of plant cell wall structures [17]. Via oxidative cross-linking of monolignols, ploysacchraides, and cell wall proteins, peroxidases can reinforce the cell wall to restrict pathogen invasion. Accumulation of certain classes of peroxidases is involved in plant cell wall lignification that occurs as stress response to prevent the spread of fungal pathogens including *Verticillium* and *Melampsora* species [5,17]. Despite the up-regulation of diverse peroxidases as well as other defense-related protein like PR proteins in the necrotrophic phase of the compatible interaction, the velocity and magnitude of activated apoplastic defenses were apparently insufficient to prevent the necrotrophic growth of Z. *tritici*.

An important aspect in studying AWF of the plant challenged with the apoplast-inhabiting pathogens is to obtain the knowledge concerning the molecular compounds produced or secreted by the pathogens, which can provide the candidates essential for the pathogenicity. Here, proteomics analysis showed that approx. half of the fungal proteins whose corresponding transcripts were not identified in the previous RNA-seq-based transcriptome dataset [4]. Similar results were observed in the global analysis of transcript and protein levels across the *Plasmodium falciparum* lifecycle [32]. The little correlation between mRNA and protein abundance has been well demonstrated due to the regulation of transcription and translation, mRNA processing, mRNA stability and degradation, and protein modifications and turnover [4]. Furthermore, the bias in methodologies of RNA-seq and proteomics can result in the discordance between detected transcripts and proteins. Combined with the poor correlation between the regulation of host apoplastic proteins and corresponding transcripts at the biotrophic stage of the compatible interaction, the results strongly suggest that the integrated 'omics' studies are required to comprehensively understand the molecular processes during *Z. tritici* colonization of host wheat.

Based on the fungal protein identifications, there is little evidence for fungal nutrient acquisition from the plant throughout symptomless biotrophic colonization by *Z. tritici* in the incompatible and compatible interactions, which further supports the previous findings from transcriptome and proteome datasets [2,10]. Different protein profiles of *Z. tritici* between two interactions were observed, suggesting specific strategies of *Z. tritici* for a successful colonization. Most notably, signaling, transport and cell wall remodeling were particularly identified in the incompatible interaction, whereas genes and proteins involved in anti-oxidative stress and modification of plant cell wall were identified and induced during the compatible interaction [4]. Fungal pathogens often mount effective responses to counter the defenses of the host plants. A well-established colonization requires circumventing and overcoming host defense responses such as generation of ROS, basically via hiding, detoxification and inhibition [33]. ROS accumulation is considered as one of the primary responses to *Z. tritici,* starting in the apoplast during the biotrophic stage and spreading to the entire leaf tissue at the late necrotrophic stage. Therefore, *Z. tritici* expresses a number of ROS-scavenging proteins and CWDEs that are known to aid invasion into plant cells by hydrolyzing the plant cell wall polymers. The increased expression of CWDEs may also play a role, although not essential, in nutrient acquisition in *Z. tritici* during the necrotrophic growth [4]. The ability to adapt to the oxidative stress, interfere with host cell structure, and overcome other host defense responses such as expression of apoplastic PR proteins probably results in compatibility with the host wheat. On the other hand, it seems that the resistant plant deploys a highly stressful apoplastic environment to put pressure on the pathogen, as evidenced by the active signaling, transport, and cell wall remodeling in *Z. tritici* during the incompatible interaction. The fungal pathogens are known to sense the stress by rapid signal transduction contributing to the regulation of stress functions including glycerol accumulation and cell wall remodeling [34].

The hypothesis of significance of anti-oxidative stress mechanisms in *Z. tritici* colonization was tested by deletion of a gene encoding YAP1 transcription factor regulating expression of antioxidants. The ability to defend against oxidative stress *in vitro* was partially compromised in *YAP1*-dirupted mutant, which, however, showed normal virulence as the wild type strain, suggesting a non-essential role of *YAP1* in *Z. tritici* pathogenicity and functional redundancy of antioxidants. Similarly, disruption of a *YAP1* homolog did not affect the virulence of the phytopathogenic fungus *Cochliobolus heterostrophus* [35] or *Fusarium graminearum* [36]. These findings highlight the sophisticated mechanisms of the involvement of YAP-mediated ROS detoxification in pathogenicity in fungi kingdom. Further investigations on anti-oxidative stress are highly required to elucidate its clear role in *Z. tritici* pathogenicity.

CONCLUSIONS

We have reported an extensive survey of leaf apoplastic proteome in resistant and susceptible wheat in response to *Z. tritici*. The results collectively demonstrate that plant resistance to *Z. tritici* is correlated with rapid activation of responses at proteome level, including enhanced carbohydrate metabolism, cell wall reinforcement and remodeling, production of PR proteins in the apoplast, and generation of a stressful apoplastic environment. On the other hand, the fungus has to interfere with host cell wall and overcome host defenses and stress at the biotrophic stage, for example, by detoxifying ROS and producing CWDEs, to achieve a successful colonization. Taken together, our work provides the valuable insights into STB resistance and *Z. tritici in planta* proteome, which form a fundamental and prerequisite step for the further research of plant apoplastic immunity and *Z. tritici* pathogenicity.

METHODS

Plant Growth, Fungal Inoculation and Extraction of AWF

Growth of wheat cultivars Sevin (susceptible) and Stakado (resistant), preparation of the inoculum of *Z. tritici* isolate IPO323, and inoculation were preformed as described [8]. Separate control plants were mock inoculated with water. Approximately 60 leaves were collected from four separate pots, serving as one biological replicate. Three biological replications were harvested separately at 5 and 14 dai prior to the extraction of AWF by water as described [37]. The remaining leaf samples were ground into fine powder in liquid nitrogen. The collected AWF and leaf powder were stored at $-80°C$ until use.

Malate Dehydrogenase Assay

Total soluble protein was extracted from the leaf powder in 50 mM phosphate buffer, pH 7.5. MDH activity assay was performed to assess the contamination of AWF by intracellular proteins. Five μL of AWF or total leaf protein extract was added to 200 μL of the reaction buffer containing 0.17 mM oxalacetic acid, 0.094 mM β-NADH disodium salt and 0.1 M phosphate buffer, pH 7.5

[14]. The change of absorbance at 340 nm was monitored for 5 min in a spectrophotometer.

Shotgun Proteomics Analysis of AWF

Protein concentration in AWF was determined by Bio-Rad Protein Assay (Bio-Rad) with bovine serum albumin as standard. Fifty micrograms protein was precipitated by 5 volumes of 10% TCA (w/v) in acetone at −20°C overnight. The protein pellet was washed three times in cold 100% acetone followed by solubilization in the buffer containing 6 M urea and 2 M thiourea. Protein samples were treated with 10 mM DTT for 30 min at room temperature and 40 mM iodoacetamide for 30 min in the dark prior to digestion with trypsin (2%, w/w) at 37°C overnight. The resulting peptides were purified on Poros Oligo R3 microcolumn and vacuum-dried prior to amino acid analysis using a Biochrom 30+ Amino Acid Analyzer (Biochrom, UK). Six micrograms peptides from each biological replicate of inoculated and control samples were labeled with iTRAQ® 4-plex (Applied Biosystems) according to the manufacturer's protocol (114 for the control at 5 dai, 115 for the inoculated sample at 5 dai, 116 for the control at 14 dai, 117 for the inoculated sample at 14 dai). Labeled peptides were combined and desalted on Poros Oligo R3 microcolumn.

The labeled peptides were fractionated using hydrophilic interaction liquid chromatography (HILIC) fractionation and analyzed by liquid chromatography (LC)-MS/MS as described [10] with modifications. Briefly, isobaric-labeled peptides obtained from biological replicate 1 were fractionated on a TSKGel Amide 80 HILIC-HPLC column by using the Agilent 1200 microHPLC instrument. Samples were suspended in solvent B (90% acetonitrile and 0.1% trifluoroacetic acid), and peptides were eluted at 6 µL/min by decreasing the solvent B concentration (100 − 60%) over 26 min. Fractions were collected and lyophilized.

Peptides from biological replicate 1 were resuspended in 0.1% formic acid and separated by reversed-phase liquid chromatography on a Reprosil-Pur C18 (3 µC; Dr. Maisch GmbH - Ammerbuch, Germany) column (22 cm x 100 µm inner diameter, in-house packed). The chromatographic gradient was 0 − 34% solvent B (90% acetonitrile and 0.1% formic acid) for 90 min at a flow rate of 300 nL/min. A LTQ-Orbitrap XL mass spectrometer (Thermo Fisher Scientific) was operated in a data-dependent mode automatically switching between MS and MS/MS. A survey MS scan (400–1800 m/z) was acquired in the Orbitrap analyzer with a resolution of 30000 at 400 m/z. The top three most intense ions with a threshold of 5000 were selected for low-resolution CID at normalized collision energy of 35 and high-resolution HCD (normalized collision energy=55; 7500 resolution at 400 m/z). Peptides from biological replicates 2 and 3 without HILIC fractionation were performed using the same LC configuration with a 120 min gradient but coupled to a LTQ-Orbitrap Velos mass spectrometer. Following a survey MS scan at a resolution of 30000 at 400 m/z, the top seven most intense ions were selected for high-resolution HCD–MS/MS (normalized collision energy=48; 7500 resolution at 400 m/z). These two biological replicates were run twice using this setup as technical replicates. Raw data were viewed in Xcalibur v2.0.7 (Thermo Fisher Scientific, USA).

Raw MS/MS spectra were processed and quantified using Proteome Discoverer (Version 1.2, Thermo Fisher) software with quantification setup as ratios of 115/114 (inoculated/control at 5 dai) and 117/116 (inoculated/control at 14 dai). Peptide identification was performed with MASCOT (v2.2, Matrix Science Ltd. - London, UK) and Sequest algorithms, searching against a target and decoy TaGI wheat gene index Release 12.0 (released on 18th April, 2010; TC sequences, 93508; ESTs, 128166; ETs, 251; http://compbio.dfci.harvard. edu/tgi/) and DOE Joint Genome Institute gene index for *Z. tritici* (released on 10th September, 2008; 10933 genes; http://genome.jgi-psf.org/Mycgr3) databases. The following parameters were set for searching: 2 missed cleavages, S-carbamidomethyl-cysteine as a fixed modification, oxidation (M), deamidation (N and Q), iTRAQ® reagents (protein N-terminus and Lys side-chain), peptide mass tolerance 10 ppm, and fragment ion mass tolerance 0.5 Da for CID and 0.05 Da for HCD. False discovery rates were obtained using Percolator selecting identifications with a q-value≤ 0.01. Only high-confidence peptide sequences with a Mascot ion score≥ 23, Sequest Xcorr value> 2.2 and rank 1 were considered for the further analysis. The identified plant and fungal proteins must contain at least two confidentially identified peptides. Protein quantification data were normalized using log2-transformed median. The statistics analysis was performed on the data merged from all biological and technical replicates by using R program. The plant proteins identified in at least two biological replicates and with average ratios≥ 2 or≤ 0.5 at either time point were defined as regulated proteins. The fungal proteins identified in at least two biological replicates with quantitative data were shown in Table 1. SignalP (http://www.cbs.dtu.dk/services/SignalP) was performed to examine signal peptides of fungal and wheat regulated proteins. The differential expression profiles of host proteins at two time points were clustered by k-means algorithm using Euclidean distance.

Western Blot Analysis

Western blotting was performed with three biological replicates as described

[38]. Five micrograms proteins from AWF were separated on CriterionTM XT Precast Gels (12% Bis-Tris, Bio-Rad) followed by blotting to nitrocellulose membranes (GE Healthcare). Membranes were blocked followed by incubation with rabbit antibodies against barley PR-1, PR-2 (β-1,3-glucanase), and PR-3 (chitinase). After extensive washes, the membranes were incubated with anti-rabbit secondary antibodies conjugated to horseradish peroxidase (1:2000, Dako, Denmark) and detected using the Immun-Star HRP Substrate Kit (Bio-Rad). Quantification of the signals on the membranes was carried out by using ImageJ program.

Agrobacterium-Mediated Targeted Gene Deletion of *ZtYAP1*

Approx. 700-bp and 1400-bp of flanking fungal genomic DNA were amplified upstream and downstream of *YAP1* (Zt35076) opening reading frame by PCR, respectively, and cloned into vector pCHYG containing the hph cassette conferring resistance to hygromycin [39]. The final plasmid pCHYG-YAP1 was sequenced to confirm the correct insert. The plasmid was transformed to Agrobacterium strain Agl-1 via the freeze-thaw method. Agrobacterium transformation of *Z. tritici* was performed as described [39]. Fungal genomic DNA was isolated and targeted insertion of the T-DNA was initially confirmed by PCR on genomic DNA directed against Hph and *ZtYAP1*. All the primers used for study of fungal mutant are listed in Additional file 5.

Characterization of *ZtYAP1* Mutant

Preparation of spores of *ZtYAP1* mutant and inoculation of 14-day-old wheat cv. Sevin seedling were performed as described above to evaluate fungal pathogenicity. Three independent pots of plant (ten leaves per pot) were prepared for inoculation with the wild type, the mutant or water, serving as three biological replicates. Leaves were photographed at 10, 12 and 15 days. Infected leaves were harvested at 21 dai. The spores were washed out from leaves and counted under microscope.

Sensitivity assay to oxidative stress was conducted by applying a 5 μL droplet of fungal spore suspension onto PDA containing compounds at appropriate concentrations and incubating under constant fluorescent light. Photographs of relative colony densities were taken after 6 days.

ACKNOWLEDGEMENTS

This work was funded by a postdoctoral grant from the Danish Research Council for Technology and Production (11–105997) to Dr. Fen Yang.

AUTHORS' CONTRIBUTIONS

FY designed the experiments, performed plant inoculation, protein extraction, cloning, and fungal mutant characterization, coordinated the study, and drafted and finalized the manuscript. WL performed bioinformatics analysis of the proteome data. MD and JJR performed agrobacterium-mediated transformation and fungal mutant generation. MRL and GP performed LC-MS/MS analysis. All authors have read and approved the manuscript.

REFERENCES

1.   Keon J, Antoniw J, Carzaniga R, Deller S, Ward JL, Baker JM, et al. Transcriptional adaptation of *Mycosphaerella graminicola* to programmed cell death (PCD) of its susceptible wheat host. Mol Plant Microbe Interact. 2007;20:178–93.

2.   Rudd JJ, Kanyuka K, Hassani-Pak K, Derbyshire M, Andongabo A, Devonshire J, et al. Transcriptome and metabolite profiling the infection cycle of *Zymoseptoria tritici* on wheat (*Triticum aestivum*) reveals a biphasic interaction with plant immunity involving differential pathogen chromosomal contributions, and a variation on the hemibiotrophic lifestyle definition. Plant Physiol. 2015;167:1158–85.

3.   Marshall R, Kombrink A, Motteram J, Loza-Reyes E, Lucas J, Hammond-Kosack KE, et al. Analysis of two in *planta* expressed LysM effector homologs from the fungus *Mycosphaerella graminicola* reveals novel functional properties and varying contributions to virulence on wheat. Plant Physiol. 2011;156:756–69.

4.   Yang F, Li WS, Jørgensen HJ. Transcriptional reprogramming of wheat and the hemibiotrophic pathogen *Septoria tritici* during two phases of the compatible interaction. PLoS One. 2013;8, e81606.

5.   Pechanova O, Hsu CY, Adams JP, Pechan T, Vandervelde L, Drnevich J, et al. Apoplast proteome reveals that extracellular matrix contributes to multistress response in poplar. BMC Genomics. 2010;11:674.

6.   Lee WS, Rudd JJ, Hammond-Kosack KE, Kanyuka K. *Mycosphaerella graminicola* LysM effector-mediated stealth pathogenesis subverts recognition through both CERK1 and CEBiP homologues in wheat. Mol Plant Microbe Interact. 2014;27:236–43.

7.   Rudd JJ, Keon J, Hammond-Kosack KE. The Wheat mitogen-activated protein kinases TaMPK3 and TaMPK6 are differentially regulated at multiple levels during compatible disease interactions with *Mycosphaerella graminicola*. Plant Physiol. 2008;147:802–15.

8. Shetty NP, Kristensen BK, Newman MA, Møller K, Gregersen PL, Jørgensen HJ. Association of hydrogen peroxide with restriction of *Septoria tritici* in resistant wheat. Physiol Mol Plant Pathol. 2003;62:333–46.

9. Shetty NP, Mehrabi R, Lütken H, Haldrup A, Kema GH, Collinge DB, et al. Role of hydrogen peroxide during the interaction between the hemibiotrophic fungal pathogen *Septoria tritici* and wheat. New Phytol. 2007;174:637–47.

10. Yang F, Melo-Braga MN, Larsen MR, Jørgensen HJ, Palmisano G. Battle through signaling between wheat and the fungal pathogen *Septoria tritici* revealed by proteomics and phosphoproteomics. Mol Cell Proteomics. 2013;12:2497–508.

11. Kema GH, van der Lee TA, Mendes O, Verstappen EC, Lankhorst RK, Sandbrink H, et al. Large-scale gene discovery in the septoria tritici blotch fungus *Mycosphaerella graminicola* with a focus on *in planta* expression. Mol Plant Microbe Interact. 2008;21:1249–60.

12. Toone WM, Morgan BA, Jones N. Redox control of AP-1- like factors in yeast and beyond. Oncogene. 2001;20:2336–46.

13. Lin CH, Yang SL, Chung KR. The YAP1 homolog-mediated oxidative stress tolerance is crucial for pathogenicity of the necrotrophic fungus *Alternaria alternata* in citrus. Mol Plant Microbe Interact. 2009;22:942–52.

14. Husted S, Schjoerring JK. Apoplastic pH and ammonium concentration in leaves of *Brassica napus* L. Plant Physiol. 1995;109:1453–60.

15. Mattsson M, Schjørring JK. Senescence-induced changes in apoplastic and bulk tissue ammonia concentrations of ryegrass leaves. New Phytol. 2003;160:489–99.

16. Floerl S, Druebert C, Majcherczyk A, Karlovsky P, Kües U, Polle A. Defence reactions in the apoplastic proteome of oilseed rape (*Brassica napus* var. *napus*) attenuate *Verticillium longisporum* growth but not disease symptoms. BMC Plant Biol. 2008;8:129.

17. Floerl S, Majcherczyk A, Possienke M, Feussner K, Tappe H, Gatz C, et al. *Verticillium longisporum* infection affects the leaf apoplastic proteome, metabolome, and cell wall properties in *Arabidopsis thaliana*. PLoS One. 2012;7, e31435.

18. Delaunois B, Colby T, Belloy N, Conreux A, Harzen A, Baillieul F, et al. Large-scale proteomic analysis of the grapevine leaf apoplastic fluid reveals mainly stress-related proteins and cell wall modifying enzymes. BMC Plant Biol. 2013;13:24.

19. Witzel K, Shahzad M, Matros A, Mock HP, Muhling KH. Comparative evaluation of extraction methods for apoplastic proteins from maize leaves. Plant Methods. 2011;7:48.

20. Djordjevic MA, Oakes M, Li DX, Hwang CH, Hocart CH, Gresshoff PM. The glycine max xylem sap and apoplast proteome. J Proteome Res. 2007;6:3771–9.

21. Soares NC, Francisco R, Ricardo CP, Jackson PA. Proteomics of ionically bound and soluble extracellular proteins in *Medicago truncatula* leaves. Proteomics. 2007;7:2070–82.

22. Casasoli M, Spadoni S, Lilley KS, Cervone F, De Lorenzo G, Mattei B. Identification by 2-D DIGE of apoplastic proteins regulated by oligogalacturonides in *Arabidopsis thaliana*. Proteomics. 2008;8:1042–54.

23. Köllner TG, Schnee C, Li S, Svatoš A, Schneider B, Gershenzon J, et al. Protonation of a neutral (S)-β-bisabolene intermediate is involved in (S)-β-macrocarpene formation by the maize sesquiterpene synthases TPS6 and TPS11. J Biol Chem. 2008;283:20779–88.

24. Schmelz EA, Kaplan F, Huffaker A, Dafoe NJ, Vaughan MM, Ni X, et al. Identity, regulation, and activity of inducible diterpenoid phytoalexins in maize. Proc Natl Acad Sci U S A. 2011;108:5455–60.

25. Agrawal GK, Jwa NS, Lebrun MH, Job D, Rakwal R. Plant secretome: unlocking secrets of the secreted proteins. Proteomics. 2010;10:799–827.

26. Chivasa S, Simon W, Yu XL, Yalpani N, Slabas A. Pathogen elicitor-induced changes in the maize extracellular matrix proteome. Proteomics. 2005;5:4894–904.

27. Cheng FY, Blackburn K, Lin YM, Goshe MB, Williamson JD. Absolute protein quantification by LC/MSE for global analysis of salicylic acid-induced plant protein secretion responses. J Proteome Res. 2009;8:82–93.

28. Yang F, Jensen JD, Svensson B, Jørgensen HJ, Collinge DB, Finnie C. Secretomics identifies *Fusarium graminearum* proteins involved in the interaction with barley and wheat. Mol Plant Pathol. 2012;13:445–53.

29. Bolton MD. Primary metabolism and plant defense-fuel for the fire. Mol Plant Microbe Interact. 2009;22:487–97.

30. Underwood W. The plant cell wall: a dynamic barrier against pathogen invasion. Frontier Plant Sci. 2012;3:85.

31. Song Y, Zhang C, Ge W, Zhang Y, Burlingame AL, Guo Y. Identification of NaCl stress-responsive apoplastic proteins in rice shoot stems by 2D-DIGE. J Proteomics. 2011;74:1045–67.

32. Le Roch KG, Johnson JR, Florens L, Zhou Y, Santrosyan A, Grainger M, et al. Global analysis of transcript and protein levels across the *Plasmodium falciparum* life cycle. Genome Res. 2004;14:2308–18.

33. Doehlemann G, Hemetsberger C. Apoplastic immunity and its suppression by filamentous plant pathogens. New Phytol. 2013;198:1001–16.

34. Brown AJ, Budge S, Kaloriti D, Tillmann A, Jacobsen MD, Yin Z, et al. Stress adaptation in a pathogenic fungus. J Exp Biol. 2014;217:144–55.

35. Lev S, Hadar R, Amedeo P, Baker S, Yoder OC, Horwitz BA. Activation of an AP-1-like transcription factor of the maize pathogen *Cochliobolus heterostrophus* in response to oxidative stress and plant signals. Eukaryot Cell. 2005;4:443–54.

36. Montibus M, Ducos C, Bonnin-Verdal M-N, Bormann J, Ponts N, Richard- Forget F, et al. The bZIP transcription factor Fgap1 mediates oxidative stress response and trichothecene biosynthesis but not virulence in *Fusarium graminearum*. PLoS One. 2013;8:e83377.

37. Shetty NP, Jensen JD, Knudsen A, Finnie C, Geshi N, Blennow A, et al. Effects of β-1,3-glucan from *Septoria tritici* on structural defence responses in wheat. J Exp Bot. 2009;60:4287–300.

38. Yang F, Jensen JD, Svensson B, Jørgensen HJ, Collinge DB, Finnie C. Analysis of early events in the interaction between *Fusarium graminearum* and the susceptible barley (*Hordeum vulgare)* cultivar Scarlett. Proteomics. 2010;10:3748–55.

39. Motteram J, Küfner I, Deller S, Brunner F, Hammond-Kosack KE, Nürnberger T, et al. Molecular characterization and functional analysis of MgNLP, the sole NPP1 domain-containing protein, from the fungal wheat leaf pathogen *Mycosphaerella graminicola*. Mol Plant Microbe Interact. 2009;22:790–9.

CITATION

CHAPTER 1

Chung et al., Resistance loci affecting distinct stages of fungal pathogenesis: use of introgression lines for QTL mapping and characterization in the maize - Setosphaeria turcica pathosystem BMC Plant Biology 2010, 10:103.

CHAPTER 2

Raquel González-Fernández, Elena Prats, and Jesús V. Jorrín-Novo, "Proteomics of Plant Pathogenic Fungi,"Journal of Biomedicine and Biotechnology, vol. 2010, Article ID 932527, 36 pages, 2010. doi:10.1155/2010/932527.

CHAPTER 3

Richard J O'Connell, Michael R Thon, Stéphane Hacquard, Stefan G Amyotte, Jochen Kleemann, Maria F Torres, Ulrike Damm, Ester A Buiate, Lynn Epstein, Noam Alkan, Janine Altmüller, Lucia Alvarado-Balderrama, Christopher A Bauser, Christian Becker, Bruce W Birren, Zehua Chen, Jaeyoung Choi, Jo Anne Crouch, Jonathan P Duvick, Mark A Farman, Pamela Gan, David Heiman, Bernard Henrissat, Richard J Howard, Mehdi Kabbage, "Lifestyle transitions in plant pathogenic Colletotrichum fungi deciphered by genome and transcriptome analyses," Nature Genetics 44, 1060–1065 (2012) doi:10.1038/ng.2372.

CHAPTER 4

Walters DR, Havis ND, Paterson L, Taylor J, Walsh DJ and Sablou C (2014) Control of foliar pathogens of spring barley using a combination of resistance elicitors. Front. Plant Sci. 5:241. doi: 10.3389/fpls.2014.00241.

CHAPTER 5

Asghar Heydari and Mohammad Pessarakli, 2010. A Review on Biological Control of Fungal Plant Pathogens Using Microbial Antagonists. Journal of Biological Sciences, 10: 273-290.

CHAPTER 6

Shaowu Meng, Trudy Torto-Alalibo, Marcus C Chibucos, Brett M Tyler and Ralph A Dean, "Common processes in pathogenesis by fungal and oomycete plant pathogens, described with Gene Ontology terms," BMC Microbiology20099(Suppl 1):S7, DOI: 10.1186/1471-2180-9-S1-S7.

CHAPTER 7

Kalunke RM, Tundo S, Benedetti M, Cervone F, De Lorenzo G and D'Ovidio R (2015) An update on polygalacturonase-inhibiting protein (PGIP), a leucine-rich repeat protein that protects crop plants against pathogens. Front. Plant Sci. 6:146. doi: 10.3389/fpls.2015.00146.

CHAPTER 8

Fen Yang, Wanshun Li, Mark Derbyshire, Martin R Larsen, Jason J Rudd and Giuseppe Palmisano, "Unraveling incompatibility between wheat and the fungal pathogen Zymoseptoria tritici through apoplastic proteomics," BMC Genomics201516:362, DOI: 10.1186/s12864-015-1549-6.

INDEX